The Darkest Days of the War

CIVIL WAR AMERICA

Gary W. Gallagher, editor

The Darkest Days of the War

The Battles of Iuka & Corinth

Peter Cozzens

The University of North Carolina Press

Chapel Hill & London

© 1997
The University of
North Carolina Press
All rights reserved
Manufactured in the
United States of America

The paper in this book meets
the guidelines for permanence
and durability of the Committee
on Production Guidelines for
Book Longevity of the Council
on Library Resources.

Library of Congress
Cataloging-in-Publication Data
Cozzens, Peter, 1957–
The darkest days of the war : the battles of Iuka and
Corinth / by Peter Cozzens.
p. cm. — (Civil War America)
Includes bibliographical references and index.
ISBN 0-8078-2320-1 (alk. paper)
1. Iuka (Miss.) — History — Siege, 1862. 2. Corinth
(Miss.), Battle of, 1862. 3. Mississippi — History —
Civil War, 1861–1865 — Campaigns. 4. United
States. Army — Drill and tactics — History — 19th
century. 5. Confederate States of America. Army —
Drill and tactics. I. Title. II. Series.
E474.42.C69 1997
973.7'33 — dc 20 96-9613
CIP

01 00 99 98 97 5 4 3 2 1

Para mi esposa, Issa Maria

Ya cuando la fe me había desamparado,

tu creíste en mi

CONTENTS

MAPS

ILLUSTRATIONS

People have asked me why I chose to write on the relatively obscure northern Mississippi campaign that culminated in the Battle of Corinth, on October 3–4, 1862. I tell them I selected this topic precisely because of its obscurity, which derives solely from a want of scholarly attention — the campaign itself was far from unimportant.

Glance at any map of the Confederacy depicting its railroads, and your eye is naturally drawn to Corinth. It stood at the junction of two of the best railroads in the South. Take and hold Corinth, and Union armies would sever the most viable Confederate line of communications and supply between the eastern seaboard and the vast trans-Mississippi region. Maj. Gen. Henry Halleck recognized this, and he counted the capture of Corinth more important than the destruction of the Confederate western armies.

Corinth was also of great importance to Federal plans of conquest. The town lay between the two strategic invasion routes into the Deep South: the Mississippi River Valley corridor leading to Vicksburg, and the Nashville, Chattanooga, and Atlanta avenue into the interior of Georgia. With Corinth in their possession, the Federals would be able to transport supplies and reinforcements to armies operating along either route.

The autumn 1862 campaign had consequences for the war in the West not generally appreciated. The utter defeat at Corinth of the Confederate forces (comprised of the combined armies of Earl Van Dorn and Sterling Price, the whole under Van Dorn's leadership) eliminated the only mobile Southern command standing between Ulysses S. Grant's Union army and Vicksburg. After Corinth, the way was clear for Grant to proceed on his great march of conquest.

Southern defeat in northern Mississippi also contributed to the collapse of Braxton Bragg's invasion of Kentucky. Bragg himself said it was a determining factor in his discomfiture. Undoubtedly Bragg exaggerated the impact of events in northern Mississippi to deflect criticism from his own errors, but there is much truth to his assertion. Bragg had counted on Van Dorn and Price to prevent Federal reinforcements from crossing the Tennessee River to oppose him and to strike north themselves to protect his strategic left flank. Van Dorn and Price disappointed him on both counts. And while Bragg already had fought and lost the Battle of Perryville by the time he learned of the defeat at Corinth, the news convinced him he must abandon Kentucky. The ruination of Van Dorn's army left

Bragg's as the only significant Confederate command in the West and exposed Chattanooga to capture, compelling Bragg to withdraw to protect the Southern heartland.

The battles of Iuka, Corinth, and even Davis Bridge are worthy of study for reasons apart from their larger significance. All were vicious fights, even by Civil War standards. Iuka stands as a textbook example of a meeting engagement gone tragically awry. A Federal and a Confederate brigade collided in march column, and the ensuing three-hour struggle was over before either army commander understood what had happened. It is hard to imagine a more bitter clash. Nearly 1,000 of 3,200 Southerners engaged, and some 800 of 3,000 Northerners fell fighting for a single ridge. Perhaps a quarter of the Union losses were caused by the fire of other frightened Federals, a pointed reminder of the hazards of exposing raw troops to battle without the support of veterans.

Iuka also demonstrated the difficulty, given the uncertain communications of the time, of coordinating a joint tactical operation of two forces separated by more than a few miles. Grant tried and failed to crush Price between the jaws of a pincer made of Edward O. C. Ord's and William Starke Rosecrans's commands. Price's escape from Rosecrans's front occasioned a falling-out between Grant and Rosecrans that ultimately wrecked the military career of the latter.

I cannot think of any other Civil War battle in which the attackers fought under more oppressive conditions than at Corinth. Van Dorn roused his troops before daybreak, marched them eight miles, and then threw them immediately into battle. For eight hours the Confederates attacked in 100-degree heat, without food and with scarcely a drop of water. The Confederate assaults on Batteries Robinett and Powell the following day stand among the fiercest of the war and among the few in which the fighting became hand to hand.

The forgotten fight at Davis Bridge on the Hatchie River, in which the Federals bungled one of their best chances to obliterate a Southern army, also was unusually cruel. In two hours 500 Yankees were gunned down on a half-acre-wide tuck of river bank.

I also found the 1862 northern Mississippi campaign fascinating for the personalities involved. Overblown egos played a large part in the results. A more ludicrous character in high command than Earl Van Dorn is hard to imagine, and his preference for solo performances over cooperation doomed the Confederates from the start. Remarkable, too, is the squabbling between Grant and Rosecrans that emerged from victory; egged on by their staffs, they engaged in a sometimes comic battle of wills that poisoned relations between them, the two principal Northern generals in the West, during the middle period of the war.

A final word on Grant. At the time of Iuka and Corinth he was still under a shadow for his near-defeat at Shiloh. General in Chief Henry Halleck mistrusted and would have been pleased to shelve him, and President Lincoln had not yet come to appreciate his talents. Had the North lost at Corinth, Grant may have been returned to the obscurity of a Galena, Illinois, leather-goods store.

ACKNOWLEDGMENTS

With each book that I write, I find myself more deeply in debt, and to a greater number of people, for help, suggestions, and encouragement. It is a debt I am pleased to acknowledge.

Moving from country to country with the Foreign Service, I would find it impossible to complete a book of this sort without the help of librarians and archivists. I would like to thank those who were particularly gracious in providing me with materials from their institutions. Many pointed me to sources unknown to me; this book is the richer for their help. My special thanks go to Bryon Andreasen of the University of Illinois, Champaign-Urbana; Yvonne Arnold of the University of Southern Mississippi, Hattiesburg; Martha Clevenger of the Missouri Historical Society, St. Louis; Richard Himmel of the University of North Texas, Denton; James Holmberg of the Filson Club, Louisville, Kentucky; Marcelle Hull of the University of Texas, Arlington; Karen Kearns and Lisa Libby of the Huntington Library, San Marino, California; Kathy Lafferty of the University of Kansas, Lawrence; Grace Linden of the Sioux City, Iowa, Public Museum; Robert McCown of the University of Iowa, Iowa City; Cassandra McGraw of the University of Arkansas, Fayetteville; Gail Redmann of the Western Reserve Historical Society, Cleveland; Randy Roberts of the University of Missouri, Columbia; Margaret Rose of the Public Library of Corpus Christi, Texas; and Ken Tilley of the Alabama Department of Archives and History, Montgomery.

William Erwin and the staff of the Special Collections Library at Duke University made my visit there both pleasant and productive, as did Richard Shrader and the staff of the Manuscripts Department of the University of North Carolina, Chapel Hill.

Gary Arnold of the Ohio Historical Society went out of his way to help me wade through the society's many fine manuscript and printed records of Ohioans who fought at Iuka, Corinth, and Davis Bridge.

My continuing thanks go to Carley Robison, curator of manuscripts and archives at my alma mater, Knox College, for making the library's fine Ray D. Smith Collection of Civil War books readily accessible to me.

I would like to acknowledge three people who helped me immensely during my visit to the Corinth area. Hugh Horton shared with me his vast knowledge of the battle and led me over the battlefield. Margaret Rogers, executive director of the Northeast Mississippi Museum Association, Corinth, opened the museum's archives to me. Herbert Wood of the

Davis Bridge Memorial Foundation, Bolivar, Tennessee, took me over the battlefield of Davis Bridge. Without his guidance on the ground, I could not have re-created the clash with any degree of certainty.

Stacey Allen, historian at the Shiloh National Military Park, gave me copies of an outstanding series of interpretive maps he had done on the three battles. They were of great help to me in tracing the movements of the armies.

Terrence Winschel, historian at the Vicksburg National Military Park, read the manuscript and shared with me his knowledge of the battles. He saved me from some embarrassing mistakes, such as pointing out to me that the Confederates could not possibly have dragged the captured cannon of Sears's battery to the Iuka town square, as I had mistakenly supposed, because Iuka did not have a town square. Any remaining errors are strictly mine.

I would also like to thank Gary Gallagher, editor of the Civil War in America series, and Kate Torrey, director of the University of North Carolina Press, for both their encouragement and most constructive criticism.

To George Skoch of Cleveland Heights, Ohio, many thanks for the outstanding maps that grace the book.

To my mother, my most sincere thanks for your countless trips to the public library gathering journal and magazine articles for me. To the staff of the periodicals department of the Wheaton, Illinois Public Library, my thanks for your filling the innumerable requests for esoteric articles.

Three dear friends deserve special mention. Robert Girardi of Chicago shared with me materials from his extensive Civil War library and made many keen observations on the manuscript. Sue Bremner, of the Foreign Service, tirelessly helped me gather sources and made numerous grammatical and stylistic improvements to the manuscript. And to Keith Rocco, once again, my especial thanks for the outstanding illustrations.

A sound of alarm. Stores and parking lots have obliterated the Iuka battlefield. Little of the Corinth battlefield remains. Battery F, a fine example of Civil War earthen lunettes, stands among a tract of modern homes outside Corinth; it owes its existence to a generous landowner. A good portion of the Beauregard line of Confederate earthworks still exists but is on private land. So, too, is Oliver's hill. The danger that both will be bulldozed is very real. Only the Davis Bridge battlefield site has been preserved intact, and that because of the initiative of members of the Bolivar Chapter of the Sons of Confederate Veterans.

The work of battlefield preservation is far from over.

The Darkest Days of the War

1

Six and a Half Feet of Missouri Soil

There was a festive air in Memphis, Tennessee, the second week of April 1862, a gaiety that seemed to mock at Southern misfortune. The week before, 11,000 Confederate soldiers had been lost at Shiloh, and the Army of the Mississippi, which was to have swept the Mississippi Valley clean of Yankees, nearly wrecked. The army had marched into Tennessee from Corinth, Mississippi, to attack Federal forces encamped nineteen miles away at Pittsburg Landing, Tennessee. The resultant Battle of Shiloh was to have been the first encounter of a lightning offensive leading to the Ohio River. Instead the Confederates found themselves back in Corinth. "Home to Tennessee" would come to be their rallying cry. But for the moment the stunned and sullen survivors of Shiloh could see no farther than the head logs of the earthworks they were constructing around Corinth.

Memphis was just ninety-three miles west of Corinth, and the victorious Northern army of Maj. Gen. Ulysses S. Grant was hardly more distant. But the war was young and Memphis was still free of Federals. What was lost did not seem irretrievable, and much might yet be won. Of more immediate concern, the town was playing host to a celebrity: Maj. Gen. Sterling Price. Shiloh had killed Albert Sidney Johnston and humiliated Pierre G. T. Beauregard, making Price one of the South's few heroes.

A large crowd was on hand when the steamer carrying Price and a regiment of his Missouri division docked late on the afternoon of April 11.

Poorly armed, roughly clad, and rougher looking, the Missourians shuffled down the gangplank ahead of their commander. The general stepped out onto the hurricane deck amid cheers of "Price! Price! Price!," making a splendid contrast with his troops. Towering above most of them at six feet, two inches tall and weighing more than 200 pounds, Price carried his bulk gracefully. His complexion was ruddy and his countenance benign, and his hair and thick sideburns were silver white; had he not been clad in a resplendent new dress uniform of Confederate gray, complete with sash and sword, one might have mistaken him for a visiting English gentleman. A band struck up "Dixie," and local dignitaries escorted Price to the town's best hotel. At a reception in his honor that night, Price pledged to repay the city's hospitality on the battlefield.[1]

Price's Memphis welcome swelled his enormous ego but could not palliate an inner dread. Like the Tennesseans at Corinth who worried over the fate of their homes, Price was troubled to be on the east bank of the Mississippi River, rather than in Missouri, the state he had joined the Confederacy to defend.

Price was born and raised in Virginia; but all he had accomplished in life he owed to Missouri, and his gratitude ran deep. As a young man Price had come west with his father, a tobacco planter whose fortunes had plummeted after the price of the Virginia leaf fell sharply in the 1820s. The younger Price moved easily among expatriate Virginians who, like him, had resettled in the Boonslick area of Missouri. Sharing with his father a dedication to republicanism and a burning ambition, within five years of settling in Missouri Price had become a prosperous planter and a respected leader of the state Democratic Party. In 1844 he resigned as speaker of the Missouri legislature to accept a seat in the United States House of Representatives.

Price saw wide service in the war with Mexico, first as commander of a regiment of Missouri volunteers and later as military governor of New Mexico. He emerged from the war a brevet brigadier general, having gained a higher rank in Mexico than any other future Confederate military leader except Gideon Pillow. His Mexican War service brought him fame and glory, but it also revealed disturbing traits that should have raised doubts about his fitness for command.

A penchant for independent action, bordering on insubordination and aggravated by a deep thirst for glory, was the most serious flaw to surface, and it drove Price to invade the State of Chihuahua. The Mexican governor tried to intercede with the invaders, claiming that peace had been declared, but Price dismissed him and attacked, capturing the capital and

Maj. Gen. Sterling Price (Louisiana Historical Association Collection, Howard-Tilton Memorial Library, Tulane University)

killing 200 Mexican soldiers at a loss of only four dead. Six days earlier the United States Senate had ratified the Treaty of Guadalupe-Hidalgo.

In the fanfare that followed, the fact that Price had waged an illegal campaign was forgotten. President James K. Polk considered the operation one of the most complete victories of the war, and the Missouri press predicted Price would emerge "with a higher military fame and with laurels more imperishable than any other officer who has been or is now connected with the army of the west." Which is precisely what Price had hoped for.[2]

Fame brought new office, but with it came great turmoil. Passage of the Kansas-Nebraska Act during his second year as governor tore asunder the state Democratic Party and placed Price in a personal quandary: he sympathized with those who would make Kansas a slave state, yet he was still a Union man. Price refused to ally himself with the growing secessionist element, preferring compromise in a time of seething radicalism.

On the strength of his reputation as a conditional Unionist, in February 1861 Price was chosen president of a state convention that voted against all proposals advocating secession. Only after Fort Sumter, when President Abraham Lincoln issued a call for troops to suppress the rebellion, did Price despair of peace. And not until the hotheaded Federal commander in Missouri, Nathaniel Lyon, provoked a confrontation with the state militia that left twenty-eight civilians dead did Price side with the South, accepting secessionist Governor Claiborne Jackson's offer of command of the militia. Still hoping for a peaceful compromise, Price cautioned his subordinates not to incite trouble and promised Federal authorities he would keep Confederate troops out of Missouri.

Price's policy of armed neutrality failed, as it was bound to, and the late summer of 1861 found him locked in a struggle with Federal forces for control of Missouri, a contest in which Price was sorely handicapped. By initially cooperating with Union authorities, he had gambled away his good standing with many Southern leaders. Governor Jackson distrusted him, and President Jefferson Davis took Price's efforts at compromise as proof of the Missourian's unreliability.[3]

Price had failed to prevent war and had offended the president of the Confederacy in the bargain. To restore both his reputation and the integrity of his beloved Missouri, Price joined Confederate forces under Brig. Gen. Benjamin McCulloch in an offensive against Nathaniel Lyon. At Wilson's Creek on August 10, Lyon was killed and his command routed. Price seemed on the verge of liberating Missouri, but McCulloch refused to join him in a pursuit, arguing that a move deep into Missouri could not succeed without the cooperation of the Confederate army in Tennessee, which was falling back.

Price went on without McCulloch, capturing 3,500 Yankees at Lexington. The triumph brought him thousands of recruits and the adulation of the Southern press, but it was strategically barren. Lacking assistance from McCulloch, Price was forced to fall back to the southwest corner of Missouri. There he spent the waning days of the year brooding. As the weather turned cold and rations became scarce, recruits slipped away by the thousands. Price pleaded for their return: "Are Missourians no longer true to themselves? Are we a generation of driveling, sniveling, degraded slaves? I will ask for six and a half feet of Missouri soil in which to repose, but will not live to see my people enslaved. Do I hear your shouts? Are you coming?"

Few came. Price's despair deepened. He railed against McCulloch for his lack of cooperation and nursed a private anger with President Davis for having neglected Missouri. In better moments his passion for freeing the state overcame his resentment, and he appealed to both McCulloch and Davis for help. Both disappointed him: McCulloch in part because he was away in Richmond answering charges arising from his failure to march with Price after Wilson's Creek; Davis because of his continued suspicions about Price's loyalty and his contempt for any general not schooled at West Point.

Davis offered nothing and demanded much. He told Price that he would entertain a request for help only if Price's Missourians were mustered into the Confederate service. And Davis told a Missouri congressional delegation requesting Price's appointment as commander of the Trans-Mississippi District that he could offer Price no command until his troops were part of the Confederate army. It was a disingenuous reply; Davis had no intention of granting Price so important a command under any circumstances. Privately he deprecated the "irregular warfare" that Price had waged in Missouri and implied that it was time to send a West Point graduate to the theater.[4]

Price did little to redeem himself with the president. He told all who cared to listen, the press included, that Richmond not only had failed to succor the Missouri troops but also was sowing disaffection in their ranks. Not surprisingly, Davis refused to entertain the renewed demands of Price's congressional allies that he appoint him to command in Missouri and Arkansas. "Gentleman, I am not to be dictated to" was all he had to say to the congressional delegation that came calling on Price's behalf.[5]

Davis already had resolved to begin 1862 with a commander for the Trans-Mississippi District able to end the discord there. Finding a general willing to take the assignment proved a problem. His friend and fellow West Point graduate, Col. Henry Heth, declined the offer after apprising himself of conditions in Missouri and Arkansas. The president next offered the command to Maj. Gen. Braxton Bragg, also a West Point graduate and friend. Bragg wanted no part of the job; the Missouri army, he said, was a "mere gathering of brave but undisciplined troops, coming and going at pleasure, and needing a master mind to control and reduce it into order and to convert it into a real army."

The third candidate was Maj. Gen. Earl Van Dorn. Unlike Price, who confined his ambition to the borders of Missouri, Van Dorn would take glory wherever he could find it. He accepted the command eagerly, and the Confederate Congress endorsed his assignment. Even Price's friends concurred in the appointment of the charismatic Mississippian.[6]

Van Dorn certainly looked a leader. Handsome and dapper, with a head of full, wavy hair, he seemed the quintessential Southern cavalier. Dabney Maury came to know Van Dorn during the Mexican War. A distinguished veteran of the Regular Army and normally a good judge of men, Maury was completely taken with the Mississippian: "He was one of the most attractive young fellows in the army. His figure was lithe and graceful; his stature did not exceed five feet six inches; but his clear blue eyes, his firm-set mouth, with white strong teeth, his well-cut nose, with expanding nostrils, gave assurance of a man whom men could trust and follow."

His classmates at West Point, where he had graduated near the bottom of his class and had come within a hair of being dismissed for bad conduct, were less impressed. Cadet John Pope thought him feminine in appearance—"he had the soft, sweet face of a girl . . . blue eyes and long, curly, auburn hair"—and possessed of only modest ability and limited intellect. But Van Dorn loved to fight, and he emerged from the Mexican War with a wide reputation for courage and daring. He also won the attention of Jefferson Davis, then a colonel of Mississippi volunteers, who recommended him for brevet promotion to captain. Later, as secretary of war, Davis secured Van Dorn a choice combat assignment fighting Indians on the Texas frontier.

Van Dorn served five years in Texas, taking a near-fatal arrow wound while chasing a band of Comanches alone. Unlike Sterling Price, Van Dorn saw no moral dilemma in secession; shortly after the election of

Lincoln, he resigned his commission to serve his native Mississippi. Experienced officers were few in the Deep South, and Van Dorn was named senior brigadier general of the state troops, second only to Maj. Gen. Jefferson Davis. Davis welcomed him into his plantation home, and the two worked closely to forge a state army. When Davis went east as president, Van Dorn succeeded him in command of the Mississippi forces. He was inducted into the Confederate service in June 1861 and transferred to Virginia that fall. Van Dorn was promoted to major general in October 1861, three months before being assigned as commander of the newly created Trans-Mississippi District of Department Number Two.[7]

Van Dorn's dashing manners and courage led most people to overlook his faults, which were egregious. Bold beyond prudence, he had no patience with reconnaissance, staff work, logistics, or anything else that might keep him from closing quickly with the enemy. A "rash young commander with a one-dimensional mind" is how his biographer described him. Disdainful of death, he showed little concern for the lives of others.

Insatiable ambition propelled Van Dorn. "He craved glory beyond everything," said Dabney Maury. The possibilities the war offered thrilled Van Dorn, and he wrote with immodest anticipation, "Who knows but that out of the storm of revolution . . . I may not be able to catch a spark of the lightning and shine through all time to come, a burning name! I feel a greatness in my soul. I am getting young once more at the thought that my soul shall be awakened again as it was in Mexico."[8]

Van Dorn would have a hard time getting anyone east of the Mississippi River to notice his exploits. The huge area of the Confederacy west of the Appalachian Mountains played hard on President Davis and the War Department in Richmond; its vastness and remoteness precluded centralized control from the capital. Then, too, there was Davis's understandable preoccupation with the enormous Federal army of George B. McClellan that was marshaling on his doorstep. Finally, his choice of Albert Sidney Johnston, in whom he had unbounded faith, to exercise supreme command in the West led him to pay even less attention to the theater than he otherwise might have.

The reality of affairs west of the Appalachians was quite different from what Davis imagined. Although the new Department Number Two included Tennessee, Kentucky, Arkansas, Missouri, Kansas, western Mississippi, and even the far-flung Indian Territory, Johnston had no time to spare on matters beyond the Mississippi River. A lack of navigable rivers beyond Mississippi and an absence of railroads meant that any Union

Maj. Gen. Earl Van Dorn (Alabama Department of Archives and History)

offensive in the trans-Mississippi would of necessity be overland and thus comparatively slow. East of the Mississippi, however, the situation was decidedly different. Leaving the Ohio River near Paducah, Kentucky, were the Cumberland and Tennessee Rivers, aquatic daggers into the Confederate heartland. The former flowed southeastward to Nashville; the latter, south to Florence, Alabama. Then there was the Mississippi itself, defensible only from Fort Pillow, fifty miles above Memphis, northward; south of the fort, low banks and swampy bottomland prevented the construction of strong points.

Good railroads also led from Federal depots into the Deep South. Among them were the Mobile and Ohio, which ran from Cairo, Illinois, through Corinth, Mississippi, and on to the Gulf Coast, and the Louisville and Nashville line, which split at the Tennessee capital, continuing on in two lines to Decatur, Alabama, and Chattanooga, Tennessee.

Cavalry and partisans could dismantle railroads, but only a strong forward defense could protect the river routes of invasion. To these Johnston gave his nearly undivided attention. In the waning months of 1861 he shuttled between his forces under Leonidas Polk at Columbus and William Hardee at Bowling Green, Kentucky, trying to make them strong enough to prevent the Federals from starting south at all. He ordered work pushed forward on forts to guard the rivers along the line of the Kentucky-Tennessee border: Island Number Ten on the Mississippi, Fort Henry on the Tennessee, and Fort Donelson on the Cumberland. When in January 1862 Van Dorn visited him at Bowling Green on his way to assume command in Arkansas, Johnston had on hand 43,000 men to defend a 400-mile front from the Mississippi River to the Cumberland Gap. Opposing him were 90,000 Federals, whose numbers were growing rapidly.

The diminutive Mississippian must have struck Johnston as slightly unbalanced. Far from contenting himself with the Federals in southern Missouri, Van Dorn had concocted a scheme for taking the war to Illinois. To Johnston he explained how he would flank the Yankees out of St. Louis and cut the supply lines of Brig. Gen. Ulysses S. Grant at Cairo, thereby easing pressure on Johnston and allowing him to carry the war across the Ohio River with Van Dorn.[9]

What, if any, encouragement the harried Johnston may have given Van Dorn is unknown, but conditions in Van Dorn's new department in any case rendered his plan moot. Nothing approaching the number of troops needed to launch a major offensive was on hand. While Van Dorn thought Price had gathered 10,000 eager recruits to fill out his state militia, the Missourian actually counted fewer than 7,000 tired and dispirited soldiers in his camp. Van Dorn had expected to find 8,000 Indians under Albert Pike, but Pike commanded only 2,000 warriors of dubious reliability. Admitting another 5,000 troops under McCulloch, Van Dorn's army of invasion, which he had estimated would be 45,000 strong, numbered fewer than 15,000 of all arms. The Federals were closer and stronger than Van Dorn had imagined. Thirty miles away, in the thinly populated, mountainous country of northwestern Arkansas, were 12,000 troops under Maj. Gen. Samuel Curtis.

Van Dorn responded to the situation with characteristic aggressiveness. Wisely deciding to leave Illinois in peace, he instead set about fashioning plans to attack Curtis, who had dispersed his army after food and forage. On March 4, under a wet and blowing snow, the combined armies of Price and McCulloch set out to meet the Yankees. Three days later the two sides clashed in the shadow of an Ozark range called Pea Ridge.

Van Dorn had intended to open the action with a flank march into the Union rear, followed by an attack on the Federal lines from behind. But Van Dorn neglected to scout the route of march beforehand, and the vanguard under Price had to stop to clear felled trees from the road, which alerted the Federals to their approach. Impatient of trailing Price, McCulloch attacked alone. Instead of striking Curtis an overpowering blow from behind, Van Dorn's small army delivered two weak jabs.

The results were predictable. Price had some success on his front but was unable to break the Federal lines. McCulloch was killed and his wing shattered. Nightfall found the Rebel army "staggering with fatigue and half-dead with cold and hunger." Artillery and cavalry horses were "beaten-out," and ammunition was dangerously low. Displaying a quixotic detachment, Van Dorn chose to stay and fight.

The Federals attacked at daybreak on March 8 under the cover of a barrage that blew terrible holes in the Rebel lines. With his right arm in a sling from an infected flesh wound, Price rode among his Missourians and urged them to stand firm, risking his life in what Van Dorn had conceded to be a losing proposition. Inevitable though it was, the order to retreat stunned and then angered Price's Missourians. They knew nothing of McCulloch's debacle and were holding their own against the enemy. Van Dorn himself did little to win back their confidence. From euphoria he fell into hopelessness, so shaken that he told Brig. Gen. Martin Green to destroy the army's wagon train — an order that Green ignored.

The Confederate retreat lasted a week. As they marched, the men in ranks took the measure of their commanders. Most agreed that Price had done well. While he had not overwhelmed the enemy on the first day, he had pushed them back. More important from the soldiers' perspective was his personal courage in battle and his concern for their well-being during the retreat. Stories of his kindly attentions made the rounds, and Price's standing grew both with his own Missourians and with McCulloch's veterans. By contrast, Van Dorn's stock plummeted. The men muttered threats of mutiny and ridiculed Van Dorn openly. Among the

Missourians the taunts were especially barbed. The night after the battle, so one story went, some of the Missourians found themselves encamped on a damp hillside covered with flat, white rocks that seemed a good surface on which to cook cornpone. But the rocks, when heated, burst with a loud report. The cornpone scattered, but no real damage was done. The soldiers likened the effect—an impotent bang—to Van Dorn's braggadocio, and they named the rocks "Van Dorn skillets."

Van Dorn halted at Van Buren, on the Arkansas River. There he penned Albert Sidney Johnston a dissembling explanation of the campaign: "I was not defeated, but only failed in my intentions." Van Dorn concluded with a promise of coming glory and spoke again of marching on St. Louis.[10]

Johnston had no time for such schemes. Since Van Dorn's bizarre interview with him in January, matters east of the Mississippi River had taken a decided turn for the worst. Although the Federals had no plan for concerted action against Johnston, the commander of the Department of Missouri, Maj. Gen. Henry Halleck, had grudgingly authorized Grant to attack Fort Henry. The fort fell on February 6. A week later, after two days of hard fighting, Grant took Fort Donelson as well. Grant wanted to keep on, but Halleck checked him, preferring that Grant wait until Maj. Gen. Don Carlos Buell, commander of the Department of the Ohio, joined forces with him.

With the Tennessee and Cumberland Rivers open to Federal gunboats and troop transports, the entire Confederate position in western and central Tennessee was untenable, and Johnston began an evacuation. He ordered Polk to abandon Columbus while he fell back from Nashville with Hardee's command. Few sympathized with Johnston's predicament, and he was mocked both in the press and within the army as a defeatist.

Control of the Army of Tennessee slipped from Johnston's grasp. Johnston yielded to the demand of Gen. Pierre G. T. Beauregard, commander of the district comprising western Tennessee, that the two generals unite to protect the Mississippi River Valley, instead of Middle Tennessee as Johnston had intended. In compliance, Johnston transferred his army to the northern Mississippi railroad junction town of Corinth, where Beauregard had moved his command.

At Corinth Beauregard officially became Johnston's second in command. In fact, he dominated the relationship. Beauregard used fear to subvert Johnston's authority over the Army of Tennessee, bombarding both him and Richmond with dire warnings of an invasion of West Tennessee. Richmond responded as Beauregard calculated, sending large reinforcements directly to his district, rather than to Johnston as department commander. Four regiments came to Beauregard from New Orleans. Recruits intended for him were marshaled in Alabama and Mississippi. Braxton Bragg reported to him with 10,000 troops gathered from the garrisons of Mobile and Pensacola.

Only Earl Van Dorn's small army eluded Beauregard's grasp, until Johnston conceded even that to him. On March 23 Johnston directed Van Dorn to report with his command to Corinth, there to join him and Beauregard in an offensive against Grant at Pittsburg Landing (where Grant's army lay encamped) designed to regain the initiative and retake Tennessee.[11]

Anxious to redeem himself, Van Dorn answered Johnston's summons at once. On March 29 he left to confer with Beauregard at Memphis. Before going, he told Price to march his Missourians to Des Arc, a village on the White River. There they would board steamers for the eighty-mile trip to Memphis.

Delays occurred. Price languished a week at Des Arcs while transportation was arranged. There, on April 8, he resigned from the Missouri State Guard to accept a major general's commission in the Confederate service. The Missourians were organized into a two-brigade division under his command. Word of their impending transfer east of the Mississippi River sparked talk of mutiny among the men—they had enlisted, they declared, to fight for Missouri. Price had misgivings equally strong, but he exerted his influence to reconcile the soldiers to their new duties. The day he accepted his Confederate commission, Price asked the Missourians to "go with us wherever the path of duty may lead, till we shall have conquered a peace and won our independence by brilliant deeds upon new fields of battle. Now is the time to end this unhappy war. Soldiers! I go but to mark a pathway to our homes. Follow me!"[12]

They did, almost to a man. Few would see their homes again.

2

Fifth Wheel to a Coach

Brig. Gen. Ulysses S. Grant fought three enemies in the spring of 1862, only one of which wore gray. Any of the three might prove his undoing. The obvious foe was Albert Sidney Johnston at Corinth. Only somewhat less dangerous was Grant's immediate superior, Maj. Gen. Henry Halleck. Jealous of what more Grant might accomplish after Fort Donelson, Halleck stood ready to exploit Grant's third enemy—his reputation among officers of the Regular Army as an unreliable drunkard.

Halleck had plenty of company. Grant had come a long way since the days when he shuffled about his father's leather goods store in Galena, Illinois, a disgraced ex-army officer bereft of a future. Officers who had known Grant in the Regular Army resented his rapid elevation to high command, and they watched for the slightest slip from sobriety.

Fort Donelson was the first significant Union victory of the war, and it elevated Grant from an obscure western general into a national hero. Newspapers called him "Unconditional Surrender" Grant and told their readers what brand of cigars he smoked. The Senate applauded him. Secretary of War Edwin Stanton enjoyed the victory immensely, telling Halleck that "the brilliant results of the energetic action in the West fills the Nation with joy."

The nation, perhaps, but not Halleck. In his mind, Grant's gain was his loss. Halleck did not even feign pleasure at the triumph. Instead he denied its importance and tried to discredit its author. Just days after the fall of

Fort Donelson, as a demoralized Army of Tennessee abandoned Nashville, Halleck wired Washington of a "crisis in the war in the west" and warned that "we are certainly in peril."[1]

Of course only Halleck's future was in peril. Union military command in the sprawling western theater was exercised through a series of departments. The vast country between the Missouri River and the Cumberland Mountains was divided into two departments: the Department of Missouri, commanded by Halleck from St. Louis, and the Department of the Ohio, commanded by Don Carlos Buell, whose headquarters initially were in Louisville. It was an imprecise division. Neither Halleck nor Buell knew where his departmental authority ended. Both reported directly to Washington and, as coequals, were jealous of their perquisites. Neither could plan a major offensive without the consent and cooperation of the other. In such a situation the gain of one was the other's loss, as was any freelance victory by a subordinate, such as Grant.

So when downplaying their importance failed, Halleck tried to claim credit for Grant's victories and use them as proof of his own fitness for overall command in the West, promising General in Chief George B. McClellan he would "split secession in twain in one month" if he were placed in command.

When McClellan refused him, Halleck tried another tack. At a time when Washington improvised strategy, he correctly adjudged the importance of conquering the Mississippi River in order to, as he had put it, split secession in twain. Having cleared the northern flanks of the Mississippi, that is to say Missouri and Kentucky, Halleck was anxious to press on farther down the river itself. To this end he instructed Maj. Gen. John Pope to reduce Island Number Ten and the nearby garrison at New Madrid, Missouri, in mid-February. Halleck also wanted to push the Federal advantage in Tennessee with a broad-front sweep southward. For this he needed Buell, or at least his army. Halleck tried to induce Buell to come into his department, which would place him under Halleck's orders, but Buell demurred, instead marching south to take Nashville on his own.[2] Already angry over Buell's act, Halleck was incensed to learn that Grant had gone to Nashville to confer with his rival. To McClellan Halleck complained that Grant had "left his command without my authority and went to Nashville. . . . It is hard to censure a successful general immediately after a victory, but I think he richly deserves it. I can get no returns, no reports, no information of any kind from him. . . . I am worn out and tired with this neglect and inefficiency." McClellan agreed that Grant should be reigned in, and he told Halleck to "arrest him at once if the

Maj. Gen. Ulysses S. Grant (Library of Congress)

good of the service requires it" and replace him with his second in command, Charles F. Smith.

But Halleck was not content with disciplining Grant. He wanted Grant's ruin, and to accomplish it, he resorted to innuendo. In early March Halleck wired McClellan: "A rumor has just reached me that since the taking of Fort Donelson General Grant has resumed his former bad habits. If so, it will account for his neglect of my often-repeated orders. I do not deem it advisable to arrest him at present, but have placed General Smith in command of the expedition up the Tennessee." Halleck ordered Grant to remain at Fort Henry and turn over command to Smith. Despondent but with his wit intact, Grant told Halleck he would have a hard time staying at Fort Henry, since "the water is about six feet deep inside the fort."

Contrary to Halleck's expectations, Grant's allies did not desert him. Congressman Elihu Washburne demanded that Secretary Stanton restore Grant to command. Nine of Grant's generals signed and placed at his disposal a letter of support "for such use as you may think proper to make of it." The efforts of Washburne and Grant's generals succeeded, and Lincoln told Stanton to have him reinstated.

In the end, both Halleck and Grant came out of the affair with enhanced responsibility. Grant was promoted to major general, and Halleck got the western command. Secure now in his authority, Halleck grew more tolerant. Also, he had learned that a pro-Confederate telegraph operator at Cairo had prevented many of his dispatches from reaching Grant. After Smith was incapacitated from an injury, Halleck restored Grant to command of the Army of the Tennessee, enjoining him to "lead it on to new victories."[3]

But Grant embarrassed himself badly in command. On April 5, while his army rested in camp near Pittsburg Landing, waiting for Buell to form a junction with him, Grant reported to Halleck that he had "scarcely the faintest idea of an attack being made upon us." The next morning Johnston's Confederates pushed his unprepared army to the banks of the Tennessee River. Grant, who himself arrived late on the field, fed reinforcements into action on April 7 and drove the Confederates off, but the thin luster had been wiped from his new reputation. People spoke again of Grant the drunkard, and Halleck hastened to Pittsburg Landing.[4]

Halleck had intended to take command of the united armies of Grant and Buell and lead them against Johnston at Corinth. The Battle of Shiloh had upset his timetable but not his resolve to seize Corinth as the first step in his sweep south through the Mississippi River Valley. The

Confederate withdrawal to Corinth added the prospect of destroying an army in the bargain.

For an advance on Corinth Halleck assembled at Pittsburg Landing the largest force ever seen on the continent. Three armies—more than 125,000 men—made camp near the putrid battlefield of Shiloh: the Army of the Ohio under Maj. Gen. Don Carlos Buell, Grant's Army of West Tennessee, and the Army of the Mississippi under John Pope, whom Halleck had summoned after the fall of Island Number Ten. Halleck organized the armies into a left wing, center, right wing, and reserve. This time Halleck showed greater finesse in dealing with Grant. Rather than remove him outright, Halleck elevated Grant to the meaningless position of second in command of the army, with direct responsibility for the right wing and reserve.

Early in the siege of Corinth Grant learned that his authority over even the right wing and reserve was illusory. "I was little more than an observer," he recalled. "Orders were sent direct to the right wing or reserve, ignoring me, and advances were made from one line of entrenchments to another without notifying me."

Among Grant's friends, Pope in particular felt his humiliation. Together he and Grant had organized Illinois volunteers in Springfield, and as a brigadier general in Missouri, Pope had been among the first to speak out on behalf of Grant, at a time when most were still trading stories of his Regular Army drinking. With nothing to do, Grant came often to Pope's camp outside Corinth. "A more unhappy man I have seldom seen," said Pope. Watching Grant pass the hours lying on a cot in his tent, silent and brooding, saddened Pope. "I never felt more sorry for anyone," he averred. Grant uttered not a word against Halleck, said Pope; when he spoke at all, it was to talk of resigning.[5]

The days wore heavily under the sultry Mississippi sun, and Grant found it hard to maintain a dignified silence. He told a colonel from Galena that he felt as useless as "the fifth wheel to a coach." Twice he asked Halleck to relieve him. Once in May Grant suggested a movement that might accelerate the pace of Halleck's painfully slow advance. "I was silenced so quickly that I felt that possibly I had suggested an unmilitary movement." Grant withdrew to his tent, banished, as his biographer William McFeely expressed it, "to an island in the midst of his men."[6]

*   *   *

Why Corinth? What made this small Mississippi town so vitally important to two huge, contending armies? Railroads. Corinth sat squarely at the junction of the Memphis and Charleston Railroad and the Mobile and Ohio Railroad. Good railroads were scarce in the Confederacy, and these were two of the best and most strategically critical. The Mobile and Ohio Railroad penetrated the Deep South all the way to the Gulf of Mexico. The Memphis and Charleston Railroad, which former Confederate secretary of war Leroy Walker called "the vertebrae of the Confederacy," was the only direct line from the Atlantic seaboard to the Mississippi River.

The loss of Corinth would play havoc with Confederate lateral communications, cutting off Southern forces operating in the upper reaches of the trans-Mississippi region, particularly Arkansas, from the Confederate heartland. A Federal army astride the Memphis and Charleston Railroad at Corinth would compromise the security of Chattanooga and render Southern control of the track west of the East Tennessee bastion meaningless. Corinth itself was the only practical base from which to support Confederate operations aimed at regaining western Tennessee.[7]

Oddly, Secretary Walker seemed to have been one of the few Confederate leaders to have appreciated the strategic value of Corinth. Absorbed in the battles before Richmond, the president pursued what might most charitably be called a laissez-faire policy toward the West. Davis had given Johnston broad authority because he admired him. Beauregard, who assumed command after Johnston's death at Shiloh, he distrusted. Yet during Beauregard's first six weeks in command neither the president nor the War Department gave him any "instructions . . . relative to the policy of the government and the movement of the armies of the Confederacy." Gen. Robert E. Lee, formally in charge of overall Confederate military operations, assured Beauregard that Richmond was concerned about his army, explaining its apparent neglect by saying that "full reliance was felt in your judgment and skill . . . to main the great interests of the country." Revealing Richmond's ignorance of the true state of affairs in the West, Lee added hopefully that Beauregard would follow up his "victory" at Shiloh with an advance northward.[8]

In contrast to the distracted leaders of the Confederacy, to whom Corinth seemed less important than it should have, General Halleck exaggerated its significance. The primary purpose of his offensive was not the destruction of Beauregard's army but the capture of Corinth. Enamored of eighteenth-century theories of "strategic points" as the keys to victory and of Baron Henri Jomini's emphasis on movement over annihilation, Halleck hesitated to give battle. He believed "Richmond and

Corinth are now the great strategic points of the war, and our success at these points should be insured at all hazards."[9]

There was little remarkable about Corinth besides the railroads that passed through it. The town lay in Tishomingo County, an area that had been home to the Chickasaw Indians. The story of white encroachment was a familiar one. Settlers drifted down the Natchez Trace to hack small farms from the forest, compelling the Indians to sign the Pontotoc Treaty, in which they ceded their remaining land east of the Mississippi River to the State of Mississippi in 1832. The vast tract was carved into ten counties, of which Tishomingo County was the largest.

Settlers swarmed into the tract, a region of sharp contrasts and sinister beauty. Swampy bottomlands choked with scrub oak, cypress, beech gum, and chestnut trees alternated with gentle hills of wildflowers, persimmon trees, and muscadine vines in an oddly scenic patchwork. The soil was rich, and crops were excellent. Corn, peas, potatoes, and wheat, with a little cotton for homespun clothing, all grew well. Surplus produce was shipped from Eastport on the Tennessee River, bringing in thousands of dollars annually. By 1851 Tishomingo County was one of the more prosperous sections of the South.[10]

What Tishomingo County lacked was potable water. Streams and creeks were sluggish, and swamps checkered the land. On the outskirts of Corinth was a two-mile-long bog, delightfully named Dismal Swamp. During the summer the brackish water of Tishomingo County bred mosquitoes, making malaria a real hazard. In times of drought, wells often had to be bored 300 feet deep to reach potable water. Besides the mosquitoes they bred, the streams and creeks were a nuisance to travelers. Their soft beds and steep clay banks impeded fording. Annoying to settlers, the swampy bottomland and lack of good water was to prove "of controlling importance in moving and handling troops" when war came to the region. As a Federal general remarked, "Men and animals need hard ground to move on and must have drinking water." But, said Confederate Colonel Lawrence Ross, Corinth was a "sickly, malarial spot fit only for alligators and snakes." The inhabitants, added a Minnesota lieutenant, were as degenerate as their natural surroundings. The men were ignorant and the women were "she-vipers," with figures like "shad-bellied bean poles. The principal products are wood ticks, chiggers, fleas, and niggers."[11]

The town that Colonel Ross and the Northern lieutenant traduced was only a few years removed from nature, born of the avarice of the Tishomingo County board, which had invited the Memphis and

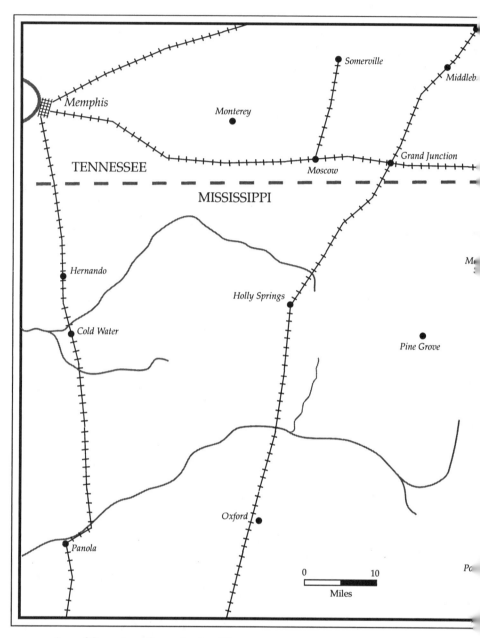

MAP 1. *Area of Operations, September 1, 1862*

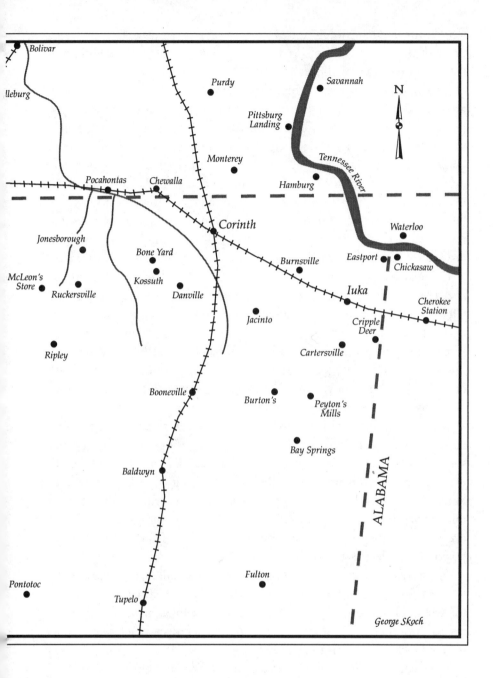

Bolivar

Purdy

Savannah

N

Pittsburg
Landing

Tennessee River

Monterey

Pocahontas   Chewalla

Hamburg

leburg

Corinth

Waterloo

Jonesborough

Bone Yard

Burnsville

Eastport   Chickasaw

McLeon's
Store

Kossuth

Iuka

Cherokee
Station

Ruckersville

Danville

Jacinto

Cripple
Deer

Ripley

Cartersville

Booneville

Burton's   Peyton's
Mills

Bay Springs

ALABAMA

Baldwyn

Fulton

Pontotoc

Tupelo

George Skoch

Charleston and the Mobile and Ohio railroad companies to lay their track through the county, and of local speculators who had raised buildings at the junction of the two lines.

The railroads nourished the town's growth. By 1861, only five years after the first building went up, Corinth boasted a population of 2,800 people, many fine homes, and an institution of higher learning called Corona Female College. Boosters lauded the college, which stood on a knoll southwest of the railway junction, as "a magnificent building surmounted by a lofty dome." Doric columns added a neoclassical flair to the brick structure, which cost a lofty $40,000 to erect. However it may have looked when built, the building aged badly, moving a soldier who visited it in 1862 to write, "It is three stories high — the design is somewhat tasteful but the workmanship poor. . . . It has a dilapidated appearance."

Corinth had a hotel that aspired to equal grandeur. Tucked in the southeastern angle of the two railroad lines and facing the Memphis and Charleston tracks was the Tishomingo Hotel, a long, two-story brick building. Despite its pretensions, the hotel wore a seedy aspect, and the second-story wooden porch looked as if it had been rattled loose by passing trains. The streets, too, were distressing. They were simply dirt lanes of varying width — pasty annoyances during rains, and the source of a miserable, fine clay dust on dry summer days.[12]

It was the Corinth of bad water and dusty streets that the Confederate Army of the Mississippi came to know in May 1862. It had been a dry spring, and potable water was unusually scarce. Thirst drove soldiers to drink from contaminated pools. Rank odors rising from Dismal Swamp hung about town. Earl Van Dorn arrived with the Army of the West in mid-April, boasting that he felt like a wolf ready for prey. But almost everyone else was ill from the water or heartsick from defeat. Of the 80,000 Confederates crowded into Corinth, nearly 18,000 were sick.[13]

Price was among the ill. Cholera had nearly killed him during the Mexican War, leaving him susceptible to severe intestinal disorders for the rest of his life. The rigors of the Pea Ridge campaign and the shock of his arm wound weakened Price badly and triggered an attack of diarrhea, so that he came to Corinth already sick.

Price had little to say about matters at Corinth, and his only public remarks on the subject were in poor taste. Shortly after Price arrived, General Beauregard took him on an inspection of the fortifications outside town. The Creole was proud of the entrenchments, which Braxton Bragg and his engineers had laid out. They began a mile and a half east of

Corinth, running first to the north and then to the west, a seven-mile-long semicircle that terminated on the Memphis and Charleston Railroad northwest of town. Bragg's engineers had done well, constructing the works on ridges whenever possible and anchoring them in forests or creek bottoms. Heavy siege guns, protected by earthwork epaulements, were mounted at vulnerable points. But Price was unimpressed. "Well, these things may be very fine; I never saw anything of the kind but once, and then I took them," he told Beauregard and Bragg. Price's impolitic refer-ence to his triumph at Lexington sat poorly with his hosts, who came to question his competence as well as his tact (Bragg eventually dismissed Price as singularly unsuited for command), but it boosted his reputation among Southern journalists. Weary of the stalemate at Corinth, the press lauded Price as a rough-and-ready general who scorned "pick and shovel warfare." [14]

That stalemate was, in any case, about to end. On May 25 Beauregard called together his chief lieutenants — Van Dorn and Price included — to discuss the propriety of evacuating Corinth. There really was little to debate; all present knew that it was only a matter of time before Halleck's anaconda tactics would crush their badly outnumbered forces. Lt. Gen. William J. Hardee urged that Corinth be abandoned at once and the army withdrawn along the Mobile and Ohio Railroad. Beauregard agreed. Four days later the Confederates slipped away. With them went all but a handful of the townspeople.

Halleck made a halfhearted pursuit with a fragment of his force, and by June 9 the Confederates held a new line at Tupelo, fifty-two miles south of Corinth. Broken in health, Beauregard went off to Mobile on June 16 to recuperate. He neither asked for nor received permission to take leave but simply transferred command to Bragg and left. President Davis had long wanted to shelve the Creole, who opposed him politically, and he used Beauregard's unauthorized leave of absence as a pretext to replace him with Bragg. Assuming command "with unfeigned reluctance," Bragg set about to reorganize and restore discipline to the army. [15]

Sterling Price also left Tupelo, although unlike Beauregard he had the good sense to secure a leave of absence. Accompanied by his chief of staff, Thomas Snead, Price traveled to Richmond to speak with President Davis. His purpose in seeking an audience with the president was three-fold. First, he wanted command of the Trans-Mississippi District. Arkansas and Missouri congressional delegations had been lobbying since Pea Ridge for Van Dorn's removal and Price's appointment. Governor

Thomas Moore of Louisiana had added his influence on the Missourian's behalf. For his part, Van Dorn wanted nothing more to do with the region and heartily endorsed Price's application. In a private letter to Davis, he wrote that "the love of the people of Missouri was so strong for General Price, and his prestige as a commander so great there, wisdom would seem to dictate that he be put at the head of affairs in the West. I drop whatever glory there may be in it on the brow of General Price, than whom there is no one more worthy to wear it and than by whom I would rather see it worn."

Of course Van Dorn was being disingenuous. Although an independent command anywhere suited his taste, as a Mississippian he preferred one close to home. So, when after the evacuation of Corinth Van Dorn was offered the command of Department Number One, which embraced Vicksburg, he resigned command of the Army of the West and eagerly set out for the river citadel.[16]

Price's second reason for going east was to get his Missouri brigades ordered back to Arkansas. And—his third reason for seeing Davis—Price believed the time was ripe for a new expedition into Missouri, an operation he was anxious to lead.

Price's journey to Richmond was a triumphal procession. His victory at Lexington and courage at Pea Ridge had made him one of the more popular generals in the South. Crowds gathered at train stations to pay their respects, and Price's arrival in the capital on June 16 was greeted with wild celebration. Exclaimed the *Daily Richmond Whig*, "The Washington of the West is now in Richmond—All hail!" The Virginia general assembly hosted a formal reception for Price, and many of the town's leading citizens opened the doors of their homes to him.

Not so the president. Davis was little disposed to hear the demands of a general whom he later described as "the vainest man he had ever met." Nor was Davis any more willing now than he had been in 1861 to assign a non–West Pointer to a major command. Too, Price's popularity was a threat. The string of Confederate defeats in early 1862 and the presence of the Union Army of the Potomac at the gates of Richmond had emboldened Davis's detractors. One cabal of disgruntled politicians from Missouri had begun speaking of deposing the president and installing Price as "generalissimo." Price probably knew nothing of this talk of impeaching Davis, which never went beyond parlor banter, but his appearance in Richmond coincided with it.[17]

Price's first interview was brief but cordial. Davis heard out Price's request for reinforcements from Beauregard and for command of the Trans-Mississippi District but was noncommittal. He told Price to submit his proposal to the secretary of war in writing and promised a second meeting.

Price dutifully propounded his proposal regarding "the proper conduct of the war west of the Mississippi" in a letter to Secretary of War George Randolph. He advocated turning the Trans-Mississippi District into a separate department, "under the command of an officer enjoying enough of the confidence of the Government to be left untrammeled by specific instructions," and nominated himself for the job. Price proposed an immediate advance from Arkansas into Missouri, which he said would divert Federal troops from Beauregard's front.

Randolph forwarded Price's letter to the president. Davis sent for Price a few days later. Both were in bad temper. After submitting his letter to Randolph, Price had learned that Maj. Gen. John Magruder, an eccentric West Point-trained Virginian, had been appointed to command the Trans-Mississippi District. Davis, in turn, was put off by the imperious tone of Price's letter. A clash of wills was inevitable.

After some perfunctory remarks, Davis told Price that he had decided not to allow him or his Missourians to return to the trans-Mississippi. They were needed where they were, the president explained. As for command of the district, he continued, that belonged to Magruder.

Price fumbled for a face-saving exit. "Well, Mr. President," he stammered. "Well, Mr. President, if you will not let me serve *you*, I will nevertheless serve my *country*. You cannot prevent me from doing that. I will send you my resignation, and go back to Missouri and raise another army there without your assistance, and fight again under the flag of Missouri, and win new victories for the South in spite of the Government."

Thomas Snead watched the president closely for his reaction. He knew Davis's reputation for eviscerating those who dared to defy him. "His eye flashed with anger as he glanced at the general's flushed face, and his tone was contemptuous," remembered Snead. Davis chose his words carefully and uttered them slowly, for greater effect: "Your resignation will be promptly accepted, General, and if you go back to Missouri and raise another army, and win victories for the South, or do it any service at all, no one will be more *pleased* than myself, or," he added after an emphatic pause, "more *surprised*."

Price slammed his fist on the table. "Then I will surprise you, sir," he shouted.

Price and Snead returned to their lodgings at the Spottswood Hotel. Price stormed inside to write out his resignation; Snead remained outdoors to harangue passersby. The president was a scoundrel and an ingrate. Ripping the insignia of Confederate rank from his frock coat, Snead told the crowd that he and Price would go to Missouri and fight under the "Bear Flag." [18]

Their threat was never tested. The next morning President Davis returned Price's resignation. As much as he wanted to see Price go, Davis knew that Price's Missourians would follow him out of the army in a mass desertion that might tear the fabric of the western armies. And his resignation might give impetus to a rumored effort by Governor Hiram Rector of Arkansas to establish a separate Confederacy beyond the Mississippi River. Davis included a conciliatory (for him) note with his disapproval of Price's resignation. He promised to instruct General Bragg to send Price and his Missourians back to the trans-Mississippi "as soon as it could safely be done." However, Magruder would remain in command of the district. Mollified a bit, Price consented to return to duty in Mississippi.

Notwithstanding his popular reception, Price's trip to Richmond had proved a fiasco. He had earned the open enmity of a president notorious for bearing grudges and had come away with an executive pledge that events were to render as empty as the spirit in which it had been given. [19]

*   *   *

William Starke Rosecrans was a man of rare ability and high principles who had an unfortunate knack of irritating his superiors. His was a world of absolutes; to Rosecrans, mendacity and mediocrity were intolerable, particularly in those who pretended to leadership.

Strength of character ran in the Rosecrans family. William's father was a stern man of unflinching integrity. Neighbors called on him to arbitrate local disputes and spoke of his "iron will and hot temper," traits William inherited. His mother imparted to William a gentle streak that, while not preventing his outbursts of temper, at last made him regret them. She also passed on to her son a deep and genuine religiosity, a passion for the truth, and a love of learning. What William lacked was a sense of humor. Too often he came across as ridiculous or insulting when he was merely being earnest.

At eighteen Rosecrans was admitted to West Point, a member of what became known as "the brilliant class of 1842." Among his classmates was Earl Van Dorn; Grant entered the academy a year later. Rosecrans

Maj. Gen. William Starke Rosecrans (U.S. Army Military History Institute)

graduated fifth in his class and won a commission in the Engineer Corps, but his army service was undistinguished. In 1853 he became seriously ill and, said his wife, "to secure rest and make a choice of a civil career more likely to support his family," Rosecrans resigned his commission.

"The next seven years," Ohio journalist Whitelaw Reid wrote, "were to Lieutenant Rosecrans years of more varied than profitable activity." He drifted into the coal oil business and with two partners built a refinery in Cincinnati. While Rosecrans was testing an experimental oil one evening, a safety lamp exploded in his face. He was badly burned and remained bedridden for eighteen months. The scars that lingered distorted his features, leaving a slight but perceptible smirk to his face.

As soon as he recovered, Rosecrans returned to the refinery. Business was poor, but he succeeded in his experiments, becoming, as he later boasted, "the first to obtain a good, odorless oil from petroleum." He also developed a new process of manufacturing soap, invented the first kerosene lamp to burn a round wick, and designed a short chimney lamp. Rosecrans formed a company to manufacture them. By April 1861 his businesses had begun to turn a profit.[20]

War fundamentally changed Rosecrans's fortunes. The soldier turned engineer found himself again in uniform, as a brigadier general of Ohio volunteers on the staff of Maj. Gen. George B. McClellan, then commander of the Department of the Ohio. In May 1861 Rosecrans accompanied McClellan into western Virginia as a brigade commander, playing an instrumental part in a string of victories that secured the region for the Union and elevated McClellan to command of the Army of the Potomac. Rosecrans stayed in western Virginia. He took charge of the Department of the Ohio and cursed the duplicity of "that damned little cuss, McClellan," who had taken credit for the climactic victory at Rich Mountain that Rosecrans had won.

Having deprived Rosecrans of due recognition, McClellan went on to rob him of most of his troops. Then President Lincoln, looking for a back shelf on which to put Maj. Gen. John C. Frémont, converted Rosecrans's department into the Mountain Department and placed Frémont in command. Rosecrans became second fiddle to a military misfit. Bitter and bored, in April 1862 he journeyed to Washington in search of a new assignment.

Before the end of May he was in Corinth, under orders to report to Halleck. Halleck sent him on to John Pope's Army of the Mississippi. Rosecrans and Pope had been classmates at West Point, and Pope was pleased to receive him. He gave Rosecrans command of his right wing, which consisted of two divisions.[21]

It was quite a fall for Rosecrans, to be reduced from department command, with direct access to the president, to leadership of two divisions in a small army, but at least he had a purpose and a place in a chain of command. General Grant, on the other hand, was still suffering the embarrassment of his nebulous role at headquarters. And he was ill. A cold he caught three months earlier had lingered and settled in his chest. Severe, recurring headaches plagued him. On a night reconnaissance after Shiloh, Grant's horse slipped on a loose boulder and fell, taking the general down with it. Grant's leg swelled badly and went numb, and he paced before his tent outside Corinth with a noticeable limp.

Repeatedly during the siege of Corinth, Grant had asked to be relieved of duty under Halleck; each request was returned disapproved. Grant renewed his demand after the town fell. His desire to leave, when Halleck was at last disposed to let him go, was irrational. As his biographer William McFeely observed, Grant "was about to make the blunder of requesting a new assignment, a move which almost surely would have been interpreted as petulance and very likely would have sent him into professional oblivion."

At that moment a friend prevailed on him to reconsider. William T. Sherman had dropped by Halleck's headquarters one afternoon in early June to pay his respects and exchange gossip. Halleck mentioned offhandedly that Grant was leaving the next morning. Halleck assured Sherman that he did not know why Grant was going but said that the Ohioan had applied for and been granted thirty days' leave. "Of course we all knew that he was chafing under the slights of his anomalous position, and I determined to see him on my way back," remembered Sherman.

Sherman found Grant at his camp—four or five tents partially hidden in a forest off the main road—seated on a camp stool, aimlessly sorting and bundling letters to pass the time until he left. Sherman begged him to stay. His own changed circumstances proved how quickly one's fortunes might turn, Sherman contended. Before Shiloh the press had disparaged him as crazy. But "that battle had given me new life, and now I was in high feather; and I argued with him that, if he went away, events would go right along, and he would be left out; whereas, if he remained, some happy accident might restore him to favor and his true place." Grant promised to stay with the army awhile longer. But he could not bear another day in Corinth and so removed his headquarters to Memphis.[22]

Sherman had rescued Grant from almost certain oblivion and, by his compassion, had cemented his friendship with Grant, who later would stand by Sherman when other commanders might have dismissed him. And Sherman's prediction came true. Grant's fortunes — and those of Rosecrans — were about to take a turn for the better.

3

The Darkest Days of the War

Henry Halleck did nothing to exploit the bright strategic situation during Grant's absence from the army. The first week of June found him in Corinth with 137,000 soldiers who were "elated at their [bloodless] success" and anxious to press on. Opposing him were 50,000 discouraged Confederates at Tupelo. In abandoning Corinth the Confederates had yielded what little remained to them of western Tennessee. With Federals astride the railroad to Memphis, Fort Pillow was indefensible, and the Confederate garrison pulled out the day after Beauregard yielded Corinth. The Federal river fleet steamed to Memphis, where on June 6 it obliterated a small Confederate flotilla. Memphis surrendered, the last river obstacle to the Federals until Vicksburg, 200 miles to the south.

To Grant and most other generals it seemed a moment of unparalleled opportunity. The Confederate interior lay open to invasion from two directions. Halleck might bypass the Rebel army at Tupelo and continue down the Mississippi River Valley to Vicksburg. Or he could move laterally along the Memphis and Charleston Railroad, concentrate his forces at Chattanooga, and then push on to Atlanta.

But after a halfhearted pursuit of Beauregard that ended on June 9, Halleck opted against an offensive. "The major object now," he told John Pope and Don Carlos Buell after reining in Pope from too vigorous a chase, "is to get the enemy far enough south to relieve our railroads from danger of an immediate attack. There is no object in bringing on a battle

[31]

if this can be obtained without one. I think by showing a bold front, the enemy will continue to retreat, which is all I desire."[1]

Grant was discouraged and later wrote in his memoirs, "The possession of Corinth by the National troops was of strategic importance, but the victory was barren in every other particular. . . . It is a question whether the morale of the Confederate troops engaged at Corinth was not improved by the immunity with which they were permitted to remove all public property and then withdraw themselves."[2]

The grumbling of his generals left Halleck unmoved. As he saw it, military logic dictated a halt at Corinth. Were he to take his huge army to Chattanooga, he risked opening the Mississippi River Valley to a Confederate counteroffensive. If he were to march on Vicksburg, he would uncover Louisville and Cincinnati. To his cautious mind, it was far more important to hold these cities than to overrun the Confederacy.

Then, too, there were the wishes of the president. Profoundly moved by the suffering of the Unionist inhabitants of East Tennessee, the preceding winter Lincoln had urged Halleck to seize the region; with Corinth in Union hands and the Mississippi River open to Memphis, he again called for the relief of East Tennessee. The president wanted Chattanooga occupied and the railroad into the region secured, tasks he thought "fully as important as the taking and holding of Richmond."[3]

Halleck acceded to the president's proposal, and during the first week of June he began to disperse his armies to effect it. He ordered Buell to take 30,000 men and "with all possible energy" open communications with Brig. Gen. Ormsby Mitchel in northern Alabama. Together they were to march aggressively against the enemy at Chattanooga. But in subsequent orders Halleck reverted to his habitual caution, admonishing Buell not to move so fast as to jeopardize repairs on the Memphis and Charleston Railroad. With no other offensive designs, Halleck dispersed the rest of his huge command along the railroads in western Tennessee. He dispatched Maj. Gen. John McClernand with two divisions forty miles northwest of Corinth to Bolivar, recalled Pope to Corinth, and sent Sherman with two divisions to repair the Memphis and Charleston Railroad to Memphis.[4]

Halleck also had been reluctant to venture far into Mississippi for fear that operations in the swampy, disease-ridden interior of the state would prostrate the army. That may have been true, but Corinth was hardly salubrious. By mid-June, quipped an Iowa soldier, "General Summer" commanded at Corinth, and a more miserable summer few Corinthians could remember. Afternoon temperatures topped 100 degrees. Dust settled six inches deep over the streets of town and swirled about marching troops in dense clouds. Two hundred thousand soldiers, blue and gray, and their animals had inhabited Corinth during the preceding three months. Their waste polluted the soil. Flies bit at men and horses, and mosquitoes swarmed about the camps. Streams dried up. Buckets dropped twenty feet into wells before reaching water. The celebrated mineral springs of nearby Iuka went dry for only the third time in a century.

With the bad water and mosquitoes came disease. Diarrhea and dysentery swept through the ranks, taking a hard toll on regiments depleted from their losses at Shiloh. Sherman was "quite unwell" until he left for Memphis, and Halleck spent much of June confined to quarters in the Curlee mansion with what he called "the evacuation of Corinth." By month's end 35 percent of the troops in town were sick.[5]

Compounding their misery was the labor to which the soldiers were subjected. One of Halleck's first acts upon occupying Corinth was to construct a chain of fortifications that rivaled the efforts of Beauregard and Bragg. The Confederate earthworks ran in a semicircle north of town. Halleck put his men to work erecting forts on every piece of commanding ground south of Corinth for a distance of three miles. From Memphis Grant marveled at the wasted effort. The works, he said, were built "on a scale to indicate that this one point must be held if it took the whole National army to do it. They were laid out on a scale that would have required 100,000 men to fully man them." For the moment, Grant could only mutter his dissent.

Rosecrans unexpectedly found himself in a position to do more. General Pope had left the army the second week of June to visit his family in St. Louis. He never returned to Corinth; a telegram from Stanton summoning him to Washington cut short his leave. On June 26 Rosecrans assumed command of the Army of the Mississippi.

After Halleck's fortifications were built, there was nothing pressing to occupy Rosecrans's attention. Nonetheless, the Ohioan kept busy. He moved the army to Clear Creek, a more sanitary campsite six miles outside Corinth, had a convalescent hospital erected, and organized an antiscorbutic and antifever diet for the troops. The army's sick list fell from 35 to 12 percent.

As the health of the men returned, Rosecrans turned to matters of command. During July he reorganized the army, which consisted of five divisions of infantry and one of cavalry, shuffling generals from unit to unit. The marginally competent Eleazer Paine led the First Division until July 15, when command passed for thirty days to its promising senior brigade commander, James D. Morgan. Brig. Gen. David Sloane Stanley, a capable but petulant officer, led the Second Division. Brig. Gen. Charles Smith Hamilton commanded the Third Division; the hot-tempered and proud Jefferson C. Davis headed the Fourth; Hungarian émigré Alexander Asboth led the Fifth; and the testy but talented Gordon Granger commanded the cavalry.[6]

A more contrary lot seldom served an army commander, but Rosecrans seems to have gotten on well with all of his generals except Hamilton. A member of Grant's West Point class, Hamilton was touchy, pugnacious, a schemer, and a liar. In Mexico he had earned a brevet for gallantry at the Battle of Molino del Rey. Hamilton resigned his commission in 1853 to farm and to manufacture flour in Fond du Lac, Wisconsin. At the outbreak of the war, he was appointed colonel of the Third Wisconsin Infantry. Six days later he was commissioned a brigadier general. Hamilton's first real test came as a division commander on the Peninsula. He failed miserably, and less than a month into the campaign McClellan relieved him. Political pressure failed to persuade McClellan to restore Hamilton, whom he declared "not fit to command a division." But Hamilton had enough friends in high office to get a second chance as a division commander. Somewhat chastened by his humiliation, Hamilton was for the moment behaving himself.[7]

Rosecrans kept occupied through the hot, quiet summer weeks, inspecting his camps incessantly and ruminating on ways to improve the fortifications around Corinth. In late July he shared his views on the subject with the new commander of the District of West Tennessee, Ulysses S. Grant.

*   *   *

Grant's return from exile had béen abrupt. It also had come as a surprise to Grant, who had no idea why Halleck wanted him in Corinth. Still bedridden, Halleck had received a telegram from the president directing him to assume the post of general in chief, with headquarters in Washington. Lincoln took the decision less from a regard for Halleck's abilities than from frustration over McClellan's failures on the Virginia peninsula. Although Halleck's personal contributions to western successes had been minimal, he at least had not failed, and that was recommendation enough for Lincoln.

Halleck lingered at Corinth for six days, but he may as well have left the day he read Lincoln's telegram. Before starting from Memphis, Grant asked Halleck why he wanted to see him. "I was not informed by the dispatch that my chief had been ordered to a different field and did not know whether to move my headquarters or not," said Grant, so he posed the question to Halleck. "This place will be your headquarters. You can judge for yourself," Halleck wired back tersely. Grant left Memphis at once and reached Corinth on July 15. Halleck remained closeted in the Curlee mansion until the seventeenth. "He was very uncommunicative, and gave me no information as to what I had been called to Corinth for," Grant recalled. Nor, apparently, did he tell Grant of a note from Rosecrans, dated July 5, relating rumors of Confederate troop transfers toward Chattanooga. Halleck went on his way, with Grant still wondering what he was doing in Corinth.[8]

By the time Rosecrans visited him, Grant had learned the nature—and limits—of his assignment. Halleck had been commander of the Department of the Mississippi with control over an area as far east as Chattanooga. Grant, however, assumed command only of the District of West Tennessee, which embraced those counties of Tennessee and Kentucky west of the Cumberland River. The Army of the Ohio, operating near Chattanooga, was outside Grant's authority, and its commander, Don Carlos Buell, reported directly to Washington. So, too, did Grant, meaning that "practically I became a department commander, because no one was assigned to that position over me and I made my reports direct to the general-in-chief."[9]

It was an empty honor. Grant had at his disposal only two badly depleted armies. The Army of the Tennessee, which he continued to command in person, numbered just 38,485 at the end of July. Rosecrans counted only 25,224 men fit for duty in his Army of the Mississippi at Camp Clear Creek. New recruits destined for Grant's district were still at Northern depots, and his veterans were a tired, sick, and dispirited lot. Oppressive heat, bad water, and inactivity blackened their thoughts. "A cloud of darkness and distress pervaded" their camps, said an officer of the Sixty-fourth Illinois. "It was without doubt the darkest days of the war."

Grant agreed. His 63,709 troops were far too few to launch an offensive, particularly as the Army of the Tennessee was scattered across the district guarding railroads. Grant could only protect what had been won, and even that posed grave challenges. Guerrilla bands and Southern cavalry cut telegraph lines and tore up track as fast as the Federals repaired them. They disabled the railroad between Chewalla and Grand Junction, and Grant abandoned the track from Grand Junction and Memphis for lack of rolling stock. Memphis and Corinth were no longer in direct communication. Messages from Grant to Sherman had to go by railroad to

Columbus, Kentucky, and then downriver to Memphis by boat. Troops were compelled to move from one town to another by way of Jackson, which left a huge expanse over which Confederate cavalry might raid with relative impunity.

And there was Halleck in Washington threatening to strip Grant's already slender command to reinforce Buell. It was no wonder Grant later wrote that "the most anxious period of the war, to me, was during the time the Army of the Tennessee was guarding the territory acquired by the fall of Corinth and Memphis and before I was sufficiently reinforced to take the offensive." War correspondent William Shanks, a fixture at headquarters, said Grant seldom joked, rarely laughed, and "whittled or smoked with a listless, absorbed air." But he and Rosecrans were on friendly terms. They dined together regularly, and Grant asked Halleck to see that Rosecrans was promoted to major general, a rank Grant thought "equal to his merit."

So Grant was disposed to hear out Rosecrans on his thoughts for the defense of Corinth. Rosecrans conveyed to him the recommendations of Capt. Frederick Prime, a gifted engineer officer on Rosecrans's staff. Prime considered the "Halleck Line," as the forts south of town had come to be called, too extensive to be defended by the small force then available to Grant. Prime suggested that a system of redoubts be built at key points closer to the town to protect the railroad depot. Rosecrans seconded Prime's proposal, but Grant demurred. Rather than abandon the works constructed under Halleck, he set his men to improving them.

Rosecrans's conversations with Grant fell into a pattern.

"How are you getting along with the line?" Rosecrans would ask after the usual greetings.

"Well, pretty slowly, but they are doing good work," Grant would answer.

"General, the line isn't worth much to us, because it is too long," Rosecrans would insist. "We cannot occupy it."

"What would you do?"

"I would have made the depots outside of the town north of the Memphis and Charleston road between the town and the brick church, and would have enclosed them by field-works, running tracks in. Now, as the depot houses are at the cross-road, the best thing we can do is to run a line of light works around in the neighborhood of the college up on the knoll," offered Rosecrans.[10]

Rosecrans's persistence paid off. Grant at last relented, and work began at once on Prime's line, designed to defend the town and the railroad depot against attacks from the west.

Five earthen lunettes were constructed within a half-mile of the railroad junction. Each had high parapets, ten-foot-wide ditches in front, and embrasures for cannon. Prime placed the first lunette 600 yards southwest of the depot (a large frame building filled with army supplies that had been erected beside the Tishomingo Hotel) and named it Battery Lathrop. He laid out the second lunette 300 yards west of Battery Lathrop on a high knoll south of Corona College and named it Battery Tanrath. The third was built 200 yards north of the college grounds and named Battery Phillips. Four hundred yards northeast of Battery Phillips, on a knoll just south of the Memphis and Charleston Railroad, Battery Williams went up. Named for Capt. George A. Williams, commander of the army's siege artillery, Battery Williams accommodated a battery of thirty-pounder Parrotts. Across the track, 200 yards north of Battery Williams and 675 yards west of town, was Battery Robinett, named for Lt. Henry Robinett of Company C, First United States Infantry. Battery Robinett mounted three twenty-pounder Parrotts, two of which were aimed west; the third, north.[11]

Raising dirt fortifications, foraging for food, and chasing marauding guerrillas were the principal distractions for the Federals at Corinth. With Grant too weak to take the offensive, the prospects for battle depended on the Confederates at Tupelo.

4

This Will Be Our Opportunity

Returning from Richmond on July 2, Price handed Bragg the only fruits of his journey: a War Department letter authorizing, at Bragg's discretion, the return of Price and his Missouri division to the trans-Mississippi.

But Bragg, intending to replace Van Dorn with Price as commander of the Army of the West, refused to release the Missourians or to let Price go. Not that Bragg had any immediate need for Price or his troops. Beyond a vague desire to strike Buell's army sometime, he gave no thought to taking the offensive. A lack of transportation and the summer drought ruled out extended marches across northern Mississippi, and the nearest possible objective, Corinth, was too well fortified to attack.

Disappointment had become part and parcel of Confederate service for Price, and he bore this latest setback well. His Missourians welcomed the news of his elevation to army command, and Price was careful to incite them to no greater antipathy than they already felt toward Richmond. Like Price, the Missourians resented fighting for the homes of others when their own were in enemy hands, and an alarming number had deserted to join guerrilla bands in Missouri.

On July 4 the First Missouri Brigade marched to Price's headquarters and demanded a speech. With a few intemperate words Price could have prompted a wholesale mutiny. Instead he calmed the waters with false assurances, telling his men he had President Davis's permission to take the Missouri division back to Missouri but was "going to stay here awhile to see if a battle was to come off." If it did not, he "would have one of his own for he was going to Missouri anyhow." The Missourians were delighted. It was a short speech, recalled one, "but what he said was to the point and just what we wanted to hear."[1]

Price made the best of duty in Mississippi. For the first time in his career, he paid real attention to matters of organization and discipline. Price inherited an army that had lost a third of its strength in late June, when Maj. Gen. John McCown's division was sent to Chattanooga. The remainder he divided into two divisions. Henry Little led the first division, which contained the seven regiments Price had brought with him from Missouri. The second division, made up largely of Arkansans and Texans, he gave to Van Dorn's former chief of staff, Dabney Maury. Price united the five understrength regiments and three independent battalions that comprised his cavalry into a brigade of 1,000 troopers. To lead them, Price selected Col. Frank Armstrong of the Third Louisiana Infantry.[2]

Price relied heavily on Little, whose promotion to major general he urged at every opportunity. They had served together since the summer of 1861, when the forty-four-year-old Marylander had resigned his Regular Army commission to help Price train his Missouri volunteers.

The Missourians took to Little at once. Little's diary entries confirm their impression of him as a "quiet, unassuming, affable man" of self-deprecating humor. Despite his reticence, Little was "a thorough soldier and an excellent disciplinarian." His efforts were appreciated. General Hardee pronounced his division "the most efficient, best drilled, and most thoroughly disciplined body of troops" in Mississippi, and Braxton Bragg thought it "as fine a division as he had ever seen."

Little needed the praise, if only to sustain his nearly broken health. Service in Confederate gray had been hard on him. At Corinth in May he contracted malaria. After the acute chills and fever passed, Little was visited with recurring headaches, nausea, and persistent boils. His condition deteriorated as the summer dragged on. Diarrhea seized him in early August, and a few days later he contracted dysentery.[3]

Little's presence about camp became sporadic. Fortunately he was ably seconded by his senior brigade commander, Brig. Gen. Louis Hébert. A forty-two-year-old native of Louisiana, Hébert had graduated third in the West Point class of 1845. Family illnesses compelled him to resign his commission after only one year of service and return to Louisiana. Hébert parlayed his military education into an appointment as state engineer and a colonelcy in the militia. When war came, he helped organize the Third Louisiana Infantry, which he turned into the best-drilled unit in the

trans-Mississippi. As colonel of the Third, said Willie Tunnard, Hébert was "genial and kind in manner and conversation." But like Little, he was a "strict disciplinarian, punctilious in enforcing a rigid adherence to all orders." Little, then, had no cause to worry that discipline would suffer while he was sick; Hébert was as exacting with the division as he had been with his own brigade.[4]

Before taking ill, Little had moved the division ten miles north, from Tupelo to Saltillo. Maury remained at Tupelo. Both places had good water, and the Confederates enjoyed better health than the Federals at Corinth. Idleness might have lowered their resistance to disease, but Hébert kept his troops occupied with incessant drilling, cleaning, and parading—a numbing routine that left the men busy but restless.[5]

For a time it had looked as if the monotony might be broken. Braxton Bragg had decided to take the offensive and in late July began moving the Army of the Mississippi out of Tupelo. His immediate objective was neither the Federals at Corinth nor Buell's Army of the Ohio but, rather, the city of Chattanooga. The agent of change in Bragg's thinking was Maj. Gen. Edmund Kirby Smith, commander of the Department of East Tennessee. Smith had the unenviable task of defending one of the most difficult positions in the Confederacy, with a paltry force of 9,000. It was his duty to keep open the East Tennessee and Virginia Railroad and the East Tennessee and Georgia Railroad (the latter a continuation of the Memphis and Charleston line) through the mountainous, sparsely settled, pro-Union counties of East Tennessee. Guerrillas ravaged the railroads in the eastern reaches of the department, and Buell menaced them from the west. To meet both threats, Smith dispersed his small command along a 180-mile front, from Chattanooga to the Cumberland Gap.

By July it seemed the march of Buell's army along the Memphis and Charleston Railroad toward Chattanooga would overwhelm the attenuated Confederate defenses. Smith's options were limited. To abandon Chattanooga was unthinkable. Its loss would sever rail communications between Virginia and the West and open the door to a Federal advance on Atlanta. Four of the South's eight arsenals would be vulnerable to capture, as would much of the Confederacy's munitions and raw materials. In late June Bragg sent McCown's division from Price's army to help hold Chattanooga. That calmed Smith, and to contemplate the fate of the department he went on leave in the Smoky Mountains. There his thinking became as thin as the mountain air. Smith had despaired to Bragg, and to anyone else who cared to listen, that he could not defend his department; now he suddenly concocted a plan to enlarge it. In early July he told Bragg that he intended to advance. Where, or against whom, he had not decided—perhaps against the Federal garrison at Cumberland Gap or into the fertile farm country of Middle Tennessee. The raider John Hunt Morgan helped clarify Smith's intentions for him. Morgan had been raiding with impunity in Kentucky—ample proof, he wrote back, that the state could be Smith's for the taking.

While Buell drew closer to Chattanooga, and before Bragg had committed himself to the city's defense, Smith began to shift troops away from Chattanooga for a Kentucky expedition. Impatient to start, Smith tried to force a decision on Bragg. On July 19 he warned Bragg that "Buell with his whole force, is opposite Chattanooga, which he is momentarily expected to attack." Then, neglecting to mention he had stripped the city's defenses, Smith urged Bragg to hurry along, for "the holding of Chattanooga depends upon your cooperation."

Nobody needed to remind Bragg how important Chattanooga was. He acceded to Smith's disingenuous demand less out of credulity than from ambitions of his own. Bragg still entertained thoughts of attacking Buell from behind, but he had given up on Tupelo as too remote to make an effective staging area. Chattanooga, on the other hand, offered not only a strong base but also several routes from which to strike at Buell and perhaps regain Middle Tennessee, a move that would be immensely popular with the army and with Richmond. Playing at diplomacy, Bragg and his generals speculated that the recapture of Nashville — or perhaps even a drive through Kentucky to the Ohio River — might induce England and France to recognize the Confederacy.

Bragg waited until his troops were boarding trains before informing Richmond of his plans and of the reasoning behind them. On July 23 he wrote the War Department that it was imperative to frustrate the Federal thrust into East Tennessee. To do so he would concede the initiative in northern Mississippi. If the chance offered itself, he would "in conjunction with Major General Smith, strike a blow through Middle Tennessee, gaining the enemy's rear, cutting off his supplies and dividing his forces, so as to encounter him in detail."[6]

Richmond had little to say in response. In late June Secretary of War Randolph had suggested to Bragg that he simply "strike the moment an opportunity offers." Randolph had nothing more to add now, and President Davis merely expressed his confidence that Bragg and Smith would be able to work together. Randolph and Davis did, however, concede the need to rationalize Bragg's department. Bragg had complained that the division of Mississippi into two departments was "exceedingly inconvenient. The only communication for me east or west passes through Van Dorn's command." Taking this as an implied request for greater authority, Richmond reconfigured Department Number Two to include the entire states of Mississippi and Alabama, northwestern Georgia, that portion of Louisiana east of the Mississippi River, and the Florida panhandle.

Bragg in turn reorganized the department he was about to abandon, dividing it into three districts. He gave Brig. Gen. John Forney command of the District of the Gulf, with orders to defend Mobile. He kept Earl Van Dorn and his 16,000 troops in Vicksburg and created the District of the Mississippi around them. Deprived of his department, Van Dorn had to content himself with a district. Bragg elevated Price to an equal footing with Van Dorn. He gave him the District of the Tennessee less out of regard for his abilities than from an absence of other suitable candidates. It was an expansive command, comprising northwestern Alabama and northeastern Mississippi as far south as Tupelo, and one that Price lacked the resources to defend adequately; he had only his Army of the West, augmented by a few hundred men from outlying garrisons.

Perhaps because he considered northern Mississippi "infinitely less important" than East Tennessee, Bragg failed to give clear and precise orders to his district commanders or to leave one of them in charge in his absence. The closest he came to naming an interim successor was to tell Van Dorn that his rank would give him command whenever he and Price might join forces. Where two generals as willful as Van Dorn and Price were concerned, such negligence as Bragg displayed was inexcusable — and bound to produce friction.

Bragg had no real instructions for Price beyond exhorting him to hold the line of the Mobile and Ohio Railroad and keep Grant and Rosecrans from reinforcing Buell. How he might accomplish that was left to Price to divine. Bragg also told Price that he and Van Dorn might later conduct a joint offensive into western Tennessee to complement his own and Smith's drives farther to the east, but he left Price no specific orders on the subject.

Bragg had little to offer to Van Dorn, either. The day before he left Tupelo, Bragg asked the Mississippian to "consult freely and cooperate with Major General Price. It is expected that you will do all things deemed needful without waiting for instructions from these headquarters. General Price will be instructed to the same effect." Even this minimal guidance was expressed as simply "the wish of the commanding general."[7]

General Hardee, whom Bragg had placed temporarily in command of the Army of the Mississippi, anticipated problems. Before leaving Tupelo, he warned Thomas Snead to watch for trouble. Having just learned that Van Dorn had staged an expedition against Baton Rouge, Hardee "feared it would lead Van Dorn into other adventures which would over task his strength, and that Van Dorn would then call on General Price to help him." This Price must not do, as "the success of General Bragg's movement into Tennessee and Kentucky depends greatly upon his (Price's) ability to keep Grant from reinforcing Buell. . . . Say to General Price that I know that General Bragg expects him to keep his men well in hand, and ready to move northward at a moment's notice." Hardee's furtive words of caution to Snead were the only meaningful orders given the defenders of Mississippi.[8]

Problems arose at once. Price interpreted Bragg's desire that he keep Grant and Rosecrans occupied to mean he should move against them. On July 29, as the last of Hardee's army departed Tupelo, he called in the scattered outposts of his district and set about assembling his army for a march against Corinth or Grand Junction, Tennessee. Price recognized that even a diversionary movement would fail without the combined strength of his and Van Dorn's army, and so on July 31 he wrote Van Dorn soliciting the cooperation Bragg had said he could expect. The Federals, Price told Van Dorn, had weakened their forces in northern Mississippi and western Tennessee in order to reinforce Buell. Price thought

Grant had only 15,000 men at Corinth and fewer than 10,000 at Bolivar, Tennessee. If that were true, then he and Van Dorn not only could re-take Corinth but might also unite with Bragg somewhere in Middle Tennessee. "This will be our opportunity," he exhorted Van Dorn, "and I am extremely anxious that we shall avail ourselves of it."[9]

Price was mistaken about the number of Federals he faced. Scouts and patrols from Frank Armstrong's cavalry had grossly miscounted the enemy. Rosecrans had spread his army across northeastern Mississippi and into Alabama to defend the Memphis and Charleston Railroad and protect Buell's flank, but he nonetheless could gather his forces rapidly. Paine's First Division was then at Tuscumbia, Alabama, near the Tennessee River; Stanley's Second Division lay encamped at Clear Creek; Hamilton's Third Division had been advanced to Jacinto, eleven miles southeast of Corinth; Davis's Fourth Division was guarding the track between Corinth and Tuscumbia; and the Fifth Division, now led by Gordon Granger, was twelve miles south of Corinth at Rienzi. Grant, in turn, had parceled out to garrison Corinth, Bolivar, Jackson, and Memphis more than three times the number of troops Price calculated.

Bragg, too, was misinformed regarding Federal troop strength and dispositions, and he based his advice to Price on poor intelligence. Thinking Grant had sent nearly all of Rosecrans's army to Buell, Bragg decided Price might do more for him than merely hold the remaining Yankees in check. An aggressive drive by Price would threaten Buell's rear and perhaps induce him to fall back into Middle Tennessee. From Chattanooga on August 2 he exhorted Price to action: "Rosecrans commands Pope's army. Nearly the whole force at Corinth should be moved this way. The road is open for you into Western Tennessee."[10]

Price was in a quandary. Although he felt compelled to act, Price was certain he could not move alone. And as Hardee had predicted, Van Dorn answered Price's call that he come north with a request instead that Price send him a brigade to support his badly stalled campaign against Baton Rouge.

Van Dorn had arrived at Vicksburg on June 28 to find a Federal brigade at work on a river canal that would bypass the city's heavy batteries. He took command with typical dash, telling his men, "Let it be borne in mind by all that the army here is defending the place against occupation. This will be done at all hazards, even though this beautiful and devoted city should be laid in ruins."

The Mississippian's gallant intentions went untested. The Federals gave up on canal digging and retired to Baton Rouge. Neither the river flotilla of Commodore Charles Davis nor Adm. David Farragut's fleet, which had effected a junction above the city, could move against Vicksburg without an army to support them. District commander Maj. Gen. Benjamin Butler at New Orleans had no troops to spare, and so they retired.

Not content to watch the Federals leave, Van Dorn sent Maj. Gen. John C. Breckinridge to retake Baton Rouge. It was another quixotic Van Dorn enterprise. Without naval support, the Confederates never had a chance. After losing half his command to extreme heat and bad water on the march, on August 5 Breckinridge attacked and was repelled with heavy loss by fire from Federal warships. Frustrated in his efforts against Baton Rouge, Van Dorn instead fortified Port Hudson, forty miles upriver.[11]

On August 4, the day before Breckinridge came to grief at Baton Rouge, Price told Van Dorn he could not reinforce him because Bragg expected him to advance into western Tennessee. "Every consideration makes it important that I shall move forward without a day's unnecessary delay," said Price. "I earnestly desire your cooperation in such a movement, and will, as I have before said, be glad to place my army and myself under your command in that contingency."

Price's anxiety grew. On August 4 he wrote Bragg's chief of staff apologetically: "I am extremely impatient to begin a forward movement, and am bending every energy to do so without any unnecessary delay. I am ordering forward the entire disposable force in the district. I expect to begin my march within a week or ten days." Then Price learned he would be moving alone. Having ignored Price's appeal of July 31, Van Dorn responded to his telegram of August 4 with an emphatic "No"—his army was in no condition to leave the District of the Mississippi. Price implored Bragg to intervene. He did, in a manner, but stopped short of ordering Van Dorn to cooperate with Price.

Bragg and Smith had worked out their differences in a meeting at Chattanooga and agreed on a basic plan for the coming campaign. While Bragg waited in Chattanooga for his artillery and wagon trains to come up, Smith would eliminate the small Federal garrison holding the Cumberland Gap. Together they would coax Buell into battle on ground of their choosing, then march north and occupy Kentucky. With their flank thus turned, any Federals remaining in western Tennessee and northern Mississippi would be compelled to retreat, presumably with Van Dorn and Price in pursuit.

The success of such a grandiose design depended partly on preventing further accretions to Buell's army. Consequently, Bragg told Van Dorn it was "very desirable to press the enemy closely in West Tennessee." Despite the obvious importance of at least a diversionary movement in northern Mississippi, Bragg only suggested to Van Dorn that he join Price, which was tantamount to allowing the Mississippian a free hand. Perhaps to induce Van Dorn to cooperate, Bragg reminded him that "of course when you join Price your rank gives you command of the whole force." Bragg wrote Price the next day, enclosing a copy of his letter to Van Dorn. "The details of your movements I must leave to your own judgment and intelligence, relying on your patriotism for a cordial co-operation." [12]

Cooperation, no matter how cordial, was no substitute for unified command. Nor were polite suggestions. Unlike Price, Van Dorn had no enthusiasm for an attack on Corinth. Anxious to clear the Mississippi Valley and unconcerned with protecting Bragg's flank, Van Dorn preferred to launch an attack farther to the west. And he was not above lying to get his way. Apparently unaware that Bragg had shared with Price his letter to him, Van Dorn distorted its content, telling Price that Bragg wanted him to take the offensive "toward Grand Junction and Memphis." But even that would have to wait. Breckinridge was languishing outside Baton Rouge, his force "too feeble to make a decisive result." Perhaps Price could spare a brigade to help him? In any case, Van Dorn concluded, it would be at least two weeks before he would be ready to move.

Van Dorn's message was delayed almost as long in reaching Price. Hearing nothing from Van Dorn in the interim and learning that earlier reports of Rosecrans's departure from his front were in error, Price grew despondent. When Bragg, who was still in Chattanooga without any clear timetable for his own offensive, counseled him not to "depend much on Van Dorn; he has his hands full," Price shelved his plan. "Believing I could not advance successfully without [Van Dorn's] cooperation, I determined to await either that or the weakening of the enemy's force in front of me and to meanwhile perfect my preparations to move." [13]

Although he entertained no hopes of assaulting the Federals without Van Dorn, Price did what he could to harass them. He called upon Frank Armstrong to take his recently organized cavalry brigade and slash at Grant's line of communication and bring back information about his dispositions.

On August 22, at the head of 1,600 troopers, Armstrong rode from Baldwyn to Holly Springs, where he was joined by the Seventh Tennessee Cavalry from Van Dorn's army. Together they pushed up the Mississippi Railroad past Grand Junction to threaten Bolivar. Along the way

Armstrong detached Col. William Falkner with 400 Mississippi partisan rangers to distract Rosecrans with a raid on Chewalla, ten miles northwest of Corinth.[14]

A nasty skirmish near Bolivar between Armstrong and the Federal garrison threw the Federal high command into a panic, taxing Grant at a time when he had little interest in the work at hand. Expecting trouble from the Confederates, he had sent his wife, Julia, and the children home two weeks earlier. Now he missed them terribly. "I wish I could be there or any place else where I could be quiet and free from annoyance for a few weeks," he wrote Julia on August 18. "From present indications you only left here in time. Lively operations are threatened and you need not be surprised to hear of fighting going on in Grant's army." Six days after his family left, Grant sent his ailing friend and chief of staff, John Rawlins, home on convalescent leave.

Alone now, Grant was tired and genuinely perplexed. Falkner's presence near Chewalla led Grant to fear an imminent attack on Corinth. Rumors spread that Price himself was marching on the town. Grant called in his outposts and evacuated the large hospital at nearby Farmington. Overnight the sick and wounded were loaded into wagons for the bone-grinding ride to Corinth. So great was the haste that nonessential supplies were abandoned.

Grant's alarm was understandable. He knew of Bragg's departure for Chattanooga but was uncertain how many troops had remained behind with Price at Saltillo and Tupelo. Grant thought 20,000; Rosecrans credited reports that gave Price 12,000 men.[15]

In northeastern Alabama Don Carlos Buell also knew Bragg was concentrating against him, and he importuned Grant ceaselessly for reinforcements. Grant pleaded that the weakened and dispersed condition of his forces precluded him from helping, but Buell persisted. Not that he could be expected to do otherwise. Halleck had told him plainly that the administration was "greatly dissatisfied" with the slowness of his march toward Chattanooga. It was common knowledge that Smith had struck out for Kentucky, and Halleck and Buell feared that Bragg was poised at Chattanooga to do similar mischief. With the majority of Confederate forces in the West threatening Buell, Grant was compelled to yield to Buell's demand, endorsed by Halleck, that he loan him two divisions.

Grant took Paine and Mitchell from Rosecrans, who was left with three divisions to cover a front fifty miles wide. Clearly that was more ground than he could defend, and Rosecrans warned that "a speedy remedy must be applied or a bad result must be expected."

Grant agreed. He tried to convince Halleck to allow him to consolidate his forces, but the general in chief refused. Instead Halleck wanted the line extended to Decatur, Alabama — 100 miles from Corinth by road — to keep open communications with Buell. In a measure Grant was responsible for Halleck's posture; on August 16 he injudiciously told Halleck cavalry patrols had revealed the country to his front "to be so dry that an attack on [Corinth] is hardly to be apprehended."[16]

Armstrong's raid caused Grant to rue that telegram. While he and Rosecrans tried to comprehend the meaning of the sudden Confederate activity, Halleck importuned Grant to send still more reinforcements to Buell. Now Grant was forced to admit he had been wrong. "I am weak and threatened with present forces from Humboldt to Bolivar, and at this point would deem it very unsafe to spare any more troops," he told Halleck on September 1. Only if Halleck permitted him to abandon the railroad at the Alabama border and shorten his front could he possibly comply. Halleck agreed to that and more — Grant was free to terminate his line at Corinth, but he must release another division. Bragg had left Chattanooga, and the War Department was hurrying all available troops to succor Buell. Again it was Rosecrans to whom Grant turned. Gordon Granger departed for Kentucky on September 4, leaving Rosecrans with two divisions — those of Stanley and Hamilton — to meet what might be an advance on Corinth by Price's entire army.[17]

Armstrong kept the Federals guessing. He bypassed Bolivar and continued north along the railroad toward Jackson, burning bridges and tearing up track as he went. On September 1 he clashed with a strong Federal detachment near Denmark. Armstrong drew off after burning a wagon train and taking 200 prisoners. The Federal commander at Jackson, Brig. Gen. Leonard Ross, panicked. Rumors spread that Bolivar had surrendered and that the Jackson garrison was about to be surrounded by 30,000 Confederates. Ross pleaded with Grant to send heavy reinforcements to save the junction town. But Grant at Corinth had none to spare. "This point," he told the frightened Illinoisan, "besides its importance, is very weak and should be reinforced rather than drawn from." Ross would have to fend for himself.[18]

Unbeknownst to Grant and Ross, the danger had passed. After the skirmish at Denmark Armstrong withdrew, delighted with the results of his expedition. The weak Federal response convinced him Bolivar could be easily taken "whenever an advance of our army is made." Armstrong's losses had been small, and morale was high. "Our commands were in the saddle for nearly forty-eight hours, and some of them longer, without food, but neither hunger nor fatigue could daunt them. They are ever ready when an opportunity offers to punish the insolent invaders."[19]

A second opportunity was to present itself far sooner than Armstrong could have imagined.

5

Things Were Beginning to Wear a Threatening Aspect

Bragg and Van Dorn were pushing Sterling Price to the limits of his patience — the former by insisting he prevent a junction of Buell and Rosecrans, and the latter by refusing to cooperate with him. On August 25, three days before starting from Chattanooga, Bragg exhorted Price to strike the moment the Federals to his front appeared vulnerable. "In the mean time harass your opponent by all your cavalry. Buell is falling back toward Nashville. We must keep them moving."

Price tried. He sent Armstrong on his raid and beseeched Van Dorn to help drive the enemy from Corinth.

No sooner had Price written Van Dorn than another message came from Bragg. Smith was on the move toward Cumberland Gap, Kentucky, and Bragg himself was about to depart Chattanooga for Middle Tennessee, where he hoped to take "the enemy's rear, strike Nashville, or perhaps . . . strike for Lexington and Cincinnati." Bragg expected Price to do his part to ensure the success of the grand offensive. "Sherman and Rosecrans we leave to you and Van Dorn, satisfied that you can dispose of them," said Bragg, who cleverly concluded with an appeal to Price's desideratum: "We shall confidently expect to meet you on the Ohio and there open the way to Missouri."

Four days passed with no word from Van Dorn. Price followed the fortunes of Armstrong and pushed his own preparations. On the night of September 1 he read a dispatch that was the closest thing to a direct order that Bragg had yet issued to Price or Van Dorn: "Buell's whole force in full retreat upon Nashville, destroying their stores. Watch Rosecrans and prevent a junction; or if he escapes you follow him closely."

Price concluded to tarry no longer. With or without Van Dorn he would march against Rosecrans's army. On September 2 Price informed the Mississippian of his decision, saying he felt Bragg's "order requires me to advance immediately. . . . I hope nothing will prevent you from coming forward without delay with all your disposable troops."[1]

It was a gracious exhortation, but Price held no real hope for a favorable reply. Instead he concentrated on his own plans. He assigned Little the mission of advance guard and on September 5 edged his division forward fifteen miles from Saltillo to Baldwyn. Maury took his place at Saltillo.

Before pushing on, Price paused to gather wagons and to write Van Dorn one final, indignant letter. The Mississippian had replied to Price's letter of September 2 in a manner that mocked Bragg's desires. Van Dorn told Price he would not be ready to march from Holly Springs, where he was concentrating his army, until September 12. More concerned with keeping Grant and Rosecrans separated than with preventing them from reinforcing Buell, he declined to attack Corinth with Price. Instead he suggested Price meet him at Grand Junction to maneuver Rosecrans out of Corinth, which he intimated was too strong to assault. Bragg's immediate needs were of little moment. "Do not leave me and go east if you can avoid it," he begged. "We can do more together west of the Tennessee, for awhile at least. We should try and shake them loose from all points in West Tennessee; then march to join Bragg, if necessary."

Price was incredulous; Bragg's orders were at last clear and unambiguous, and he intended to follow them. A move toward Holly Springs would uncover the Memphis and Ohio Railroad, upon which he depended for supplies, and take him far from Rosecrans, whom Bragg wanted him to watch.[2]

Price correctly placed Rosecrans at Iuka. Although Halleck had permitted Grant to abandon the country east of Corinth, Grant felt obliged to maintain a presence near Tuscumbia until Paine and Mitchell completed their crossing of the Tennessee River and Rosecrans was able to draw off the excess property they left behind. At the same time both

Grant and Rosecrans recognized that, given the vulnerability of the rail-roads to depredations by Rebel cavalry, the most reliable line of supply for the Army of the Mississippi was the Tennessee River. The best river landing was thirty-five miles east of Corinth at Eastport. Rosecrans considered it imperative to protect Eastport as long as possible, and so he deployed a brigade from Stanley's division at Iuka to shield the river town. The remainder of the division he hoped to draw off to Corinth. Rosecrans also decided to leave outposts at Jacinto, Rienzi, and Danville to warn of a Rebel approach.[3]

An urgent telegram from Bragg, dated September 6, removed any doubt Price may have had about the wisdom of refusing Van Dorn. Reporting the Federals had evacuated Alabama and were falling back fast toward Nashville, Bragg was sure Rosecrans would follow; consequently, Price "should move rapidly for Nashville" to keep Rosecrans from uniting with Buell.

Price made his final preparations. Bragg, of course, was wrong about Rosecrans, mistaking the crossing of Paine and Mitchell for a transfer of the entire Army of the Mississippi. Price acted on the assumption that Bragg was right, but he nonetheless was prepared to do battle in Mississippi should Rosecrans remain at Iuka. On September 8 he advanced his headquarters to Guntown. Weary of his futile exchange of letters with Van Dorn, Price told Major Snead to inform the Mississippian's adjutant that he "expected to move immediately against Iuka in accordance with order just received from General Bragg, who again instructs him to follow Rosecrans."[4]

*   *   *

Thursday, September 11, dawned bright, hot, and dry. Reveille sounded long before daylight, and the troops spent the morning loading wagons. The effort was badly organized. It took six hours to pack down the munitions and stores, leading General Little to complain to his diary that there was "great inefficiency somewhere." His ire, however, ranged wider than mere frustration with the quartermaster corps. Little was dead set against a march on Iuka. That morning he had risen, sick and angry, and tried one last time to try to dissuade Price from his plan, which Little argued would leave the interior of Mississippi open to a Federal counterthrust. But Price was not to be persuaded, not even by Little.[5]

Perspiring and caked with dust, at noon the soldiers of the Army of the West filed onto a confusing web of trails that laced together the wild, uncultivated country between Baldwyn and Corinth. Fourteen thousand strong they marched in the direction of Marietta, a small village eight miles east of Baldwyn. Armstrong's cavalry screened them. The roads converged and then branched unexpectedly, and by late afternoon many of Little's regiments were badly entangled.

Little reached Marietta an hour and a half after sunset. He was in passable health, but his men were exhausted. Most camped in a swamp near Twenty Mile Creek with only their blankets to protect them from the rank air and mucky ground; others simply lay down on the road, and a fortunate few made camp beside a clear stream. A storm came up during the night, and by morning all were drenched and equally miserable. It rained off and on all day, and the infantry covered only eleven miles. At Price's behest Armstrong rode on toward Iuka. Little encamped a mile east of Bay Springs, a decrepit village with only a gristmill to brag on. There he was joined by Maury, who had marched his division over the Saltillo–Bay Springs road. Sick again, Little retired to an ambulance. In his diary that night he speculated that a battle seemed likely at Iuka.[6]

Grant and Rosecrans also expected a fight. Civilian scouts from Charles Hamilton's division, which was posted at Rienzi, had detected the movement of Little from Saltillo to Baldwyn. Two days later, on September 8, patrols from Col. Edwin Hatch's Second Iowa Cavalry ran into troopers from Armstrong's brigade along Twenty Mile Creek. Deserters began to drift in bearing news that Price and Van Dorn had combined for a strike toward Kentucky. An escaped prisoner from an Ohio regiment told Rosecrans he had overheard two angry civilians, whose farms had been sacked, threaten to bushwhack the Yankees responsible. Price had admonished them to go home; he "would rectify matters in a week or two."

Rosecrans was concerned but confident. As "things were now beginning to wear a threatening aspect," he removed his headquarters from Iuka, where he had gone to supervise the crossing of Paine and Mitchell, to Camp Clear Creek. Grant had given him the division of Thomas A. Davies and two brigades of Charles McArthur's division from his own army to help hold Corinth, and Rosecrans counted himself secure there. He doubted Price could muster 12,000 men and thought the Rebels were "playing a game of bluff." Grant was less certain. "For two days now I have been advised of the advance of Price and Van Dorn on this place," he wired Halleck from Corinth on September 9. "Should the enemy come I will be as ready as possible with the means at hand. I do not believe a force can be brought against us that cannot be successfully resisted." Yet Price's movements baffled him. "With all the vigilance I can bring to bear I cannot determine the objects of the enemy," Grant confessed to Halleck the same day. "Everything threatens an attack here but my fear is it is to cover some other movement." Rosecrans's cavalry reported Price's departure from Baldwyn promptly, but the knowledge did little to clarify the uncertainty in the Federal high command. Grant became more convinced that Corinth was about to be attacked; Rosecrans was just as certain the whole thing was a heavy-handed demonstration.

Perhaps, but Grant was not inclined to take chances. On September 11 he ordered Rosecrans to gather his forces at Corinth in anticipation of an attack Grant thought would come within forty-eight hours. Rosecrans returned from Iuka on September 12, bringing with him Col. John Fuller's Ohio brigade of Stanley's division. Col. Robert Murphy's brigade was left behind to protect the stores at Iuka.[7]

At Bay Springs Price was still unsure where Rosecrans was or what he was doing. He predicated his marching orders for September 13 on the assumption that Rosecrans's army, rather than just Murphy's lone brigade, was at Iuka and that Rosecrans was about to cross the Tennessee River at Eastport. Not wanting to lose him, Price exhorted Little to make better time in the morning. Armstrong would ride on ahead before dawn, and General Maury was to march over the old Natchez Trace, east of Little's division. The day's objective for both divisions was Peyton's Mill, a hamlet twelve miles south of Iuka.

Price had his troops on the road at 3:00 A.M. The recent rain had settled the dust but not broken the hot spell. The soldiers shuffled past houses sacked and burnt by the Yankees and forded streams long dry. By mid-morning the temperature reached ninety degrees, and only a low ceiling of dark clouds and a light breeze made the march bearable.

The infantry converged on Peyton's Mill at 5:00 P.M. Wagons were still several miles behind, so the men set about cooking what rations they carried with them. Watching his troops huddled about their campfires, General Little grew hopeful. He liked their spirit and good humor the night before what he was certain would be a battle at Iuka. "God grant us victory," he wrote in his diary before retiring to his ambulance.

Little's wagons rolled into camp at 8:30 and parked. The men threw themselves on the bare ground to sleep, and the camp fell silent. At 10:30 P.M. the warm stillness of the night was broken by the rattle of beating drums. Dirty and exhausted, the men grounded their knapsacks and fell into line. Word spread that Frank Armstrong's cavalry had met the enemy at Iuka but needed infantry support to take the town.[8]

The news cheered the men. Back onto the road they filed at 11:00 P.M., "old men and young boys, rich planters on blooded horses and Negro laborers on foot; farmers and clerks; grizzled hunters and tough keelboat men; prosperous merchants and plain backwoodsmen," remembered a Texan. "On they came, glutting narrow roads, overflowing into the forest; undulating, talking in smooth drawls or emitting shrill, terrifying cries — as strangely assorted and colorful an army as ever human eye rested upon."

As the excitement wore off, the men became aware of their fatigue and discomfort. Few had done any real marching since abandoning Corinth in May. Several hundred of Maury's men had drawn new shoes just before leaving Tupelo, and their feet were now painful welts. Troops by the score slipped from the ranks and melted into the forest to sleep.

Somewhere near the head of the column was Henry Little. His stomach was weak and fever gripped his body. Fleas had made a home in his clothing. Little had expended his last reserves of strength mounting his horse. Now the gentle rocking of the saddle lulled him into a deep sleep, and the horse walked on, guided by the light of a full moon and the silhouettes of the marching soldiers.[9]

*　　*　　*

Soldiers of both armies agreed on the merits of Iuka, Mississippi. It was "a pretty little village," recalled a Missourian. A member of the Eighth Wisconsin remembered Iuka as "the first place we had seen in the South that looked anything like a business town. Wealth, affluence, and southern grandeur were plainly visible. Houses built in the most improved style. Gardens beautifully arranged and blooming. It seemed a pity to see such a beautiful village become the prey of contending armies."[10]

Iuka enjoyed a national reputation for its abundant mineral springs. Invalids came to bathe in the restorative, sulphur-tinged waters and relax in surrounding hotels. Named for a Chickasaw chief who had made his village beside the springs, Iuka grew rapidly after white settlers took the Indian lands, expanding in a long strip on either side of the Memphis and Charleston Railroad, which laid track through the town in 1857. Merchants flocked to Iuka with the tourists, and by 1860 the permanent population of Iuka was nearly 1,500.

Besides the railroad four good wagon roads served the town. The Eastport and Fulton Stage Route entered Iuka from the northeast. The Iuka and Corinth Stage Route connected Iuka and Corinth from the west. Two roads led into town from the south: the Jacinto road and, running parallel to it a mile and a half to the east, the Fulton road. Like that surrounding

Corinth, the country around Iuka consisted of vast swamps, rolling hills, sharp ravines, indifferently plowed fields, bottomland pastures, and forests of oak and pine.[11]

* * *

Frank Armstrong relished a good fight. Born in the Indian Territory to a Regular Army father, Armstrong accepted a direct commission after graduating from Holy Cross Academy in Massachusetts. He fought for the North at First Bull Run but resigned three weeks later to serve on the staff of Ben McCulloch. After Pea Ridge the soldiers of the Third Louisiana Infantry elected him colonel; it was as commander of the Third that the twenty-six-year-old Armstrong attracted the notice of General Price. His performance at Bolivar two weeks earlier affirmed Armstrong's worth, so that Price had no misgivings about turning him loose against the garrison of Iuka, a day's march ahead of the infantry.

Armstrong struck at 8:00 A.M. on September 13. His troopers captured thirty pickets before Col. Robert Murphy mustered enough men to drive them off. All morning Armstrong's men probed the Federal defenses for weaknesses. Three times they attacked, capturing a handful of prisoners on each occasion. The Yankees counterattacked and took two prisoners themselves, a captain and a private whom they hustled back to headquarters for interrogation. Both confessed that Price was one day away with the rest of the army — "a very strong hint to us," said an officer of the Eighth Wisconsin, "that it was best to retire in good order."

Murphy agreed. Rosecrans had ordered him to defend Iuka until the stores could be removed; nothing had been said about holding at all cost. Murphy tried to telegraph Rosecrans of the attack, but the wires between Iuka and Corinth were dead. Murphy sent a staff officer with a three-man escort to take a dispatch to Rosecrans; they were never heard from again. All day Murphy waited for a train that was to transport the thirty carloads of stores at Iuka to Corinth. Convinced that Armstrong had torn up the track, he sent repairmen toward Burnsville to find the break; they never returned. Murphy assumed the worst and made his preparations accordingly. He sent a single courier on a fast horse to Burnsville with a dispatch explaining his predicament, then told his brigade quartermaster to gather all the wagons in town, load them with supplies, and be ready to move.

The courier returned at 8:00 P.M. still bearing the dispatch. Burnsville had been evacuated; townspeople told him that the last Yankee had left by train for Corinth that morning. For the next six hours Murphy agonized over the commissary stores and waited for some word from Rosecrans. At

2:00 A.M. he gave orders to evacuate Iuka and detailed a company of cavalry to set fire to the stores. The wagon train left for Farmington under cover of darkness. The infantry followed at daybreak.[12]

Out on the Fulton road, two miles from Iuka, Price's infantry prepared for battle. Armstrong had reported Murphy's presence in town, and Price was eager to capture the Federals along with their stores. It was Price's desire to close on Iuka rapidly that had driven Little and his weary soldiers back onto the road on the night of September 13. Progress had been slow. Between 11:00 P.M. and daylight Price's infantry covered fewer than ten miles, and at dawn the army was three miles short of Iuka. A courier from Armstrong's cavalry intercepted Price along the road. The enemy were in Iuka in strength, he said, and reinforcements were on the way to them from Burnsville. Price's soldiers sunk into the road to sleep while the Missourian pondered the intelligence. Having no reason to doubt the courier, Price ordered the troops roused. The Third Texas Dismounted Cavalry and the Third Louisiana were brought forward and told to move on the town at the quick time.

Armstrong's troopers led the way. They galloped into Iuka at 7:00 A.M., brushing away a few Union cavalrymen who had lingered to burn the stores. Armstrong detailed a detachment to pursue the Yankees toward Burnsville, then sent back word to Price that the town was empty.[13]

Murphy's escape disappointed Price; but his tired soldiers were pleased to have taken the town quietly, and the wealth of abandoned stores that greeted them palliated any regret over a missed chance to fight Federals. The two officers whom Murphy left behind to superintend the destruction of the stores had failed miserably, running at the first faint echo of beating hooves. They left behind $30,000 worth of commissary and quartermaster stores, a long train of railroad cars laden with supplies, and dozens of well-stocked sutlers' shops.

The Rebels broke ranks and assaulted the stores. Remembered a Texan, "It was a sight to gladden the heart of a poor soldier, whose only diet had been unsalted beef and white leather hoecake, the stacks of cheese, crackers, preserves, mackerel, coffee, and other good things that lined the shelves of the sutlers' shops and filled the commissary stores of the Yankee army."[14]

The Southerners were determined to rectify the discrepancy. As the first regiment in town, the Third Louisiana fared especially well. Sgt. Willie Tunnard came away with an armful of condensed milk, canned fruit, lager beer, and fine wines. After depositing their spoils in a safe place, a few Louisianans amused themselves with abandoned handcars.

Tunnard watched them push the cars up a steep grade, jump on, and descend at breakneck speed, yelling like children.

Sadly for the trailing regiments of the army, order was quickly restored and guards posted over the stores. The latecomers skulked away, convinced that the officers had appropriated the delicacies for their exclusive use.

Lt. Col. Robert Bevier of the Fifth Missouri certainly profited. He had been ordered to post a guard over the Iuka Springs Hotel, every room of which "was filled with cheese and crackers, ham and hominy and molasses and whiskey. I found infinite difficulty in protecting my treasure. True to the 'old soldiers' motto, every guard would provide for his own mess and jealously protect from all others." But, confessed Bevier, "my feelings were somewhat mollified when I found that my cook had amply provided for my own larder, and I ate in silence and asked no questions."[15]

General Little was too tired and sick to partake of the spoils. While his colonels restored order, he slipped from the saddle, accepted as his headquarters a dilapidated house that fortunately was occupied by an attentive family, and after reading a bit of a captured Yankee newspaper, fell fast asleep.

Pvt. Sam Barron of the Third Texas Cavalry was equally apathetic. Laid low with dysentery at Baldwyn, Barron only recently had returned to duty. The night march to Iuka nearly killed him. When his comrades ran off to gather spoils, Barron collapsed in the road, "and as soon as we went into camp I fell down on the ground in the shade of a tree where I slept in a kind of stupor until nearly midnight."

Barron's comrades slowly drifted into camp. The excitement of the capture was wearing off, and in its place came a numbing fatigue. Those lucky enough to slip away with Yankee rations cooked and ate them, then spread out into the forests outside town to rest. By sunset nearly the entire army was asleep.[16]

Price indulged the men; his mind was given over to deeper musings. Apart from Federal stores Price's nearly bloodless capture of Iuka had gained him nothing. In fact, it was beginning to look as if the whole march had been a grueling waste of time. Reports came in that, rather than attempting to join Buell, Rosecrans had fled westward with his two remaining divisions. Assuming the Ohioan's plans indeed were to enter Middle Tennessee, Price recognized that he had at best delayed him. The Army of the West was too far south to prevent a Federal crossing of the Tennessee River north of Eastport; Rosecrans need only march north eighteen miles and cross at Pittsburg Landing. On other hand, Price could not himself push on to join Bragg without exposing his line of communications or inviting a flank attack by the superior forces of Grant and Rosecrans. Calculating that Bragg's orders for him to come to Nashville had been based on bad intelligence regarding Rosecrans's whereabouts, Price opted to remain in northern Mississippi.

The only sure way of preventing Rosecrans from slipping into Middle Tennessee, Price concluded, was to attack him at Corinth. But he dare not risk his army alone. On the evening of September 14, as his men bedded down around Iuka, Price again tried to interest Van Dorn in a combined attack on Corinth: "Rosecrans has gone westward with about 10,000 men. I am ready to co-operate with you in an attack upon Corinth. My courier awaits your answer." [17]

Van Dorn had been busy during Price's advance on Iuka, but none of his activity was directed toward cooperating with Price. To the contrary, the Mississippian was zealously availing himself of Price's preoccupation with the Federals to his front to subvert him from behind.

Van Dorn had begun his unseemly machinations on September 8, when he urged Price to ignore Bragg's orders and instead help him retake West Tennessee. After giving up on Baton Rouge, the Mississippian had placed a garrison at Port Hudson, a river strong point located between Baton Rouge and Vicksburg. Feeling that he had sufficiently provided for the security of the lower Mississippi, Van Dorn looked to take the offensive — on his own terms. Bragg's predicament elicited only his contempt. "If Rosecrans has crossed the Tennessee and got beyond your reach do you not think it would be better for us to join forces at Jackson, Tennessee, clear Western Tennessee of the enemy, and then push on together into Kentucky. . . . If Rosecrans is much ahead of you he could join Buell and meet Bragg before you could aid him," Van Dorn argued, as if Bragg's orders had been discretionary. While awaiting Price's reply, he made ready to move his own army from Holly Springs to Grand Junction, Tennessee, preparatory to an attack on Memphis.

Price was not fooled: he must obey Bragg, and if need be, he would march alone into Tennessee. Van Dorn changed his tactics. He wrote Secretary of War George Randolph and demanded that, by virtue of date of rank, he be given "command of the movements of Price, that there may be concert of action." Van Dorn also asked Randolph to grant him 5,000 recently returned, but not yet exchanged, prisoners from Fort Donelson that Bragg had promised to Price, arguing that "I can make a successful campaign in West Tennessee with them; little without them."

At a loss what to do, Randolph bucked the matter to President Davis.

"I supposed these matters would be regulated by General Bragg, and feel some hesitation in giving directions which might conflict with his plans. Something, however, should be done," Randolph endorsed the letter before passing it to the president. "Shall I order him to take command of the prisoners, subject to General Bragg's orders?"[18]

Jefferson Davis's tendency to pay less attention to affairs in the West than their importance dictated, too often intervening only after a crisis became acute, made itself felt now. Despite the chivalrous tone of their correspondence, there could be no cooperation between Van Dorn and Price so long as neither commanded the other. Someone had to step in, but the president's unfamiliarity with the situation in northern Mississippi and with Bragg's intentions meant whatever action he took would be uninformed.

Davis made a halfhearted effort at informing himself. He told Bragg about the mess in Mississippi and confessed, "I am at a loss to know how to remedy [these] evils without damaging your plans." The danger was great, Davis pointed out. "If Van Dorn, Price, and Breckinridge each act for himself disaster to all must be the probable result."

His impatience for a quick fix got the better of Davis. Having affirmed that a misstep in Mississippi might prove fatal, the president nonetheless went ahead without waiting for Bragg to answer. Davis addressed Randolph's query first, but his answer was tentative and unclear. Returned prisoners should join their respective regiments as soon as they were formally exchanged, regardless of the district in which the units were serving. On the issue of overall command, Davis said only that "the rank of General Van Dorn secures to him the command of all the troops with whom he will be operating."

What did that mean? Van Dorn was not operating with Price, and there was no prospect that, short of being in charge, Van Dorn would do so. Davis apparently recognized the inadequacy of his response, because on September 9 he wrote Van Dorn directly and decisively on the question: "The troops must co-operate and can only do so by having one head. Your rank makes you the commander."

No one bothered to inform Price or Bragg of the decision. Unaware that he was now subordinate to Van Dorn, Price settled into his headquarters at Iuka to ponder his next move.[19]

* * *

Grant and Rosecrans were baffled. Neither could discern the purpose behind Price's march on Iuka. Grant guessed it portended an attack on

Corinth, perhaps within forty-eight hours. Rosecrans speculated that the elusive Price might bypass Grant and make for the Tennessee River. "As Price is an old woodpecker, it would be well to have a watch set to see if he might not take a course down the Tennessee toward Eastport, in hopes to find a landing," he suggested to Grant.[20]

Reports from Union scouts only compounded the confusion. One report claimed that Price intended to invade Middle Tennessee and that if Grant tried to pursue him, Van Dorn would attack Corinth. Other reports repeated rumors that the Confederates were combining to attack Corinth — Van Dorn by way of Ripley, and Price from the northeast. A third report had Price over the Tennessee River and on the way to Kentucky.

All the reports agreed on one essential, that the Confederates were concentrating their forces east of the Mississippi Central Railroad. Consequently, Grant elected to draw down his scattered detachments west of the line. He ordered Maj. Gen. Stephen A. Hurlbut to reinforce the garrison at Bolivar with his division from Memphis. Next he called in the troops guarding the railroad itself, intending to bring them to Corinth. Grant countermanded the order when he learned that Hurlbut had been delayed in getting to Bolivar, and he told them to reassemble at Jackson instead. From Bolivar he took the brigade of Col. Marcellus Crocker, and he advised Maj. Gen. Edward O. C. Ord, who held the center at Jackson and Bolivar with the divisions of Ross, McArthur, and Davies, to stand ready to move to Corinth.[21]

In good spirits and good health, Rosecrans could joke about the Old Woodpecker Price, but the uncertainty was playing hard on Grant. He felt deeply the absence of Julia and the children. Rawlins had returned, which was some comfort, but he was as yet too weak to do much strenuous work. And Grant himself was far from well. "My health is still like it was when you left; a short appetite and a loss of flesh," he wrote Julia. "From being some fifteen or twenty pounds above my usual weight I am now probably below it. I am beginning to have those cold night sweats again which I had a few years ago." To his sister Grant despaired, "I have not been very well for several weeks but [have] so much to do that I cannot get sick."

Grant's patience was easily taxed. News of Murphy's retreat from Iuka "disgusted" him, as it did Rosecrans, who relieved the colonel and brought him up on court-martial charges. Absorbed with the fate of Kentucky, Halleck badgered Grant for reinforcements while dismissing his fears. "There can be no very large force to attack you," Halleck wired Grant. "Attack the enemy if you can reach him with advantage."[22]

To reach Price Grant first had to find him. On September 15 he directed Marcellus Crocker to reconnoiter east along the Memphis and Charleston Railroad toward Iuka until he found the Confederates.

It was a poorly conceived order, a reflection of Grant's anxiety. Crocker's brigade was a reliable command, but the men were nearly played out from three days' hard marching, and they had yet to reach Corinth. The usual agonies of a summertime march in Mississippi—oppressive heat and humidity, swirling dust, and no water—had knocked nearly half the men from the ranks and consigned most of the officers to ambulances. When the brigade at last stumbled into camp outside Corinth, it had dwindled to less than a regiment.

Crocker could ask no more of his men. He ignored Grant's order and permitted the brigade to pitch camp. Rosecrans seconded him, telling Grant that Crocker's brigade was in no condition to move. Instead Rosecrans gave Colonel Murphy's brigade, now led by Joseph A. Mower, a chance to redeem itself.[23]

The assignment was less ironic than it seemed; Mower was no Colonel Murphy. A scrappy, hard-drinking, hard-fighting Vermonter, he had entered the army during the Mexican War as a private. Mower found army life to his liking, and seven years after the Mexican War he was promoted to second lieutenant. He was in Missouri when Fort Sumter was fired upon, and there he accepted a commission as colonel of the Eleventh Missouri. Mower's troops liked him and nicknamed him "the Wolf," perhaps for his predatory instincts in battle. Already Mower had made an enviable record for himself as one of the most reliable regimental commanders in Rosecrans's army, when sober.[24]

At daybreak on September 16 Mower set out for Burnsville by train. Voluble townspeople told him that Price was at Iuka with his entire army, but the Vermonter was skeptical and decided to press forward on foot.

It was eight miles to Iuka over the sandy, narrow Burnsville road. The day was hot and the going hard, recalled one Illinois lieutenant: "Stifling dust, parched throats, and aching eyes irritated by sand, the smell of sweaty leather from burdening knapsack . . . gun barrels that are hot and bayonets whose glint is an irritation, the water in the canteen hot and brackish, though the cloth cover is kept well wet."

Mower pushed his men hard on the march, so hard that his brigade came to be called "Joe Mower's Jack Ass Cavalry." At 4:00 P.M., six miles from Iuka, his advance guard ran into pickets from Armstrong's cavalry brigade. The Rebel horsemen squeezed off a few harmless shots, then turned for town with news of the Federal approach.[25]

The troopers' excited arrival roused the Confederate camp. For two days Price and his lieutenants had expected the Federals to return in strength. They required the men to sleep within reach of their rifles, and the flimsiest rumor of a Yankee approach was enough to send the army out of camp and into line of battle. Already the men had been called out four times to meet phantom threats, and they were wearying of the game.

This time the crack of pickets' rifles and the low boom of cannon gave the alarm a ring of truth. Price hurried Dabney Maury's division out the Burnsville road at the double-quick time. A mile northwest of town Maury deployed his command along a ridge that offered an unobstructed view for nearly a mile in the direction from which the Yankees were expected. He pushed out the Second Texas Infantry and a battalion of Arkansas Sharpshooters as skirmishers, protecting their flanks with dismounted detachments from Wirt Adams's Mississippi Cavalry Regiment.[26]

Late afternoon shadows stretched across the road and melted into the pine forest. The air was damp and smelled of rain. On high ground a mile northwest of Maury's position, Colonel Mower formed line of battle, and the two forces contemplated each other across 1,200 yards of darkly forested ravines. Under a warm, heavy rain their skirmishers kept up a desultory fire until nightfall.

Both sides slept on the field. The temperature fell sharply and the rain kept falling. A deserter from the Second Texas slipped into the Yankees' lines and told Colonel Mower that Price was in Iuka with at least 12,000 men. He intended to draw the Union army out from Corinth, the Texan added, after which Van Dorn would attack the town. Mower believed the Texan. Under the cover of night he quietly withdrew his outnumbered command. By 11:00 P.M. on September 16 Mower was back in Burnsville.[27]

Rosecrans forwarded the results of Mower's reconnaissance to Grant. General Hamilton at Jacinto corroborated the Vermonter's intelligence. His cavalry had just captured an ordnance train; the Rebel drivers confessed that they had come from Iuka, where Price lay with "his whole force."[28]

The news awakened Grant. From doubt and vacillation he swung in the space of a few hours to an almost reckless frenzy. On the strength of Colonel Mower's report he decided to attack Price. Whatever might be Van Dorn's intentions, Grant was sure the Mississippian could not possibly reach Corinth in less than four days, giving him time to draw down the garrison there for a rapid march against Price.

Hastily composed, Grant's plan was simple enough. He would bring together the commands of Ord and Rosecrans for an attack on Iuka from the west. After leaving enough troops at Rienzi and Jacinto to prevent a surprise attack on Corinth, Rosecrans was to march with the remainder of his army to Burnsville, there to meet Ord, who had just reached Corinth with the understrength divisions of Brig. Gens. John McArthur and Thomas A. Davies. From Burnsville Ord would march over the Burnsville road, north of the Memphis and Charleston Railroad, to Iuka and then attack from the northwest. Rosecrans, meanwhile, would advance south of the railroad on Ord's right. Grant would travel with Ord. To reinforce Ord Grant summoned Brig. Gen. Leonard Ross's division from Bolivar. Assuming Ross arrived promptly, Grant calculated he would have 15,000 men — 6,000 in Ord's column and 9,000 in Rosecrans's — with which to meet Price, whose numbers he estimated to be roughly the same.[29]

Rosecrans agreed with Grant's purpose but questioned his method. He believed that the best "chances of success lay in the celerity of our movements." Since Hamilton already was at Jacinto, nine miles southwest of Burnsville, and the country from Burnsville to Iuka was, as Rosecrans described it, "full of morasses and covered with brush, [which] would be difficult to operate in," Rosecrans suggested that he be allowed to march from Jacinto to Iuka — it would save time, and the ground was better.

Actually the idea had originated with Hamilton. On the night of September 17, while reporting the capture of a Confederate wagon train, he admonished Rosecrans to "let Stanley join me here and we will slip into Price's rear, and have another like force move on him from Corinth and he will be in a tight place." There was much in Hamilton's proposal to recommend it, as Rosecrans perceived. From Jacinto he might march both his divisions along the Tuscumbia road to Barnett's Crossroads, four miles southwest of Iuka, then swing to the northeast and block both the Jacinto and Fulton roads, Price's only lines of retreat. Leaving a small force to hold the Jacinto road, Rosecrans then could move his main body along the Fulton road toward Iuka. If Ord attacked Price simultaneously, the Confederates would be snared in a four-sided trap with no means of escape: the Tennessee River to the northeast was impossible to cross without lengthy preparations; Ord would close off the northwest exit from Iuka; and Rosecrans would seal off the roads leading southwest and southeast from town.

Grant deferred to Rosecrans in his "plans for the approach" because the latter "had a most excellent map showing all the roads and streams in the surrounding country [and] was also familiar with the ground."[30]

Grant was gambling. In the Civil War simultaneous movements of two widely separated commands were, as Rosecrans's biographer aptly put it, "notoriously tricky" to orchestrate. Even a plan seemingly as simple as Rosecrans's "required close timing and careful leadership for each jaw of the pincers, for Price had sufficient strength to match either Union force." The able French student of the war the Comte de Paris thought the risks should have been obvious: "In a region the topography of which was so little known, where the roads became broken up at the first rainfall, and the streams, the swamps and the forests combined to retard the movements of armies, and communications between headquarters were extremely uncertain, such a maneuver, undertaken in the presence of so active an adversary as Price, was full of danger." But both Grant and Rosecrans judged the opportunity to catch the armies of Van Dorn and Price apart to be greater than the danger of a preemptive attack by either one. Anxious to start even before he heard Rosecrans's proposal, Grant had written General Ord on the afternoon of September 17: "We will get off all our forces now as rapidly as practicable. I have dispatched Rosecrans that all our movements now would be as rapid as compatible with prudence." Ord was to come on to Glendale before dusk, there to be ready to board trains for Burnsville at dawn on September 18.[31]

Despite a torrential rain that fell all day, Ord reached Glendale as planned. Cyrus Boyd of the Fifteenth Iowa marveled at the misery the men endured on the march. "Have traveled all day and toward all points of the compass and the rain has poured down all the time turning the dust all to mud," he complained to his diary. "We are wet to the hide and the air is very cold. We made up a fire in some old pine logs and stayed up most of the night drying our clothes and keeping warm." All night long the rain came down, pounding the roads to paste.

It was an inauspicious beginning to a movement that demanded rigid adherence to a timetable.[32]

6

Let Us Do All We Can

Earl Van Dorn was ready to fight. The president had made it clear he was to command in Mississippi and that the better part of the paroled prisoners were to be his. With them, Van Dorn boasted to Secretary of War Randolph, he could "clear the west" of Federals even without Price's cooperation.

Although he had the authority and might soon have the men needed to take the offensive, Van Dorn had trouble settling on an objective. First he told Randolph he was inclined to move against Memphis from Holly Springs. Then, on reflection, the Mississippian concluded Memphis was of minimal strategic value and could never be held against the combined land and river counterattack that would surely follow its capture. Van Dorn also ruled out Bolivar. Stripped of its garrison to reinforce Corinth, the Tennessee railroad town could be taken easily, but a march against it would expose Van Dorn's lines of communication to the Federals at Corinth.

Van Dorn at last conceded that Price had been right; Corinth was indeed the key to clearing the Federals from northern Mississippi and to recovering West Tennessee. On September 16 Van Dorn told Price he was disposed to join forces for an attack on the town. While neglecting to mention Davis had placed him in overall command, Van Dorn nevertheless ordered Price to bring his army to Pocahontas, by way of Rienzi, "so that we may join our forces and attack without loss of time. . . . Rosecrans is a quick, skillful fellow, and we must be rapid also." Regardless of how the Federals responded, he added, Corinth would be the objective and Pocahontas the assembly area. Two couriers delivered Van Dorn's message on the morning of September 19 and, rather unceremoniously, announced to Price that President Davis had placed him and his Army of the West under Van Dorn's command.[1]

Price welcomed Van Dorn's summons. His relief at the prospect of decisive action overcame any pique he may have felt at what Thomas Snead, at least, considered a high-handed indignity on the part of the president. Indeed, Price was eager to leave Iuka under any circumstances. The evening before, Frank Armstrong had warned that the Yankees were six miles northwest of town and coming on "in considerable force"; his own pickets were fast falling back before them. Maury pushed his division four miles down the Burnsville road to meet the Federals, who halted two miles short of his lines, and at nightfall Price committed Little to extend Maury's right.

The stay in Iuka had agreed with Little. His appetite had returned, and he had slept well two nights running. With his health back, Little felt "very hopeful; if they attack us we will not only repulse them but gain a victory! God grant it!"[2]

Perhaps God would grant him victory, but Price was not going to tempt fate to find out. Unlike Little, Price had found nothing in Iuka to comfort him. Each day the army lingered in town, his fear of Grant and Rosecrans grew—and with it his calculations of their strength—so that by nightfall on September 18 Price was convinced that they had marshaled 30,000 or 40,000 men against him. Van Dorn's directive gave Price the nudge he needed to leave. He answered Van Dorn decisively at dawn on September 19: "I will make the movement proposed in your dispatch. . . . Enemy concentrating against me."

But by noon Price doubted he could get away without a fight. Daylight had revealed Leonard Ross's lead division of Ord's command, drawn up on Turnpike Hill, and Price could only assume that more divisions lay behind it. He ordered supply trains loaded and the troops readied to march the next morning. To Van Dorn Price now wrote, "I will move my army as quickly as I can in the direction proposed by you. I am, however, expecting an attack today, as it seems, from the most reliable information which I can procure, that they are concentrating their forces against me."

Price knew that the enemy he faced northwest of Iuka was commanded by Ord. What he did not know was that he was about to confront a threat from the south that not only would rule out a quick departure from Iuka but also would jeopardize the very survival of his army.[3]

*   *   *

Rosecrans was on the move, slowly and behind schedule, but closing on Iuka nonetheless. The drenching rain of September 17 had disrupted his timetable. Rosecrans had assured Grant that both his divisions would be at Jacinto before midnight, but darkness and muddy roads had prevented Stanley from reaching the rendezvous point with Hamilton at Davenport's Mill, on the outskirts of Jacinto. The rain lingered until nearly 10:00 A.M. the next morning, compelling Rosecrans to write Grant that it would be 2:00 P.M. before he had his forces concentrated at Jacinto. But, Rosecrans was quick to add, he would then immediately push on beyond Jacinto to the Bay Springs road: "If Price's forces are at Iuka the plan I propose is to move up as close as we can tonight and conceal our movements."[4]

Concerned but not alarmed, Grant concluded that Rosecrans would be near enough to Iuka by nightfall on September 18 to permit Ord safely to begin his attack the next morning. He guessed that Armstrong's troopers had been deployed on Ord's front to cover a Confederate retreat. Anxious to strike a blow before the Old Woodpecker flew off, Grant told Rosecrans Ord would advance at 4:30 A.M. and admonished him to close up.[5]

Grant's appeal found Rosecrans not south of Iuka, as he had assumed, but at Jacinto, many hours behind schedule. General Stanley again was the unwitting culprit. Whether from treachery or by accident, a local guide had led Stanley astray, so that instead of entering Jacinto at 2:00 P.M. on September 18 as Rosecrans had expected, his column stacked up on the road behind Leonard Ross's division near Burnsville. When Rosecrans learned of Stanley's predicament, he elected to bivouac at Jacinto. At 9:00 P.M., as Stanley's soldiers shuffled into camp, Rosecrans wrote Grant of the delay. He tried to put a hopeful face on matters. Hamilton was only eight miles from Barnett's Crossroads, and Stanley would start at 4:30 A.M. to make up the lost ground. Although the men were tired and Iuka was twenty miles away, Rosecrans thought he could reach the town by 2:00 P.M. on September 19 — certainly no later than 4:30 P.M. In the worst case, Rosecrans assured Grant, two and a half hours of daylight would remain to give battle. "When we come in," Rosecrans concluded, "[we] will endeavor to do it strongly."[6]

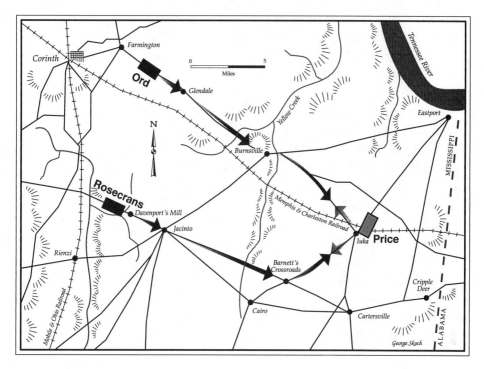

MAP 2. *Federal Advance on Iuka, September 17–19*

Rosecrans's feigned exuberance fooled no one at Burnsville. Grant was "very much disappointed" and dismissed Rosecrans's guarantees. Recalled Grant, "He said, however, that he would still be at Iuka by two o'clock the next day. I did not believe this possible because of the distance and the condition of the roads, which was bad; besides, troops after a forced march of twenty miles are not in a good condition for fighting the moment they get through. It might do in marching to relieve a beleaguered garrison, but not to make an assault." Grant altered his plans accordingly. He later claimed that he "immediately sent Ord a copy of Rosecrans's dispatch and ordered him to be in readiness to attack the moment he heard the guns to the south or south-east." No copy of any such communication from Grant to Ord has been found, and Ord in his report said he did not hear of the change of plans until 10:00 A.M. on September 19, when he was handed a message from Grant's aide, Col. Clark Lagow, that read, "I send you dispatch received from Rosecrans late in the night. You will see that he is behind where we expected him. Do not be too rapid with your advance this morning unless it should be found that the enemy are evacuating."

Why Grant would have waited until morning to inform Ord of Rose-crans's situation and risk bringing on a battle before both columns were in place is a mystery. In any case Ord did not move out aggressively at day-break, so no harm was done. Rather, Ord undertook a leisurely reconnais-sance, gently probing the Rebel lines but otherwise avoiding a fight.[7]

In the meantime perhaps Grant and Ord could bluff Price into surren-dering. A telegram just in from the War Department gave a stunning — and badly distorted — account of the recently concluded Battle of Antie-tam. There, near the banks of the Potomac River in Maryland, Robert E. Lee's invasion of the North had been turned back. Despite suffering heavy losses, Lee's army had recrossed the river into Virginia intact. But what Grant read was this: "Longstreet and his entire division prisoners. General Hill killed. Entire rebel army of Virginia destroyed." Were this true, Grant reasoned, then the war was all but over; a fight at Iuka would be a senseless effusion of blood. He asked Ord to pass the telegram through the lines and call on the Confederates "for the sake of humanity to lay down [their] arms."

Price was not intimidated. He placed no credence in the dispatch; even if it were true, Price assured Ord, the news "would only move him and his soldiers to greater exertions in behalf of their country. . . . Neither he nor [his soldiers] will lay down their arms . . . until the independence of the Confederate States shall have been acknowledged by the United States."[8]

While it failed to produce the desired effect on Price and his lieu-tenants, the erroneous news of Lee's annihilation did cheer Rosecrans's tired soldiers, who stepped onto the Tuscumbia road before daybreak on September 19 for the thirteen-mile hike to Barnett's Crossroads, the first leg of their march on Iuka. That the Rebels might stand and fight troubled few: a large number of Rosecrans's soldiers were green and knew not what to expect, and the battle-hardened preferred a stand-up fight to senseless marches and vicious clashes with guerrillas.

This day, fine weather eased the march. Said the medical director of the army, "The roads were in splendid order, hard, and entirely free from dust. The men marched . . . in fine order, none lagging and very few straggling." Morale was high. Samuel Byers and his comrades of the Fifth Iowa were "perfectly gay with anticipation of being killed. . . . We sang jovial songs as we marched along. . . . Hurrying through the woods to-wards the enemy, we saw the poetry of war."[9]

Rosecrans too was cheerful, pleased to have gotten his army off early. Col. John Mizner had two regiments of cavalry on the road to screen the advance by 4:00 A.M., and Col. Edward Hatch rode off a few minutes later with his Second Iowa Cavalry to reconnoiter beyond the army's right flank, in the direction of Peyton's Mill. Awakened at 2:00 A.M., the vanguard of the infantry, Col. John B. Sanborn's brigade of Hamilton's division, was on the march two hours later. Gen. Jeremiah C. Sullivan's brigade followed at 5:00 A.M. Even Stanley, who pulled up the rear, looked like he would keep pace.

Progress was good. The cavalry covered six miles before 6:00 A.M., the infantry three, moving Rosecrans to write Grant: "Troops are all on the way, in fine spirits by reason of news [of Lee's defeat]. Eighteen miles to Iuka, but think I shall make it by the time mentioned — 2 o'clock p.m." Unaware that Grant had cautioned Ord to do nothing until Rosecrans opened the fight, he added, "If Price is there he will have become well engaged by the time we come up, and if so twenty regiments and thirty pieces [of] cannon will finish him." Wrongly assuming that Ord would start the battle, thereby preventing Price from turning his full attention southward, Rosecrans reiterated his intention to split his command when it reached Barnett's Crossroads: Hamilton's division would go up the Fulton road, and Stanley would advance along the Jacinto (Bay Springs) road.[10]

First contact occurred when Colonel Hatch and his Iowa troopers stumbled upon a handful of Rebel cavalry pickets two miles northwest of Peyton's Mill. Hatch gave chase. At noon his Iowans rode into the camp of Falkner's First Mississippi Partisan Rangers at Peyton's Mill. The Mississippians were unprepared and badly outgunned. The Iowans dismounted, deployed in the woods near the mill, and poured a rapid fire from their carbines into Falkner's men. Falkner shoved enough of his rangers into line to present a respectable front, and the two regiments traded volleys until the Mississippians receded into a dark swamp. Hatch remounted his regiment and continued on to Thompson's Corner, where he burned some Rebel commissary stores. Satisfied no serious threat existed to Rosecrans's right flank, Hatch returned to Barnett's Crossroads to camp for the night.[11]

Hatch's brief skirmish was largely superfluous to Rosecrans's main effort. Of far greater moment was a clash shortly before noon on the Tuscumbia road, one-half mile west of Barnett's Crossroads, between eight companies of the Third Michigan Cavalry under Capt. Lyman Willcox, charged with screening Sanborn's infantry brigade, and Rebel scouts. Willcox brushed the Southerners away and pressed on toward Cartersville, but Colonel Sanborn chose to pause at Barnett's Crossroads for orders.[12]

They came thirty minutes later. A courier from army headquarters told Sanborn to resume the march until he reached the intersection of the Tuscumbia and Fulton roads at Cartersville. There Sanborn was to "go into the strongest position I could find for defense, and to hold that crossing at all hazards against the enemy." Sanborn sought clarification of the order from his division commander, but General Hamilton could add nothing. Both he and Sanborn concluded that Rosecrans wished Sanborn merely to wait at Cartersville until the rest of the army began the last leg of the march — Stanley by way of the Jacinto road, and Hamilton along the Fulton road.[13]

They were wrong. Cocky and full of fight at daybreak, Rosecrans at noon felt the first flutters of doubt. As Sanborn pressed on to Cartersville, Rosecrans and his staff rode into Barnett's Crossroads. There, beside the recumbent soldiers of Sullivan's brigade, they huddled around their maps. Rosecrans studied the roads confidently until Colonel Mizner reported the true distance between the Jacinto and Fulton roads to be nearly five miles, more than twice what he had assumed.

Rosecrans had made a terrible error. He had cajoled Grant into allowing him to advance in two columns on the assumption that Hamilton and Stanley would be within supporting distance of each other. When he realized they would not, he changed his plan of march. Rather than move in two columns and risk defeat in detail, Rosecrans elected to send both Hamilton and Stanley up the Jacinto road. In doing so he would leave Price a means of escape from Iuka, but Rosecrans reasoned that he lacked the troops needed to block the Fulton road.

The absence of any word from Grant also constrained Rosecrans from dividing his forces. Having heard nothing from him in nearly eleven hours, Rosecrans had no idea what, if anything, had transpired on Ord's front. At 12:30 P.M. two members of Grant's staff, Cols. Clark Lagow and Theophilus Dickey, met Rosecrans at Barnett's Crossroads. They may or may not have brought a message from Grant — Dickey said they did; Rosecrans said they did not — but they were full of gratuitous advice.

"General, do you think the enemy is in force at Iuka?" Lagow asked.

"Yes," answered Rosecrans.

"Are you going to pitch into him?" Lagow wondered.

"Yes of course, that is the understanding of my movement, and we are only five or six miles from the enemy. We ought to hear Grant's opening guns on the Railroad by this time."

"Maybe he is waiting for you to begin," Lagow suggested, adding that there had been no activity on Ord's front since 3:00 A.M.

Rosecrans grew impatient with the impertinent staff officer. "Not so," he snarled. "The main attack should begin on the Railroad to attract the enemy's attention and enable me to surprise his left flank and get the roads in his rear."

Lagow persisted. Did Rosecrans not understand that he must attack at once to prevent the Rebels from retreating? Of course he did, but Rosecrans thought "it would be very bad policy to allow the enemy's attention to be first attracted towards his line of communication, to seize and secure which was the object of my movement."[14]

Despite his reservations Rosecrans issued orders for the army to press on to Iuka. He recalled Sanborn to Barnett's Crossroads and then sent him up the Jacinto road. General Hamilton rode with Sanborn at the head of the column. Captain Willcox redeployed his cavalry to screen the march. Sullivan roused his brigade and fell in behind Sanborn.

Rosecrans hastened to apprise Grant of his whereabouts. From Barnett's Crossroads he wrote, "Reached here at 12. Cavalry advance drove pickets from near here; met another stand at about 1 mile from here. Hamilton's division is advancing; head of column a mile to the front now. Head of Stanley's column is here." He concluded the note with a few trivialities, saying nothing of his decision to leave open the Fulton road.[15]

Rosecrans loitered at Barnett's Crossroads to watch the first of Stanley's infantry turn onto the Jacinto road. It was nearly 1:00 P.M. That morning Rosecrans had blithely assured Grant he would be at Iuka by 2:00 P.M., yet the town was still eight miles away. Colonels Dickey and Lagow remained with him, Lagow still carping about the need to make haste, and Rosecrans just as emphatically insisting that it was Ord who must attack first.

The crackle of rifle fire ended their unseemly debate.[16]

7

Their Delay Was Our Salvation

Captain Willcox and his Michigan cavalrymen were having a grand time. Since swinging into the saddle before daybreak, they had brushed aside every Rebel trooper who dared confront them. Riding north now along the Jacinto road, ahead of Sanborn's brigade, they had no reason to expect anything less. On the far side of Barnett's Crossroads the Michiganders made contact with a squadron of General Armstrong's Confederate cavalry under the command of Lt. E. B. Trezevant.

They drove the Southerners handily to within four miles of Iuka before matters took an abrupt turn for the worst. Drawing rein at a shallow rivulet known as Moore's Branch, Willcox's scouts scanned the horizon. Four hundred yards up the road, on a commanding knoll, stood an attractive, whitewashed house belonging to Mrs. Moore, the widowed matron of a prominent Iuka family. Around her home Rebel cavalrymen were dismounting and hurrying into line. It was nearly 1:30 P.M. Without pausing to reconnoiter, Captain Willcox sent forward at the gallop his lead element, Company K under Sgt. H. D. Cutter. Cutter's troopers splashed across the streamlet and charged up the road. The Rebels scattered into the timber behind the Moore house, but they returned—a full squadron strong. The Michiganders wheeled their horses and made for Moore's Branch. A hail of bullets hurried them along.[1]

At that moment General Hamilton arrived at Moore's Branch. With him were his staff and mounted escort. Cutter's cavalrymen scattered,

Brig. Gen. Charles S. Hamilton
(Massachusetts Commandery of the
Military Order of the Loyal Legion)

making Hamilton's entourage the principal target of the Rebels, who had dismounted around the Moore house. Several horses were hit, and a bullet struck Lt. Louis Schraum, commander of Hamilton's escort, in the chest. The volley missed Hamilton, who had worries greater than his own safety. "I was immediately in rear of the skirmishers, and taking in the situation at a glance, dashed back to the head of the column," he recalled. "If this should become enveloped by the enemy, a rout was inevitable, and our force would be doubled back on itself."

Hamilton met John Sanborn at the head of his brigade. Messengers from the Third Michigan Cavalry had apprised Sanborn of the skirmish at the Moore house, and he agreed with Hamilton that his lead regiment, the Fifth Iowa, must sweep the farm clear of Confederates. Hamilton told Sanborn to hasten his follow-on regiments; he would see to the Fifth Iowa himself.[2]

The Fifth Iowa needed no encouragement. Its commander was every bit as eager to dislodge the Rebels as were Sanborn and Hamilton. An officer in the Prussian army before emigrating to America in 1849, Col. Charles Leopold Matthies was one of the most capable commanders in Hamilton's division. With a patriotic fervor common to European immigrants, he had offered to raise a company for Federal service in January 1861 — four months before Fort Sumter. His offer was politely rejected as premature. Five months later he was back as a captain in the First Iowa.[3]

Aggressive but prudent, Matthies had been listening to his men grumble for a chance to fight since the first shot from the Moore house echoed through the woods. He deployed three companies under Lt. Col. Ezekiel Sampson—a political intriguer whom the men despised—in a long skirmish line across the road and retained the rest of the regiment in march column. After Willcox's cavalry passed to the rear, Sampson's men advanced, shooting as they went. The Confederates fell back firing, and at 2:00 P.M. the Iowans took the Moore house. General Hamilton and his retinue were right behind them. Either from anger at the death of Lieutenant Schraum or from plain mean-spiritedness, Hamilton ordered Sampson's skirmishers to set fire to the widow's home.[4]

The sky was a cloudless blue, the day was warm and dry, and the wooden planks of the Widow Moore's home burned well. Blast-furnace heat rolled from the tall flames out over the Jacinto road, nearly suffocating the passing Federals. Lieutenant Trezevant's Southern troopers kept up their annoying fire, challenging the Iowa skirmishers to venture beyond the Moore knoll. It took the Iowans an hour to inch forward a mile; all the while the long Federal column stacked up behind them.

The Iowans emerged from the timber into a poorly plowed field, too exhausted to keep skirmishing. Colonel Sanborn relieved them at 3:00 P.M. with four companies from the next regiment in line, the Twenty-sixth Missouri. Lt. Col. John Holman deployed one company forward and one in reserve on either side of the road and started across the field. The Rebels fled, and the cat-and-mouse game of skirmishing resumed.[5]

Confederate resistance weakened. A handful of rebels took refuge around the Curtis farm and contested the progress of the Missourians a few moments. Farther up the road, part of the enemy command remounted and prepared to charge. Holman brought up his reserves. The sight of four companies of infantry arrayed in line of battle dissuaded the Rebels, and they withdrew.

A half-mile north of the Curtis farm Holman's Missourians came upon a wide, level field. On its western edge stood a large, two-story frame house belonging to a farmer named Ricks. In the farmyard was a well and a small circular signboard that told the Federals they were only two miles from Iuka. Troops from the main column would pause later to drink at the well—and pillage the abandoned farmhouse—but Holman pushed his men on, intent on closing the final two miles to town.

Four hundred yards beyond the Ricks farm, on a hill beside a log meetinghouse, Holman paused. The Confederates had disappeared from his front, but in their place was an obstacle potentially more vexing. Where the meetinghouse stood, the Jacinto road forked. Should Holman choose the wrong branch, he might fatally delay the arrival of the army before Iuka. Fortunately Holman was well served by his guides, who not only knew the ground but also were honest in what they told him. The left fork, they said, ran along a ridge. Although it was wider and looked more promising, the left fork actually trailed off to the northwest, away from town. The narrower right fork, which initially followed a low ridge toward the east, was in fact the Jacinto road—unquestionably the correct route. Holman reported his location and with General Hamilton's concurrence started his line of skirmishers forward at 4:00 P.M.[6]

The Missourians followed the fork through an open wood of young oaks to the base of a short, convex ridge. It was some 300 yards long and trended generally from the northwest to the southeast. Tall grass, dense bushes, and tangled thickets blanketed its slopes and low summit. The Jacinto road ran up the center of the ridge, then dipped sharply into a deep, tree-choked ravine that paralleled the front face. In the ravine the road again forked: the Jacinto road veered sharply to the north, and a wagon road branched to the southeast to connect with the Fulton road just south of Iuka. On the far side of the ravine, 300 yards from the ridge, stood a high knoll.

Holman swung his line up the ridge. At the top a scattered volley met the Missourians, who returned the fire, and Holman hastened up the slope to learn the cause of the shooting.

His gaze fixed upon the ravine to his front. Drawn up in line of battle, 150 yards from the ridge where his skirmishers stood, was what looked to be a Confederate regiment of infantry. The bright glint of sunlight on bronze betrayed its artillery support, and a rising cloud of dust from the forest beyond suggested that more infantry was on the way. Holman halted his skirmishers and reported to Sanborn that "he had driven the rebel skirmish line into the main line of battle; that the rebel army was in line of battle, batteries in position; and that, in his judgment, the skirmishers should be withdrawn; saying that every indication was that the rebel army was about to advance and attack." It was 4:30 P.M., two and a half hours later than the time Rosecrans boasted he would be at Iuka.[7]

*   *   *

Lieutenant Colonel Holman was guilty of hyperbole. The entire Rebel army was not about to attack him. Indeed, there was grave doubt whether the Confederates would be able to muster enough troops even to resist Sanborn's brigade.

The first report of Rosecrans's advance up the Jacinto road had reduced Price to a perfect nonplus. All his attention, and all his infantry,
had been concentrated on Ord's force northwest of Iuka. Apart from
Lieutenant Trezevant's rapidly retreating cavalry squadron, at 3:00 P.M.
not a single Confederate soldier stood between Rosecrans and Iuka, and
the head of his column was as close to town as the divisions of Maury and
Little.

Price had learned of the Federal appearance south of town thirty minutes earlier. A sweat-stained scout, whose comrade had been shot down
beside him in the initial clash with the Yankee cavalry, brought Price the
news. The dead man's blood covered the scout's jacket, proof enough that
the story was true. Other pickets brought in corroborating reports, and
Price acted. Knowing nothing of the strength or intentions of the approaching Federals, Price sent what he guessed were the minimum number of troops needed to stop, or at least to slow, them.[8]

The mission went to Hébert's brigade, the only unit not committed to
the line opposite Ord. General Little received the order to detach Hébert
at 3:00 P.M., about the time Lieutenant Colonel Holman relieved the
Fifth Iowa on the Yankee skirmish line south of town. The long roll
sounded, and Hébert's brigade stumbled into march column, caught unprepared. "All is confusion, hurrying to and fro," observed a soldier from
another brigade. "Officers away from post. Can't find commands. With a
few bottles of the stuff which screws the courage up to the sticking point
. . . their legs stick badly on terra firma and their heads set less firmly in
the direction of the field of glory."[9] Every Confederate asked himself the
same question: "How on earth, with the woods full of our cavalry, could
they have approached so near our lines?" The simple truth was that the
woods were not blanketed with Southern cavalry. The Fulton road was
unguarded between Peyton's Mill and Iuka, and the Jacinto road was pa-
trolled only as far west as Barnett's Crossroads. Armstrong's resources
were limited, but he should have pushed farther along the roads leading
into town from the south. However, the ultimate responsibility for not
ensuring that his southern flank was better screened rested with Price.[10]

Hébert's infantrymen paid in perspiration for the neglect at headquarters, setting off at the double-quick time to meet the enemy. A mile south
of town they collided with Trezevant's demoralized troopers. Hébert
urged his men toward the large, high knoll overlooking the ravine in
which the Jacinto road rejoined the branch trail from the Fulton road. Before them rose a second eminence—the low ridge upon which Holman's
Missourians had appeared. Hébert rushed his lead regiment, the Third

Texas Dismounted Cavalry, into line near the base of the knoll, on its sheltered northeastern side. While waiting for the Texans to deploy, Hébert told the commander of the Clark Missouri Battery, Lt. James Faris, to run a section of guns up the knoll and shell the Federals off the far ridge.

It was an impetuous act, deploying artillery without infantry support to silence enemy skirmishers. Lieutenant Faris chose his own section for the duty and accorded his second section the privilege of going into battery on the northwest side of the Jacinto road, where intervening timber would protect it from the Yankee skirmishers. As Faris feared, his cannoneers came under fire the instant they showed themselves on the knoll. Faris led them back down just as quickly, putting 150 yards between himself and the hill.[11]

When Hébert saw artillery alone would not frighten off the Federals, he ordered the Third Texas Dismounted Cavalry to sweep forward in a skirmish line, both to silence the Yankee fire and to tease the enemy into revealing his strength.

Col. Hinchie P. Mabry had his Third Texas Dismounted Cavalry formed and ready before Faris's guns abandoned the knoll. It was to be their first fight on foot; a few weeks earlier, the Texans had sent their horses to Holly Springs to rest and graze. Their training as infantry had been minimal, and some of the men still carried double-barreled shotguns, effective only at close range. As the Texans stepped off toward the ravine, a nervous regimental officer asked Hébert, who was watching the movement, "General, must we fix bayonets?" "Yes sir!" barked the Creole in impatient, heavily accented English: "What for you have ze bayonet, if you no fix him? Yes, by gar; fix him! fix him!" Those Texans with riflemuskets paused to fix bayonets.[12]

It was a needless bit of posturing. The handful of Missourians who had ventured into the ravine to get a closer shot at the Rebel cannoneers scampered back up the ridge when Mabry's dismounted troopers replaced Faris's cannon on the knoll. The Texans marched down the knoll and settled into the ravine. Holman's skirmishers took cover behind a growth of blackjack oak trees seven to fifteen feet high, which covered most of the ridge. As the Federal fire died out, Faris returned to the knoll, loading his guns with canister and case shot to greet the Federal main body, which Hébert expected momentarily.[13]

Apparently it was then, at 4:30 P.M., that Lieutenant Colonel Holman came onto the ridge and made his desperate report. His regimental commander, Col. George Boomer, arrived a few minutes later. He found no cause to doubt Holman's estimate of Rebel strength, nor did Colonel Sanborn, who showed up next.

By the time Sanborn arrived, a courier had returned with Hamilton's reply to Holman's note: Colonel Boomer was to advance his skirmishers aggressively. Sanborn passed Boomer the paper. The Missourian protested — to go forward off the ridge was suicide. "You have shown me your orders; under them I will assemble the skirmishers, and bring back my regiment as soon as possible," Boomer announced defiantly.

Hamilton's arrival spared Sanborn the need of reconciling what he called a blatant "clash of authority." Hamilton scanned the front as the remainder of Hébert's brigade came into line on the knoll and in the tall grass on either side of it. He rescinded the order to advance; from confident aggressiveness he lapsed into an agitated panic. That he could fashion an adequate defense struck the New Yorker as doubtful: "The ground was on the brow of a densely wooded hill, falling off abruptly to the right and left. The underbrush and timber were too thick to admit of deployments, and the most that could be done was to take a position across the road, by marching the leading regiments into position by a flank movement." Before returning to the column, Hamilton told Boomer to get his regiment into line of battle immediately. Sanborn was to place each succeeding regiment.[14]

Colonel Matthies's Fifth Iowa arrived first, and Sanborn directed it to the right side of the Jacinto road. The left companies of the regiment formed on top of the ridge in the timber, and the right companies lined up down the southeastern slope. Matthies drew in his right flank to follow the convex trace of the ridge.

General Hamilton was right behind the Iowans, too worried to remain in the rear. Had the aim of Faris's gunners been better, his anxiety might have cost him his life. "Just as the first regiment was placed, the enemy opened one of his batteries with canister," said Hamilton. "The charge passed over our heads, doing no damage beyond bringing down a shower of twigs."[15]

Next up was the Eleventh Battery, Ohio Light Artillery. The battery presented Sanborn with a dilemma similar to that which had confronted Hébert moments earlier: should he order his artillery into action on what was commanding — but exposed — ground before more of the infantry came up? The Minnesota lawyer elected to do so. He told 1st Lt. Cyrus Sears to take his battery into action on the brow of the ridge, astride the Jacinto road. A decrepit log cabin, framed by a small clearing, stood just in front of the position Sanborn chose. Young timber blanketed the rest of the ridge. Although he disliked the ground, Sanborn felt Hamilton's hectoring precluded him from looking for a better spot.

For a moment 1st Lt. Henry Neil, commander of the right section of Sears's battery, doubted he would have enough men alive to place his guns. Rifle fire swept the ridge the instant his limbers rattled up the slope. Neil whirled in the saddle to follow the flight of the bullets: "I turned my head to look for the men, expecting to see half the men and horses down. To my great joy I found all uninjured. The storm of bullets was passing just over our heads. We hastened to get into position and unlimber before they could get the range." Sears chose ground a few yards behind the spot that Sanborn had selected for him, planting the battery colors in a dense growth of hazel at an angle where the road bent to the right. There he waited for his artillerymen to clear a space large enough to accommodate his six cannon, which Sears intended to align nearly hub to hub.[16]

The next regiment of infantry on the ridge was Col. Norman Eddy's Forty-eighth Indiana, which had never fired a shot in anger. Organized nine months earlier, the regiment still numbered a respectable 434 men. Colonel Eddy hurried them into line on the left of the Eleventh Ohio Battery, fronting an old, farmed-out field. He ordered the men to lie down and hold their fire. Hamilton, who had returned to the business of placing Sanborn's regiments for him, accorded Eddy poor ground, especially for green troops. In front of the right wing of his regiment the forward slope of the ridge fell off so sharply that, as Eddy complained, "it was impossible to reach with musketry. On the left of the regiment the descent was less rapid and abrupt, and in many points offered a cover to an enemy's approach." Eddy conformed his left wing to the slope.[17]

Sears's battery was booming out its first salvos when the Fourth Minnesota came up still farther to the left, facing the same field the Forty-eighth Indiana fronted. Four hundred eight strong, it too was untested in battle. Worse yet, it was led by a captain, Ebenezer Le Gro, who was in over his head. After Sanborn was promoted to brigade command, Le Gro had wangled regimental command by convincing his fellow captains he had seen combat service somewhere before. If so, it taught him little. First Le Gro ordered the men to fix bayonets, which would make it harder for them to deploy. Next he formed line of battle too far to the rear, so that his front was masked by timber. Then he overcompensated for his mistake, marching the men forty yards forward into the open field. A few minutes later he marched the Minnesotans back to their proper place beside the Forty-eighth Indiana. Unfortunately their assigned position offered little more in the way of cover. The ridge ended just beyond the regiment's right flank, and the left wing was anchored on nothing more substantial than a lightly wooded ravine, just south of a cabin belonging to the Yow family. The men were able to lie down in a slight hollow, but the ground was bare, the timber belt being thirty yards behind them.

Le Gro's peregrinations so confused T. P. Wilson, the commissary sergeant of the Fourth Minnesota, that he rode into the ranks of the Third Texas. Lying face down in the grass to escape the canister that Sears's cannon were spewing into the ravine, the Texans failed to notice that the rider in their midst wore blue. "Where are you going," yelled one of them, surprised to find someone foolish enough to be on horseback amid the flying canister. Wilson recognized his mistake at once. "Oh, just looking around a little," he replied as he cast a glance back up the Jacinto road. "You had better not go very far in that direction," cautioned the Rebel, still too preoccupied by the roaring artillery to notice Sergeant Wilson's Union blue. "I'll be careful," said Wilson, who wheeled his horse to the left and made his way back to the Fourth.[18]

By the time Sanborn's last regiment, the Sixteenth Iowa, appeared, Hamilton had run out of defensible ground on which to place troops, and he confessed as much to Sanborn. As the veteran regiment in the brigade, the Sixteenth might have made a decisive contribution in the front line, but Hamilton, judging the right to be well protected and the left to be in little danger, had the Sixteenth Iowa wait below the ridge behind Sears's battery and the Forty-eighth Indiana.[19]

Returning from skirmish duty, Colonel Boomer re-formed his Twenty-sixth Missouri behind the Fifth Iowa. Before Boomer could look for a place in the front line, Colonel Sanborn intervened and told him to remain behind the Iowans and watch the brigade's right flank. Boomer's left companies rested behind the Fifth Iowa, and his right companies settled into a shallow ravine laced with sassafras and hazel bushes that ran between two low spurs.

It was 5:00 P.M. before Hamilton had Sanborn's 2,200 infantry formed to his satisfaction. In the thirty minutes it took him to deploy, the firing had risen from a light rattle to a steady thunder. Both sections of the Clark Missouri Battery were engaged and had the ridge nicely ranged; a single shot from Faris's section had torn apart six soldiers from the Forty-eighth Indiana while the regiment was still in column. The Confederate line of battle was formed and appeared ready to advance. Hamilton was both pleased and mystified that they had not: "Why they did not move forward and attack us at once is not understood. Their delay, which enabled us to form the nearest three regiments in line of battle . . . was our salvation."[20]

The arrival of General Rosecrans and his retinue interrupted Hamilton's musings. The New Yorker hastened to explain his dispositions. Rosecrans agreed that he had placed the Eleventh Ohio Battery on the "only available ground" and otherwise approved of his actions—or so said Rosecrans in his report. Colonel Sanborn told a different story, one decidedly more in keeping with Rosecrans's contentious temperament. The general, said Sanborn, rode up "in a fault-finding mood," utterly disgusted with the way Hamilton had formed the line.

Rosecrans's conduct certainly was more consistent with Sanborn's story than with his own report. Whatever he may have thought of Hamilton's dispositions along the ridge, he obviously concluded that the New Yorker had neglected his flanks. Both rested near roads that might provide the enemy with a means of approaching the Federal flanks undetected. On the right a narrow and sandy trail called the Mill road led southeast toward a small settlement. Rosecrans called for Colonel Mizner, whose cavalry had been relegated to the role of spectators, and ordered him to send a battalion of the Third Michigan Cavalry to ascertain whether the Mill road joined the Fulton road or simply trailed off harmlessly into the woods. Captain Willcox rode off with four companies. Observing that, in his obsession with drawing up Sanborn's command, Hamilton had forgotten he had a second brigade, Rosecrans took it upon himself to call on Brig. Gen. Jeremiah Sullivan for reinforcements to cover the road that led to the left of the line.[21]

Sullivan was pleased to oblige. Eleven days shy of his thirty-first birthday, the young Hoosier was on familiar terms with Rosecrans, having served under him in western Virginia. Shortly after receiving his commission as a brigadier general in April 1862, Sullivan was sent west to command a brigade in the Army of the Mississippi, perhaps at Rosecrans's behest. Although he enjoyed the commanding general's confidence, Sullivan was an undistinguished commander; his sobriquet of "Fighting Jerry" was attributable to his personal bravery, which the major of the Seventeenth Iowa said was the only quality that fitted him for command. For the past thirty minutes here at Iuka he had been waiting with his brigade stacked up along the Jacinto road below the meetinghouse while Hamilton fidgeted about the ridge. To Rosecrans's request that he patrol the settlement road that forked to the left at the meetinghouse, Sullivan responded with the Tenth Iowa and a two-gun section of the Twelfth Battery, Wisconsin Light Artillery.[22]

Col. Nicholas Perczel commanded the Tenth Iowa, and 1st Lt. L. D. Immell led the Twelfth Wisconsin Battery. They were an odd pair. Perczel, a gruff and gray-haired Hungarian loved by his troops, had seen service in Europe during the rebellious decade of the 1840s. Immell, a young martinet utterly despised by his men, had no combat experience and had come to command by default.

The Twelfth Wisconsin Battery had been organized in St. Louis as a company for the First Missouri Light Artillery, Wisconsin having fulfilled its first quota of volunteers. The Wisconsin men tolerated their assignment until the battery was attached to Hamilton's division, when they demanded that their affiliation with Missouri be severed and that their commander, a Missourian, be dismissed. The captain took an indefinite sick leave to save face, and command devolved on Immell, who lost no time in applying his own unique brand of discipline to his fellow Wisconsin volunteers. During the march to Iuka Immell had dragged a sick private from an ambulance at gunpoint, then drawn his saber and threatened to thrust it through the artilleryman's chest if he persisted in malingering. On another occasion he yanked a gravely ill private from his bed, shook him violently, and set him to work digging ditches until the man collapsed, near death. Those who got sick during the march he denied recourse to the battery's ambulance, offering instead to tie them spread eagle to the spare wheels of a caisson. Those who complained of moldy rations Immell had bucked and gagged.

Immell was as contemptuous of his subordinate officers as he was of his enlisted men, and he showed it now by taking charge of the section detailed to support the Tenth Iowa. However, the army's chief of artillery, Lt. Col. Warren Lothrop, had little confidence in Immell and so rode along with him to superintend matters. Immell set up the cannon north of the Yow place, on the east side of the road, facing the same played-out field that the Fourth Minnesota and the Forty-eighth Indiana fronted south of the cabin.

Colonel Perczel formed his regiment to the right of Immell's cannon. Finding his right flank in the air, Perczel dispatched three companies into a wooded ravine that paralleled the settlement road to try and make contact with Sanborn's left flank regiment, the Fourth Minnesota.[23]

After sending Perczel and Immell to reinforce the left, General Sullivan took it upon himself to strengthen the right. Two hundred yards southeast of the meetinghouse the prevailing timber gave way to a huge, oblong field of corn and cotton, 500 yards long from north to south and 600 yards wide at its widest, east to west. The eastern fringe of the field ran to the right and rear of the Twenty-sixth Missouri, which then held the extreme right of the Federal front line. Should the Rebels outflank the Missourians and penetrate the field, they would be able to sweep behind Sanborn with impunity. Mizner's cavalry had ridden too far down the Mill road to detect any such movement.

To the far end of the cotton field, then, Sullivan sent the Tenth Missouri. Col. Samuel Holmes deployed his men at the edge of the woods, fronting east, and shook out skirmishers into the trees. His position was as perilous as it was crucial. Faris's Confederate gunners on the knoll had glimpsed Holmes's moving men, and they blasted them when they crossed the field. A single shell burst in the ranks of Company D with sickening effect, killing or mangling fourteen men. The regiment stopped, dressed ranks, and lay face down amid the cotton shrubs in what one Missourian called a "most trying position." Rebel cannoneers continued to send occasional shells their way. "The smoke soon became so dense that we could scarcely see the man next to us," said a Missourian, "and though lying down we met with considerable loss, under the most trying conditions that soldiers can be placed in, that of receiving fire without the ability to return it or to move." [24]

Rosecrans had come onto the ridge, reached his conclusions about Hamilton's dispositions, and sent instructions to Sullivan to bolster the left flank, all in the space of three or four minutes. With plenty of nervous energy left to expend, Rosecrans turned to upbraid Colonel Sanborn for having placed the Eleventh Ohio Battery so far forward on the crest.

But Sanborn was spared what would have been an unwarranted dressing down. As Rosecrans began to speak, Colonel Eddy of the Forty-eighth Indiana galloped up and announced that "the whole rebel army was advancing." Sanborn excused himself and rode forward to confirm the report. "I did, and the entire rebel line opened fire with both artillery and musketry," said Sanborn. "As I came back through my own line . . . I gave the command as loud as I was able to both infantry and artillery to commence firing, and the battle opened as fiercely as it is possible to conceive. Leaves, twigs, men, horses — everything — was falling." [25]

8

Like Lightning from a Clear Sky

General Hébert regained his composure during the lull following the Third Texas Dismounted Cavalry's march into the ravine. While Hamilton and Sanborn sorted out their regiments atop the ridge, Hébert brought the rest of his brigade from column into line.

He placed the First Texas Legion of Col. John Whitfield[1] on the right side of the Jacinto road, 300 yards northeast of the Fourth Minnesota. The Fourteenth and Seventeenth Arkansas regiments, merged because of battle losses, formed on the left of the First Texas Legion. To the left of the Arkansans went the Third Louisiana Infantry. Behind the left flank of the Louisianans Hébert placed the St. Louis Artillery. The untried Fortieth Mississippi he retained in the second line.

Hébert had his brigade deployed behind the knoll a full fifteen minutes before Sanborn formed his on the ridge, but he was loath to leave his sheltered position. While contemplating his next move, Hébert ordered skirmishers from his flank regiments to augment the Third Texas Dismounted Cavalry in the ravine. Company F of the Third Louisiana drew the duty on the extreme left of the brigade. The Louisianans felt their way across the heavily wooded bottom, where the long afternoon shadows cooled the air. Near the ridge they ran into skirmishers from the Fifth Iowa. Fighting among the dark trees was highly personal, bringing out the best or worst qualities in the combatants. The Louisianans killed four or five Yankees, one quite needlessly. Recalled Willie Tunnard,

Two of the enemy took shelter behind a large tree directly in front of Company F. The tree, however, was not large enough to protect the two, one of whom was instantly killed by Private Hudson; the other begged for his life most piteously, which would undoubtedly have been granted him had he relied on the word of a rebel. He was ordered several times to come to the company and his life should be spared, but he was afraid to expose his person. During the conversation between him and the captain, Private J. Jus, it seems, became rather restless, left his position in the line, and slipped around until he came in view of the Yankee, then raised his gun and shot him through the head, at the same time remarking, "Damned if I don't fetch him." The Federal proved to be a lieutenant.[2]

Hébert had concluded he lacked the troops needed to attack the Yankees—a wise decision, as he had no idea how many Federals might be lurking behind the ridge—when General Little arrived at the head of Col. John D. Martin's brigade and took command. Price had decided to detach a second brigade south of town. To improve the odds of success, he asked Little to accompany the brigade.[3]

Little gave no hint of the afflictions that had brought him to the edge of collapse. He sized up the situation quickly and, after a few words with Hébert, decided to attack with both brigades. Little called for Colonel Martin, whose Mississippians and Alabamians were double-quicking down the road under a rain of shells one new soldier thought sounded like "buzzards passing." Little told Martin to support Hébert's left wing with the Thirty-seventh Alabama and the Thirty-sixth Mississippi; he would assume command of the Thirty-seventh and Thirty-eighth Mississippi regiments himself and support Hébert's right wing, which was drawing fire from Immell's Wisconsin battery across the open field. Unlike General Hamilton, Little recognized the importance of securing his flanks before beginning a fight with an enemy of unknown strength.[4]

Little reminded the commanders of the Thirty-seventh and Thirty-eighth Mississippi regiments that Hébert's men were to their front; they were to hold their fire until Little said otherwise. With that final word of caution, Little judged his brigades ready. At 5:15 P.M. he ordered the attack to begin. Hébert's brigade, numbering 1,774 officers and men, and Martin's brigade, some 1,600 strong, moved forward to do battle with the 2,200 Federals of Sanborn and, in close proximity, a similar number under Sullivan. All 8,000 combatants, North and South, were arrayed on a tract of rolling forest and field slightly more than one-half mile wide.[5]

Hébert's men let go a high-pitched "Rebel Yell" and marched slowly down the knoll. The Federal line was silent. In the ravine Col. Hinchie P. Mabry recalled the soldiers of his Third Texas Dismounted Cavalry. Hébert halted the brigade long enough to allow the Texans to replace the Fourteenth and Seventeenth Arkansas, which fell back to join the Fortieth Mississippi in the second line. The advance resumed as deliberately as before. Faris's cannon fell silent for fear of hitting their own men in the backs. Hébert's Confederates descended into the ravine, 150 yards short of the ridge, and the Yankee line opened fire with a deafening crash.

"I can never forget that moment—it came like lightning from a clear sky," remembered Sgt. W. P. Helm of the Third Texas Dismounted Cavalry. "The roaring artillery, the rattle of musketry, the hailstorm of grape and ball were mowing us down like grain before we could locate from whence it came. We were trapped; there could be no retreat, and certain death was in our advance. We fell prostrate to the ground." Over the din a familiar voice crying, "Steady, boys, steady," caused Helm to look up. It was his company commander, braced on one knee. Before he could repeat the exhortation, a cannonball carried away his head. A lieutenant stood up to stop a soldier from retreating, and both he and the terrified private were cut in two by canister. "Our ranks," marveled Helm, "were shattered in the twinkling of any eye."

Smoke blanketed the ravine and blinded the Rebels. "The evening was one of those damp, dull, cloudy ones, which caused the smoke to settle down about as high as a man's head," observed Willie Tunnard. The view was little better from the ridge. "The smoke hung over the battlefield like a cloud, obscuring every object ten feet off," said a Yankee defender.[6]

But the Federals needed do no more than point their rifle-muskets into the smoke; the ravine was too choked with Southerners for them to miss. To stay there meant death. The men were ordered to their feet, officers screamed the command "double-quick," and the line of Southerners lurched up the slope of the ridge. The Third Texas Dismounted Cavalry had the dubious distinction of being directly in front of the Yankee guns, so that every blast of canister told with especial savagery on its ranks. No one needed to tell them what must be done. The objective was evident: the six guns of the Eleventh Ohio Battery that had been decimating their ranks. "Unless the battery was immediately silenced the result might be most disastrous," concluded Colonel Whitfield. He turned the First Texas Legion obliquely toward the cannon and gave the order to charge. The Fourteenth and Seventeenth Arkansas followed.

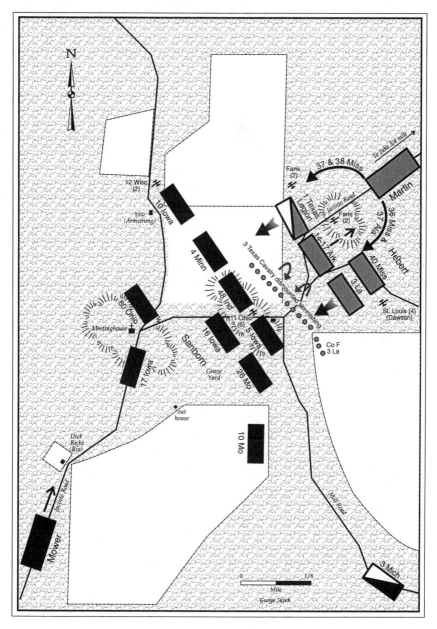

MAP 3. *Hébert Attacks, September 19, 5:15 p.m.*

Every regiment of Hébert's brigade claimed to have charged at once for the guns; had that been the case, Hébert's line would have taken on a concave aspect the moment the men climbed out of the ravine. There was a perceptible massing upon the battery; however, the regiments of Yankee infantry on either side of the battery deflected the dash on the guns, and

Like Lightning from a Clear Sky   [89]

those Confederates fortunate enough to be out of the line of fire of Sears's Ohioans adopted as their objective the battery's infantry supports.[7]

The first Federals to face the Rebel onslaught were the troops of the Forty-eighth Indiana, posted to the left of the Eleventh Ohio Battery. It was their misfortune to stand opposite the First Texas Legion and the right companies of the Third Texas Dismounted Cavalry, the latter as hardened a lot of men as the Hoosiers were green. The Indianans had suffered heavily from the well-aimed fire of Faris's battery, both sections of which were now trained on Sears's guns and the infantry on its flanks, and the declivity in front of the Forty-eighth Indiana was unusually sharp, which masked the Texans in the ravine. Unable to fire on the enemy, the Indianans grew restive. When the Texans suddenly emerged from the ravine and delivered a near point-blank volley, the regiment crumbled. First the right companies gave way, leaving the left section of the Eleventh Ohio Battery unsupported. Then the entire regiment collapsed, falling back to the far brow of the ridge. There Col. Norman Eddy and his officers had ridden, hoping to rally the regiment.

Their efforts were wasted. The Third Texas Dismounted Cavalry had lost heavily—Sam Barron said four men were killed by his side in the opening minutes of the fight, and W. P. Helm said twenty-seven of the forty-two men in his company were hit early—but the Texans were accustomed to the shock of combat. Most of the right companies of the Third Texas Dismounted Cavalry halted on the ground abandoned by the Forty-eighth Indiana, neither advancing nor retreating, but a number of the men, maddened by the horror, raised a yell and dashed after the retreating Yankees, following so closely that in some places on the ridge pursuers and pursued were only a few paces apart.[8]

Colonel Sanborn saw the Forty-eighth collapse. Beside himself with rage, he galloped into the mass of fleeing Indianans and ordered them to stand and fight. When two near him refused, Sanborn drew his revolver and shot them dead.

Colonel Eddy also rode among his men, imploring them to rally. Here and there squads answered his call, restoring enough of the line that Eddy briefly entertained a hope of returning to the front to succor Sears's cannon. But five bullets brought down the colonel and his horse. Seeing their commander fall, the Indianans broke for good, bounding down the reverse slope with the Texans only a few yards behind.

Sanborn was near Eddy when he was hit. After the Forty-eighth Indiana unraveled for the second time, Sanborn galloped into the swale behind the ridge, where the Sixteenth Iowa waited. As he neared the

Iowans, Sanborn "shouted over and over again at the top of my voice for the men to hold their fire until the Forty-eighth had passed." Sanborn assumed his order would be obeyed: "The troops of the Sixteenth were cool and looked up as if they understood the command." Sanborn reined his horse beside the regiment and waited for the Indianans to pass.

Suddenly the unthinkable happened. Perhaps it was the nearness of the enemy, many of whom were intermingled with the Indianans. Perhaps it was the murky gloom in the swale—sunset was only an hour away and already the thick smoke of battle had combined with the dark shadows of the forest to cast a deep pall over the field. Or perhaps it was simply fear bred of inexperience. In any case the Sixteenth Iowa let go a volley into the disorganized mass to its front, in that instant killing more Indianans than had died from Rebel bullets.

Sanborn at once took charge of the Sixteenth Iowa. The moment the last Indianan cleared its ranks, he led the regiment in a charge up the ridge that sent the Texans scampering back to their commands on the far slope.

Despite the defection of the Forty-eighth Indiana, the left section of the Eleventh Ohio Battery was still in action, thanks to confusion in the Rebel ranks. The Third Texas Dismounted Cavalry was paralyzed. Its right companies were regrouping behind the ridge, and the rest of the men were hugging the ground or kneeling behind trees in front of the cannon. Colonel Whitfield tried to break the stalemate by maneuvering his First Texas Legion up the ridge and to the left, over the ground vacated by the Forty-eighth Indiana. He hoped to take the artillerymen from the flank, but his men fired too soon. "We still hoped for help," said a soldier of the Third Texas Dismounted Cavalry—which had tried twice to cover the final 100 yards to the guns—"but Whitfield's raw troops sent to our relief, from our proximity to the Federals, mistook us for the foe and fired into us." Several Texans fell in the volley.[9]

At that moment the Sixteenth Iowa showed itself atop the ridge. Had anyone paused to check their watch, they would have noticed the time to be 6:00 P.M., just ten minutes before sunset. The Iowans opened fire. Whitfield's Texans returned it. Col. Alexander Chambers, commander of the Sixteenth, slipped from the saddle, shot in the neck and shoulder. His regiment stood long enough to take seventy-six casualties before fleeing from the ridge. Like the Forty-eighth Indiana, the Sixteenth Iowa was lost to Sanborn for the remainder of the battle.[10]

With the ridge to the left of the Eleventh Ohio Battery finally clear of Yankee infantry, Whitfield's First Texas Legion resumed their assault on

Sears's left section. Colonel Mabry got his Third Texas Dismounted Cavalry on its feet for a fourth try at the guns. After the volley in their backs, Mabry's men were ready to move. Explained one, "Seeing certain death between friend and foe, the order was given: 'Boys, if we are to die, let it be by Yankee bullets, not by our friends. So let us charge the cannon.'"[11]

*　*　*

The carnage and confusion that marked the first forty-five minutes of combat on the left side of the Eleventh Ohio Battery was exceeded only by the bewildering brutality of the fighting to the right of it. There the Fifth Iowa of Col. Charles Matthies met the Third Louisiana. Both Hamilton and Sanborn had great faith in the Prussian, and Matthies intended to show he merited their confidence. He greeted Hamilton's order to hold at all hazards with a positive, "That is what I calculate to do." As the Third Louisiana crossed the ravine to his front, Matthies rode the length of his line, "encouraging his gallant men, and cautioning them about keeping cool." The Louisianans marched into the ravine and momentarily disappeared from view. The Iowans heard them prepare for the lunge up the slope but could not see to shoot. Said one, "Their steady tramp comes nearer, and in a moment Colonel Matthies commands, 'Attention, battalion! Ready! Aim! Fire!' and a sheet of flame and lead is sent into the ranks of the advancing enemy, they first appearing in front about fifty yards away."

"The fire immediately opened from both lines like a sudden clap of thunder," marveled Louisianan Willie Tunnard. "At the commencement of the firing our boys dropped down on their knees, the best thing they could have done, as the greatest portion of the enemy's fire flew harmlessly over their heads, while their firing had a telling effect on the enemy. The firing was fearful—the smoke enveloped both lines, so that they became invisible to each other. The lines could only be distinguished by the flash of the guns." For fifteen minutes, perhaps longer, the two regiments traded volleys without either gaining the advantage.[12]

Matthies broke the deadlock. When his left companies wavered, the Prussian ordered a bayonet charge, as much to forestall a general collapse as to drive off the enemy. Matthies was pleased with the result: The "charge was executed in the most gallant manner, every officer and man moving up in line, cheering lustily. The enemy gave way, when we poured a deadly fire into their ranks, causing them to fall back down the hill."

The Louisianans regrouped in the ravine and in a few minutes were pushing Matthies's Iowans back up the slope. Again Matthies ordered a charge, and a second time his men drove the enemy from the ridge. In the fast gathering darkness the lines intermingled. A squad of Louisianans edged through the smoke to within an arm's length of the national colors of the Fifth. One of the Rebels reached out to grab them, saying, "Don't fire at us; we are your friends." The Iowans shot him dead. Another Rebel shoved through the ranks of Company B, intent on taking the regimental flag. He was shot and bayoneted behind the line.

The two sides untangled themselves, stepped back, and renewed their volleys. "Five of my eight messmates of the day before were shot," said Samuel Byers of the Fifth. "It was not a question who was dead, or wounded, but who was not."

The fight was going badly for Matthies. From the left side of the Eleventh Ohio Battery, no more than 150 yards away, Whitfield's First Texas opened an enfilading fire on Matthies's left. It was now probably 6:15 P.M. The Fifth Iowa had been fighting for nearly forty-five minutes and had already lost more than 100 men. Then Whitfield's volleys decimated Matthies's three left-wing companies. Nearly every officer was shot, and the men "went down like grass before a scythe." Those still alive began to give ground.[13]

Watching the fight from behind the Fifth Iowa was Col. George Boomer, whose Twenty-sixth Missouri had brought on the battle almost two hours earlier. The caissons of the Eleventh Ohio Battery were lined up to the rear of the Missourians, and artillerymen periodically passed through their ranks bearing ammunition for the guns. Boomer appreciated the exposed position of the cannon, and when Colonel Matthies called on him to help shore up his decimated left wing, the Missourian responded immediately, knowing the safety of Sears's right section was at stake. Boomer yelled to Maj. Ladislaus Koniuszeski to get on his horse and lead the four left-wing companies of the regiment — 162 men in all — up the slope to close the gap between Sears's cannon and the right wing of the Fifth Iowa, which was still largely intact. He told Lieutenant Colonel Holman to stay in the swale with the six remaining companies of the Twenty-sixth.

On the ridge all was a sickening confusion. When they reached the ridge, Koniuszeski's Missourians walked into a volley from the Third Louisiana. Capt. De Witt Brown, the commander of Company C, which was nearest the cannon, remembered the first Rebel fire: "Bushels of minie balls came thick and fast among us. My men were ordered by me to 'lie down, load,' rise and fire. In this way I saved the lives of my men." With each successive volley, the accuracy of the Louisianans improved.

Confessed Brown, "After a few rounds were fired, a command was given by a rebel officer, in a loud tone, to fire low, when a leaden hail swept through our ranks, wounding several of my men, and throwing my company into confusion." Evading Brown's gaze, men began sneaking rearward from the center of his line.

Colonel Boomer called their departure to Brown's attention. Disgusted by Koniuszeski's poor performance — he had watched the major dismount and then stumble befuddled about the ridge — Boomer rode up to take command in person. First he upbraided Captain Brown, who responded by blocking the path of his skulking soldiers: "I immediately moved from the right to the center of the company, struck up the guns of the men with my sword, commanded them to stand fast and face the enemy, and turned two men around into their places with my own hands." While Brown was thus distracted, the rest of his company fell back a few steps. Faced with the defection of all, Brown screamed, "Charge." The men obeyed. They surged back up the ridge, past the battery and "almost into the Rebel lines."

The other three companies were faring no better than Brown's. To forestall their collapse, Colonel Boomer rode off for his six reserve companies. But at the bottom of the ridge he found only trampled underbrush where Lieutenant Colonel Holman and the rest of the regiment had been. Having neither the time to look for them himself nor a staff officer to send, Boomer returned to the fight.[14]

In the few minutes Boomer was gone, the situation deteriorated beyond repair. Lieutenant Sears was shot from his horse. A moment later Sears's ranking section commander, 1st Lt. Henry Neil, came to Captain Brown, pleading with him to withdraw a few paces so he could limber up and get away with the guns of the right section. Brown did so. "At this juncture, the scene became perfectly terrible." He watched Neil, bleeding profusely from two shell wounds and two bullet wounds, call for horses from the caisson in the rear to replace the dozens of dead animals that lay tangled in their harnesses beside the limbers. It was an effort worse than futile. The replacement horses were shot as fast as they came up. Bleeding and dying, they lunged at Brown's right flank, and his Missourians dove for cover.

Colonel Boomer could offer Brown no help in rallying his broken company. About the time Lieutenant Sears fell, he had been shot through the left lung. Before he was carried off, Boomer ordered his four companies on the ridge to retreat.

There were few left to obey the order. Of the 162 Missourians who had gone into the fight, 21 were dead, and 76 were wounded or missing. Captain Brown, who had pieced together a firing line, never received Boomer's order. For Brown's company, the retreat, when it came, was spontaneous and ineluctable.[15]

Not everyone heard or understood the order to withdraw. A squad of five Missourians, congregated behind Neil's battery, suddenly found itself alone amid fallen artillery horses, with the Rebels only a few dozen yards off. Screaming over the din, Pvt. J. H. Allen asked the corporal commanding the squad what had happened to the rest of the company. The corporal put his mouth to Allen's ear and shouted, "I don't know." "What should we do?" pleaded Allen. "Die here!" replied the corporal. Bullets ripped through both Allen's thighs. As he crumpled to the ground, Allen saw the corporal fall dead.

In the wooded swale Captain Brown and the other survivors found refuge. They came together, formed into column, and marched into the large cotton field to the right of the recumbent Tenth Missouri. There they found the six derelict companies of their own regiment. Lieutenant Colonel Holman shrugged off his absence from the fight. He had not gotten Boomer's order to stay put, Holman told Captain Brown. Lacking instructions and witnessing the carnage on the crest, Holman simply had faced his men about and led them away.[16]

The defeat of Colonel Boomer's Missourians forced the Fifth Iowa to concede the ridge to the Third Louisiana. The Iowans had expended nearly all of their forty rounds and had lost an astonishing 217 men—nearly half those engaged—in seventy-five minutes. The withdrawal of the Twenty-sixth Missouri exposed the Iowans' left flank to an enfilading fire. They might have been able to respond had their right flank not been endangered by the advance of the Fortieth Mississippi, which General Hébert had committed to the left of the Third Louisiana to lengthen his badly shot up front line. Matthies saw the Mississippians march through the ravine into position beyond his flank, and he withdrew while there was time. Getting his men off in good order "by the right of companies to their rear," Matthies passed through the Tenth Missouri and halted the regiment at the far fringe of the cotton field.[17]

The destruction of Colonel Boomer's four companies also doomed the Eleventh Ohio Battery. From the moment they fired their first salvo at 5:15 P.M., the Ohioans fought as if the battle began and ended with them. "I had my hands and mind so thoroughly occupied with my own particular part of the business, that it seemed to me we were going it alone on a weak hand, with a strong prospect of being euchred," said Lieutenant Sears.

Sears was right to feel abandoned. At the start of the fight Colonel Sanborn had told him to "form in battery" astride the Jacinto road and "'await further orders.' Orders never came, but the enemy did, in force, sneaking up with their rifle-muskets at 'charge bayonets,' in plain view and at easy canister range," Sears remembered. "On the charging masses came, 150 or 200 yards. Still the battery was waiting 'further orders,' every man at his post, toeing the mark, with everything 'ready.'"

The Rebels opened fire first. Said Sears, "Their bullets were zipping among the battery with very uncomfortable frequency, and occasionally winging a two or four-footed victim."

Sears's gunners grew impatient. Grumbled a sergeant, "By God, I guess we're going to let them gobble the whole damned shooting match before we strike a lick, if we don't mind and quickly too."

"I guess we are obeying orders," answered a corporal.

"Damn the orders!" snarled the sergeant. "To wait for orders in a time like this!"

Sears heard the rumble of discontent. "This dialogue struck a responsive chord in my mind, and was, perhaps, the last straw that moved me to take a chance and shoulder the responsibility." Sears yelled, "With canister, load, aim low, and give them hell as fast as you can!" so loudly it was heard in the first rank of the Third Texas Dismounted Cavalry, before the crash of cannon shattered the air.[18]

Texans fell by the dozen. Colonel Mabry of the Third Texas Dismounted Cavalry had his ankle fractured but kept on his horse. At 100 yards' range Sears ordered the cannon double-shotted with canister, and the enemy vanished in smoke. Lieutenants Sears and Neil felt certain they could hold the foe to their front; both agreed, however, that the retreat of the Forty-eighth Indiana from their left and the Twenty-sixth Missouri and Fifth Iowa from their right condemned the cannon to capture.

First came enfilading volleys from Whitfield's First Texas Legion and the right companies of the Third Texas Dismounted Cavalry. "The guns were being worked with greater speed and smaller crews," said Neil. "Cannoneers were falling. Other cannoneers coolly took their places and performed double duty. Drivers left their dead horses and took the places of dead or wounded comrades, only to be struck down in turn."

As the cannonade slackened, the Third Texas Dismounted Cavalry sprang forward for a third and final lunge up the ridge. Whitfield's men

Going into Battery: The Eleventh Ohio Battery at Iuka *by Keith Rocco*

charged on their right, overwhelming the gunners of Sears's left section at their pieces.

Several stood by the cannon to the last, exhibiting a brand of courage that Lieutenant Sears conceded to be foolhardy. David Montgomery, an ammunition carrier who found himself the only man unhurt at his cannon, yanked the lanyard in time to discharge a double load of canister into the bodies of a squad of Rebels who were just an arm's length from the gun tube. The instant Montgomery pulled the lanyard, a tall Texan raised his rifle to brain him. Montgomery darted aside as the Texan brought down his weapon. Grabbing a round of canister, Montgomery smashed in the Texan's skull, then dodged under a limber and into the brush, where he hid safely.

Driver John Dean was less prudent than Montgomery. As Montgomery dove for cover and every other cannoneer still on his feet ran for the rear, Dean stood holding the two horses of his limber team precisely where they had stopped when his piece was taken into battery. Three of the limber's six horses were dead, and the Rebels were a stone's throw away. A battery mate screamed at Dean to save himself. Dean refused. "My sergeant ordered me to hold this team right here, and by God, I'm going to do it or die, till I get proper orders to do something else." The next morning Dean was found holding his team with a death grip on the bridles. His body lay beside his dead horses.[19]

The tenacity of the Ohioans' defense stunned the soldiers of the Third Texas Dismounted Cavalry. "I shall never forget what happened," said W. P. Helm of the moment when his regiment crested the ridge. "Sword and bayonet were crossed. Muskets, revolvers, knives, ramrods, gun swabs—all mingled in the death-dealing fray." Sam Barron confirmed the doggedness of the Federal cannoneers: "The horses hitched to the [limber] tried to run off, but we shot them down and took it, the brave defenders standing nobly to their posts until they were nearly all shot down around their guns—one poor fellow being found lying near his gun, with his ramrod grasped in both hands, as if he were in the act of ramming down a cartridge when he was killed."[20]

A similar scene was played out when the Third Louisiana converged on Lieutenant Neil's own section. By 6:30 P.M., twenty minutes after sunset, the fight for the cannon was over. The battery had gone into action 97 strong; 18 lay dead and 39 were wounded. Of the 54 cannoneers, 46 were hit. Forty-two dead limber horses were heaped near the guns; almost as many caisson horses were shot coming forward to replace them. Only 3 of the battery's 80 horses left the ridge unhurt. A few of the Ohioans were taken prisoner while swinging their sponges. In an act of remarkable restraint, the Confederates not only spared their lives but also let them go. Explained a Rebel eyewitness, "Those battery boys had so much spunk that we took pity on the few who were left."[21]

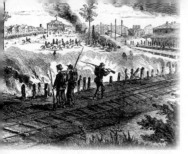

9

Night Quickly Set In

Sanborn's line was shattered. Four of his five regiments had been routed and his battery captured. Only the Fourth Minnesota on the extreme left remained intact. After seventy-five minutes of close combat the ridge belonged to the Rebels. Then was the moment for Confederate reserves to pass to the front, sweep over the ridge, swallow up Sullivan's dispersed Federal brigade, and complete the victory, or for Martin's brigade to take the remaining Yankees in the flank. Instead a lull fell over the field. No fresh Confederate regiments appeared, and those congregated around the Federal cannon went no farther.

What accounted for the sudden stop to the Southern attack? Darkness played a part. In the smoky grayness of approaching night, images were blurred beyond a few feet, so the Confederates could not know how many Federals lay beyond the ridge. Heavy losses also contributed to the halt. Three regiments had borne the brunt of the battle — the Third Louisiana, the Third Texas Dismounted Cavalry, and the First Texas Legion — and they suffered commensurately. Colonel Mabry's bleeding ankle finally compelled him to leave his regiment. Colonel Whitfield was shot in the shoulder during the final charge of the First Texas Legion, and Lt. Col. Jerome Gilmore of the Third Louisiana had taken five flesh wounds. Each regiment had lost about 100 men. More important than the loss of regimental commanders or men was the absence of a guiding hand. Among the dead was General Little.[1]

Sterling Price reached the front a few minutes after Hébert began his attack, and he at once sought out General Little. Price found him 100 yards south of the Jacinto road, behind the Third Texas Dismounted Cavalry. Little had just ordered the Thirty-seventh and Thirty-eighth Mississippi regiments of Martin's brigade to reinforce Hébert. Sitting on horseback a few feet from each other, Price and Little briefly reviewed the situation. It was about 5:45 P.M. The bitterness of the battle astounded Price: "The fight began, and was waged with a severity I have never seen surpassed." Guessing that Rosecrans had at least 8,000 men on hand, Price concluded that Hébert needed reinforcements merely to hold his own. Turning from the fight to face Little, arms akimbo, he implored the Marylander to bring up the rest of his division. At that instant a bullet passed under Price's arm and struck Little above the left eye. It dug through his brain and stopped just under the skin on the back of his head.

Little threw his arms in the air, dropped the reins, and slumped forward onto the neck of his horse. Sgt. T. J. Cellum of the Third Texas Dismounted Cavalry caught the general and eased him to the ground. Price dismounted and knelt beside Little. His friend had died instantly. Price wept "as he would for a son" until an ambulance came to claim the body. Mounting his horse, Price repeated to Capt. John Kelly of Little's staff his order that the remaining two brigades of the division be brought up, then sent word to Hébert that the division was now the Creole's. At the insistence of his staff Price rode to a less exposed position.

Galloping down the Jacinto road to join Little, unaware of his death, was the general's "guardian angel," Father John Bannon, chaplain of the First Missouri Brigade. He and Little were close friends, eating, traveling, and tenting together from Elkhorn Tavern to Iuka. Solicitous of Little's precarious health, Bannon was bringing him a canteen of water when he met Captain Kelly on the road. Bannon searched out the ambulance bearing Little's body and escorted it to the general's headquarters in Iuka.[2]

* * *

"If Little's death had not produced a pause in the Confederate assault, Hamilton's entire division might have been routed," historian Albert Castel has asserted. His claim has merit. Little fell just as the Thirty-seventh and Thirty-eighth Mississippi regiments were about to make their presence felt against the Federal left flank. Separated from their brigade commander, whom Little had sent to the left to direct the movements of the Thirty-sixth Mississippi and the Thirty-seventh Alabama, the two Mississippi colonels looked to Little for instructions. When he died, they were forgotten. Hébert was too busy with his own brigade to shoulder the responsibilities of division command, and Price neither gave the Mississippians orders nor reminded Hébert of his larger duties.[3]

Had he been a man of initiative, Col. Fleming W. Adams of the Thirty-eighth Mississippi might have contributed something to the Confederate attack. In keeping with his last orders from Little, Adams had his command advancing with its left flank touching the Jacinto road to reinforce the stalled assault on the Eleventh Ohio Battery. Just as the regiment ascended a low rise in the woods to the right and rear of Colonel Whitfield's First Texas Legion, someone commanded the Thirty-eighth to fall back. Colonel Adams was perplexed: "I asked who the command came from, but was unable to ascertain." Unfortunately he chose to obey the anonymous order. He withdrew his Mississippians sixty yards at precisely the moment when their continued presence might have hastened the capture of the Eleventh Ohio Battery and provided the crucial reserves needed to carry the assault over the ridge and into Sullivan's dispersed brigade. As it was, the Thirty-eighth Mississippi took no further part in the battle.[4]

Adams would have found little in his way, had he pressed on. The only Federal regiment near enough to have interrupted his advance, the Fourth Minnesota, was in no condition to challenge anyone. Unlike Adams, Captain Le Gro was a man of action; unfortunately the actions he took were generally wrong. When the Forty-eighth Indiana gave way early in the fight, Le Gro removed his right wing companies to a less exposed position in deep timber sixty yards to the rear.

Le Gro's order was "a terrible blunder," said the regimental historian. "A right oblique fire from us along the front of the battery and into the woods in front of it would have prevented any force [such as the Thirty-eighth Mississippi] from going through the gap, and also aided those who were engaged in a desperate struggle against far superior numbers on the right. . . . The movement made by Le Gro's order virtually drew the regiment out of the battle."[5]

Col. Robert McLain of the Thirty-seventh Mississippi found himself in a predicament far graver than the "very embarrassing situation" Colonel Adams of the Thirty-eighth later tried to explain away in his report. General Little had ordered him "to move forward on the extreme right, with instructions not to fire, as there was a brigade of our own troops between us and the enemy."

That was true when Little gave the order, but the situation had changed markedly. The First Texas Legion, which had stood directly in front of the Thirty-seventh Mississippi, had inclined toward the Eleventh Ohio Battery as it advanced, unmasking McLain's left companies. Nonetheless, McLain moved in accordance with Little's order. He followed the initial path of the First Texas Legion, evident by trampled brush and broken branches.

For 250 yards McLain's Mississippians tripped through what one veteran called "the thickest place I ever saw of vines, bushes, and briars" before striking the southeast corner of the Yow field. A high wooden fence bordered the field. McLain ordered his men over it. While they were climbing, the Mississippians were stunned by a heavy and unexpected fire delivered into their right flank from the far side of the field, some 400 yards away. At that range few were hit, and Colonel McLain fell back to re-form the regiment and await instructions from Little. None came; not a single courier appeared to inform McLain of Little's fate or to offer guidance. Unlike Colonel Adams, McLain took matters upon himself. After restoring order he elected to challenge the Federals, who were arrayed on a ridge along the field's western fringe. The Mississippian conducted a half wheel to the right and marched his regiment due west into the open, away from the rest of the division and the fight for the Eleventh Ohio Battery. McLain's decision cost him dearly. A sharp frontal volley of musketry and crushing salvo of canister met his men midway across the field. Then an irregular but well-aimed fire erupted from the forest 200 yards to their left. The accuracy and increasing volume of the Yankee volleys convinced the Mississippians that they confronted a whole brigade — Capt. Absalom Dantzler guessed they faced four times their number — and the men fell to the ground to return the fire as best they could.[6]

In the thickening twilight it was easy to exaggerate the strength of an enemy only dimly seen, even at close range. In reality McLain confronted only Col. Nicholas Perczel's Tenth Iowa, the two-gun section of Immell's Twelfth Wisconsin Battery that had accompanied the Tenth on its reconnaissance up the settlement road, and the left companies of the disorganized Fourth Minnesota.

Colonel Perczel had ample warning of the Rebel approach. The three companies he had sent into the timber south of the Yow house to find the flank of the Fourth Minnesota returned to warn him that two enemy regiments—the Thirty-eighth and Thirty-seventh Mississippi—were coming. Perczel realigned his regiment along a ridge paralleling the settlement road north of the Yow house, which gave his men an unobstructed view of the field. Immell's cannon went into battery on his right. Perczel felt ready, until Lieutenant Immell limbered his guns. Perczel drew his saber and blocked Immell's path. "You are not going to retreat, are you?" bellowed the silver-haired Hungarian. "Wait a moment and see," Immell snapped back. He recalled with satisfaction what followed: "Immediately south of the cabins was a depression just back of the crest of the ridge of open ground, a fine position for the guns: the recoil placing them below the crest to load, and then 'by hand to the front.' On understanding the situation the Colonel rode up to me, and under a severe fire from the enemy apologized."

The Rebel fire was less effective than Immell portrayed it. Lying down and firing uphill, burying their heads to dodge the canister that raked their line, the Mississippians aimed badly and hit little. And they were angry. The Thirty-eighth Mississippi had disappeared from their left at the same time McLain's men took their first volley from that direction, leading them to conclude that their fellow Mississippians had run. Fighting alone and taking fire from two directions, McLain conceded the contest as dusk settled over the field. The Thirty-seventh re-formed in the timber and delivered a few more volleys, until darkness ended the pointless exchange. McLain's futile sally into the Yow field had cost him seventy-eight men.[7]

While McLain and Perczel fought their private duel near the Yow house, the action raged with unremitting fury around the Eleventh Ohio Battery. The cannon became a vortex into which opposing lines were drawn; attack followed counterattack in the smoky darkness.

For perhaps the only time during his army service, "Fighting Jerry" Sullivan proved worthy of his sobriquet. Sitting with his staff in the swale behind the ridge when Sanborn's brigade broke, Sullivan rode back to the meetinghouse to bring up his two reserve regiments, the Eightieth Ohio and the Seventeenth Iowa. The Eightieth Ohio was drawn up on a ridge parallel to and 250 yards in behind the one on which Sanborn was fighting, the regiment's left seventy-five yards north of the old meetinghouse and its right touching the Jacinto road. The Seventeenth Iowa was in route column on the road. Sullivan yelled at its commander, Col. John Rankin, to get his men into line, while the Eightieth Ohio marched off the slope and into the wooded ravine between the two ridges.

Sullivan expected little of Rankin, so he probably was not surprised to find his regiment unprepared for battle. No one really had much confidence in Rankin's ability, not even Rankin himself. A descendant of the Scottish poet Robert Burns, Rankin was a small, quick-minded man who studied Latin and law at Washington College before coming west to practice in Keokuk, Iowa, in 1848. Despite his general popularity and sharp legal mind, Rankin was singularly unfit for military command. He never grasped the basics of the military art and in the field was constantly sick. Rankin recognized his shortcomings, which he tried to remedy by resigning two weeks before Iuka on the pretense of having pressing business back home. Rosecrans immediately accepted his resignation. Unfortunately for all concerned, the lieutenant colonel was absent, the major was under arrest, and the new colonel of the Seventeenth had not arrived, so Rankin had to stay on.

The Seventeenth Iowa was a new regiment, but the men were good. Their mustering officer and surgeon were unusually conscientious, so that "no man was passed if he had the slightest physical blemish, and no man mustered unless, in size, he more than filled the letter of the regulations." But good men can do little without experience or a good leader, and the Seventeenth had neither.

Rankin deployed the regiment in a manner that thoroughly bewildered Capt. John Young: "We were filed off to the right by Colonel Rankin, until a little more than the right wing had filed to the right, when the regiment was halted and brought to a front and the remainder of the left wing formed on the left of the road." When Sullivan upbraided him, Rankin tried to redeem himself with brave words. Riding to the front of the regiment, he shouted, "Now, if there is a coward in the ranks of the Seventeenth Iowa, I want him to step out." No one moved. One man muttered, "You old fool." Deeply impressed with himself, Rankin ordered the regiment forward at the double-quick.

Into the ravine marched the Iowans. Bullets peppered the underbrush. Through the dusky timber came the "infernal, blood-curdling, ear-splitting, hair-raising rebel yell," recalled Pvt. Daniel Spencer of Company C. "To say that I was scared is putting it mildly. That first rebel charging yell rings in my ears yet. It brought my heart into my mouth, and I thought at the time that I had swallowed it back down on the wrong side." A glance toward his colonel did little to reassure him; "Rankin was too much rattled to give a command."

Crossing the ravine, the Iowans met a cavalcade of frightened officers and cavalrymen from Rosecrans's bodyguard that had strayed too close to the enemy. They stampeded down the road toward the center ranks of the Seventeenth. Rankin managed to yell an order to "give way to the right," so that the men on the road were not trampled, but in the bedlam that followed, the regiment separated. Watching the left wing drift off and his own company waver, Captain Young called Rankin's attention to the impending breakup of the regiment. "He told me to do the best I could for them and keep them together if possible," recalled Young. "After this I saw no more of him during the engagement."

Those who saw Rankin differ on what happened next. All agree that he tumbled from his horse and hit his head against a tree. As to the cause of his fall, there was considerable debate. Some said his horse was shot and Rankin was thrown from the dying animal. Others said he simply rode into the tree in a blind drunk. Averred one veteran, "The Colonel dashed into a tangle which brought down the colonel from his horse, tearing his sword and one epaulet off him. After that the colonel returned to his law practice in Keokuk." The regiment left him lying unconscious beside the tree and struggled onward.[8]

Word of Rankin's fall reached Captain Young, and he took command. An ambitious, scheming sort, Young had his eye on the colonelcy and was gravely disappointed when an Iowa judge was named to replace Rankin, but a good showing at Iuka as temporary commander would at least boost his standing when the new colonel arrived.

The situation, however, had deteriorated beyond Young's ability to salvage it. The left companies were beyond his reach, and those on the right unraveled under a plunging fire from the Third Texas Dismounted Cavalry. Capt. Samson Archer was shot while trying to get the men of the center companies to lie down. All panicked. Young ran rearward with the Iowans to an old graveyard, 400 yards from the Rebels. There, winded and frightened, the men collapsed.

"I succeeded in rallying them and got them back to about where our line was first formed and succeeded in quieting them for a time," said Young. General Sullivan rode up. Young reported himself in command of the regiment and asked for orders. Sullivan told Young to do "the best [he could]," then hurried away. Young walked to the right of the line and led the men forward. "We had proceeded but a short distance when a tremendous volley from the enemy caused a panic in the battalion, and with all my efforts I could not rally them until they had retreated almost to the road near the old log church." There Young reassembled them. Stragglers from other shattered regiments wandered into his ranks. Two thrashings had satisfied Young's taste for command, and he held the men in place, content to wait for orders.[9]

The Eightieth Ohio gave a better account of itself than did the Seventeenth Iowa. Lt. Col. Matthias Bartilson got his men across the ravine in good order. Near the base of the ridge a concealed line of Rebels — probably skirmishers belonging to the First Texas Legion — rose from the underbrush and at a range of thirty paces fired into the Ohioans. Colonel Bartilson's horse was shot dead. As he rose, the colonel was struck in the shoulder by a buckshot. Bleeding profusely, Bartilson called together his company commanders. Before leaving for the rear, he enjoined them "to hold their position until relieved by some proper officer." They did. Returning fire, the Ohioans bested the Confederates after a ten-minute exchange of volleys.[10]

The Ohioans' stand bought Sullivan time to patch together a scratch line from soldiers of the Twenty-sixth Missouri, the Forty-eighth Indiana, the Sixteenth Iowa, the Fifth Iowa, and the Fourth Minnesota who were milling about in the ravine. At General Hamilton's behest he led them past the Eightieth Ohio and recaptured the battery.

Sullivan's twilight counterattack caught the Confederates off guard. The Third Texas Dismounted Cavalry recoiled down the ridge. The Third Louisiana held on to the brow until a volley from behind disrupted the Louisianans' resistance. While his men sought cover in the underbrush, Maj. William Russell galloped rearward to find the source of the misdirected fire. Lieutenant Colonel Gilmore had been hit five times, but he stayed with the men, riding among them until his horse was shot. A sixth bullet struck Gilmore in the shoulder. He kept on his feet and, as much to escape the friendly volleys as to drive away the enemy, led a wild bayonet charge into the Federal ranks. Major Russell returned to the regiment and "came near losing his life by giving orders to a company of the enemy, mistaking them for one of our own." Gilmore's audacity paid off. Darkness intensified the ferocity of the Confederate charge, and the Federals gave way.[11]

The seesaw struggle for the ridge continued. The First Texas Legion was momentarily thrown off the ridge. Sullivan's left wing reassembled along the brow, and those in the center re-formed in front of Sears's cannon, three of which the Rebels spiked before falling back. The Federals' stay among the guns was brief. The acting major of the First Texas Legion had concealed three companies in the brush on Sullivan's left. They went undetected by the Yankees until, a few minutes after 7:00 P.M., they

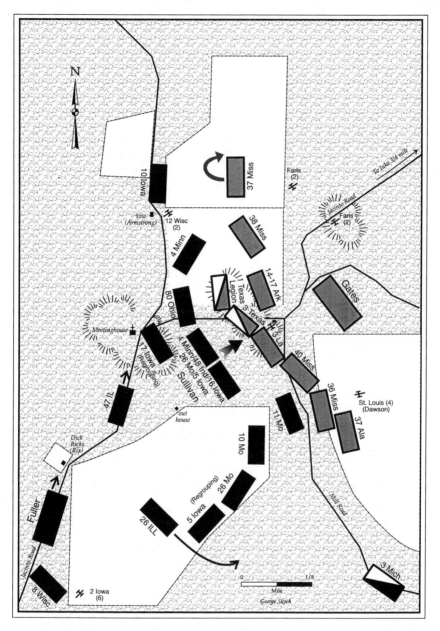

MAP 4. *Sullivan Counterattacks, 7:00 p.m.*

let go a volley, rose to their feet, and charged. Seeing only the flash of rifles beyond his flank, Sullivan ordered a retreat. The Confederates poured off the ridge in pursuit. Sullivan rallied his men and met the Rebels at close range. "It was so dark that friends could not be distinguished from foes," said Sullivan, but the fighting went on nonetheless.

Sullivan got help of a sort from an unexpected source. After waiting forty-five minutes for orders, Captain Young had found the nerve to start forward again. Badly frightened by their first two encounters with the enemy, Young's men fired randomly as they advanced. A soldier from the Fifth Iowa emerged from the dark to scold them; they were shooting into the backs of friendly troops. Young halted, restored discipline, and then edged the regiment forward, intending to draw near enough "to support our men in case they should fall back." But darkness thwarted Young's good intentions: "We had not proceeded far when some of my men again commenced firing, which was apparently answered by a tremendous volley from the direction of the enemy; but [another] soldier who was some distance in advance came rushing back and said that our own men were firing upon us. I then ordered my men to fall back in good order so as not to come in contact with them." Thirty yards to the rear, near the graveyard, he directed them to halt, face about, and kneel. Frightened soldiers from Sullivan's shattered counterattack came back onto his line. Behind them came the crash of an unseen volley. Again Young's men panicked. "In spite of all exertions they ran back about one hundred yards, when I succeeded in forming another line, and having advanced a few yards I ordered them to stand and wait for orders."

None came, but Sullivan brought his mixed command into line beside Young. Sullivan's troops opened fire and checked the Rebels. Young was about to join in when "a tremendous volley was fired by the enemy, and was immediately answered by some regiment still in our rear. We were now between two heavy fires from front and rear." Captain Young was helpless to prevent what followed. The deadly cross fire in the dark "caused a dreadful stampede among the men, and all commenced firing in all directions without regard to where their guns were aimed." Young had little trouble stopping their wild shooting after the volley from the rear ceased; his men were too exhausted to reload.

The same misdirected volley played havoc with Sullivan's line; it "killed and wounded more than my whole loss prior to that time," he complained bitterly. The Confederates were too tired to exploit the chaos in the Federal ranks, if they were even aware of it. All they knew was that the volume of enemy fire suddenly had increased, and that was enough to cause them to retire to the ridge, where they contented themselves with dragging away the cannon of Sears's battery. A few fitful volleys lit up the woods before the shooting sputtered out at 8:00 P.M.[12]

10

We'll I-uker Them Today!

The fire from behind that decimated Sullivan's ranks and terrified the soldiers of the Seventeenth Iowa was the consequence of a belated effort by General Rosecrans to strengthen the Federal center. Rosecrans's whereabouts during the battle are a matter of conjecture. Sergeant Brown of the Fourth Minnesota said Rosecrans walked behind his regiment on a quick tour of inspection shortly after the Minnesotans came into line. George Pepper of the Eightieth Ohio said, "Rosecrans was everywhere, cheering and encouraging the troops." Pepper said Rosecrans exhorted Sanborn's brigade into line with the words, "Come on, boys! We'll I-uker them today!"

Pepper's account is plausible, given Rosecrans's penchant for silly turns of phrase. But if Rosecrans remained on the ridge when the shooting started, his presence there was unknown to Hamilton or Sanborn. Perhaps, as a veteran of the Seventeenth Iowa suggested, the commanding general had been borne back with the cavalcade of staff officers that stampeded his regiment early in the fight. In any case Hamilton insisted Rosecrans was nowhere to be found when he most needed him: "Thinking General Rosecrans was in the rear, where he could hurry up the troops of Stanley's division, I dispatched an aid with the request that General Rosecrans would come forward far enough to confer with me. All the while the battle waxed hotter and more furious." Hamilton grew worried. "The dead lay in lines along the regiments, while some of our troops gave signs of yielding. I dispatched another officer for General Rosecrans. He happened to be one of General Rosecrans's staff, and at my request he started to bear the message to the general."

When the Forty-eighth Indiana crumbled, Hamilton's patience ran out. Four of his staff officers were down, badly wounded. Hamilton sent two of those still in the saddle to tell Rosecrans he "considered it imperative he should come forward to see me and should hurry forward fresh troops."

Sanborn waited with Hamilton for a reply. By then the Sixteenth Iowa had been swept from the field, and Colonel Boomer was struggling to keep his four companies of the Twenty-sixth Missouri in the fight to the right of Sears's battery. No word came from Rosecrans, but General Stanley galloped up.

"Sanborn, this is the heaviest musketry fire that I have ever heard in my life, and I was in every battle in Mexico. What can I do to assist you?" asked Stanley.

"I must have reinforcements on that portion of the line about the battery, or I shall lose it," Sanborn said.

"I will send up the Eleventh Missouri double quick," Stanley assured him, then rode off to carry out his promise. Hamilton and Sanborn waited for Stanley's return until the tide of retreating Federals carried them off the ridge. The Eleventh Missouri, lamented Sanborn, "was ten minutes late for the relief needed."[1]

That fatal delay indirectly was Rosecrans's fault. Wherever he may have been when Hamilton was searching for him, Rosecrans knew the division commander needed help. He tried to provide it by ordering Col. Joseph A. Mower, commander of the lead brigade of Stanley's division, forward from the Ricks house.

Mower moved at once. He beckoned Maj. Andrew Weber, whose Eleventh Missouri stood at the head of the brigade column, to follow him. To make better time Weber filed the Eleventh off the road and into the Ricks field at the double-quick time. Crossing the field Mower and Weber noticed the firing from Hamilton's extreme right wax louder. As the volleys from the direction of Hamilton's center had diminished—a consequence of Sanborn's line having been thrown from the ridge—Mower diverted the Eleventh Missouri to the right. Major Weber halted the regiment at the northeastern corner of the field, deployed, and advanced into the timber. When Mower turned to form the rest of his brigade, he found the field empty. Judging the moment too critical to

Brig. Gen. David Sloane Stanley, in later dress as major general (Massachusetts Commandery of the Military Order of the Loyal Legion, and the U.S. Army Military History Institute)

allow him to search for the brigade, Mower followed the Missourians into the forest.[2]

Both Rosecrans and Stanley were to blame for the breakup of Mower's command. Rosecrans erred in not telling Stanley he had ordered Mower to send in his entire brigade. When Stanley reached the brigade column, Mower had already moved out with the Eleventh Missouri. Apparently assuming one regiment was enough to relieve the pressure on Sanborn, Stanley opted to fashion a strong reserve in case Hamilton's division buckled. He halted the Forty-seventh Illinois near the meetinghouse and sent the Twenty-sixth Illinois midway into the Ricks field. Stanley placed Mower's trailing regiment, the Eighth Wisconsin, 300 yards south of the Ricks house, beside the Second Iowa Battery.

Mower was "all fight from head to foot," said one of his men, and he proved himself deserving of his nickname "Fighting Joe." He had guessed correctly in diverting the Eleventh Missouri to the right. The firing was heaviest in that direction because the unit that fought there — the Fifth Iowa — was the only regiment of Sanborn's brigade still in action. Major Weber maneuvered the Eleventh Missouri into line on the right of the Iowans. No sooner had Weber's regiment entered the forest north of the Ricks field than it came face to face with an enemy line of battle, thirty yards away.[3]

The Missourians got off the first volley. No response came from the Rebels. Instead, said Weber, a confused Confederate ran into Federal ranks screaming, "For God's sake, stop firing into your own men; you are firing into the Thirty-seventh Mississippi." The Missourians collared the man, raised a cheer, and released a second volley "more terrific than the first." The Confederates returned the fire, but their bullets largely passed over the Missourians, who were on lower ground.

Major Weber had misunderstood the bewildered Rebel who had tried to call off his men. He was not fighting the Thirty-seventh Mississippi but, rather, the Thirty-seventh Alabama and the Thirty-sixth Mississippi regiments, which Col. John Martin finally had brought to bear to the left of Hébert. Martin had been slow in coming up; perhaps the death of Little interrupted his advance. Had he moved fifteen minutes earlier, he would have outflanked the Fifth Iowa, caught the Eleventh Missouri in route column in the Ricks field, and perhaps rolled up Rosecrans's entire right flank. As it was, the two lines settled into a fierce firefight after sunset.

After the first two or three volleys blanketed the forest with smoke, neither side hit much. The Missourians overshot their targets as badly as did the Confederates. To break the impasse, Martin ordered a bayonet charge. Pvt. Sam Singletary of the Thirty-seventh Alabama recalled the moment the two lines collided. "We went right into their lines. A Yankee [officer] being the first to see our approach shot a man named Judkins through the arm with his pistol. By this time five or six of us had our muskets on him and were pressing our triggers when he said for God's sake men don't fire, which saved his life." Judkins was handed the pistol and allowed the privilege of escorting the officer to the rear. Singletary last saw him sitting astride the Federal, punching him with the pistol.

Weber repelled three charges in brutal, close-quarters fighting. Men darted forward in the dark to yank enemy soldiers from the ranks, bayonets crossed, and powder burns were common. An hour after sunset the Confederates broke off the attack. After they disappeared, the Missourians, out of ammunition, fell back slowly. In forty-five minutes of fighting, Major Weber lost seventy-five men. Colonel Martin reported seventy-seven casualties, including both the commander and the lieutenant colonel of the Thirty-seventh Alabama. Their clash was the last action of the day, but not the last killing.[4]

In an appropriate denouement to the two-and-a-half-hour, added affair, misdirected fire inflicted the last casualties of the day. On the Federal side the principal culprits were the Thirty-ninth Ohio of Fuller's brigade, brought forward by Rosecrans and Stanley to steady Sullivan's patchwork command, and the wayward Fourth Minnesota.

Maj. Alfred Gilbert of the Thirty-ninth said he had no idea what was happening. It was dark when his regiment reached the battlefield, said Gilbert, and "large clouds of smoke issuing from above the trees" obscured the fighting. General Rosecrans and his staff paused near the Thirty-ninth Ohio. On the commanding general's instructions, Stanley led the regiment forward. The Ohioans marched up the road and filed into line thirty yards behind the Seventeenth Iowa and the right wing of Sullivan's force. The timidity of Captain Le Gro had kept the Fourth Minnesota out of the battle; now his ineptitude caused a clash between the Thirty-ninth Ohio and the Seventeenth Iowa.

Shortly after the Eleventh Ohio Battery changed hands for the last time, the Minnesotans were removed from their superfluous position near the Yow house to bolster the new line near the meetinghouse.

The Minnesotans trudged down the settlement road. "The fighting at this time had entirely ceased. On our way to the front we stepped over a good many of our wounded who belonged to other regiments, several of whom begged us to shoot them and put them out of their misery," remembered Sergeant Brown. "Before arriving at the place of our destination we were halted. It was now very dark." Captain Le Gro lost control of the regiment while forming a line of battle. The left wing companies kept marching after the rest of the regiment stopped, stumbling between the lines of the Thirty-ninth Ohio and the Seventeenth Iowa. A few of the Minnesotans heard officers, whom they swore were Rebels, whispering commands. They panicked and let go a few random rounds, which set the black woods ablaze with gunfire. The Eightieth Ohio, which had regrouped to the right of the Thirty-ninth Ohio, joined in the shooting. Six soldiers in the Thirty-ninth were lightly wounded—two having been kicked by the regimental commander's startled horse—but Pvt. J. H. Van Eman said a lieutenant of the Seventeenth was killed at his side. Several more Iowans must have been shot trying to pass through the lines of the Thirty-ninth, said Van Eman, "judging by the amount of shooting done at close range."

Being on lower ground the Fourth Minnesota suffered the least. Nevertheless, the Minnesotans lost more in that brief exchange of friendly fire than they had in sidestepping the Rebels all afternoon.[5]

The Confederates suffered less from misdirected volleys because there were far fewer demoralized Southern soldiers milling about behind the front line. Hébert and Martin fought the battle alone, and no regiment in either of their brigades broke. By nightfall every regiment was committed to the fight. Not until 7:00 P.M. did the reinforcements Price had asked Little to bring forward just before the Marylander's death reach the battlefield. Col. Elijah Gates's brigade appeared "just in time to fire and receive one round when the enemy withdrew on account of darkness," said a Missouri lieutenant. Or so the lieutenant thought. Gates's volley actually struck the Third Louisiana, which was then bracing itself for Sullivan's counterattack. Major Russell of the Third succeeded in quieting the Missourians before they could reload.[6]

That the Missourians were anxious to kill Yankees is understandable. They had "double-quicked six miles through the suffocating dust and beneath a sultry sun, increasing their gait to a run as they neared the battlefield." As they drew near the fighting, stragglers from the front told them of the death of Little. "Sadness and sorrow" greeted the news. "Their leader and first brigadier had fallen. . . . In Henry Little our brigade lost its main stay and support," attested Lt. Col. Robert S. Bevier, commander of the Fifth Missouri. As word of Little's death spread through the ranks, General Price appeared. With hat in hand and a wild look in his eyes, Price exhorted his "brave Missouri boys" to "stand up to the work." Cried Price, "Move up, boys, we are whipping them—we have already driven them back a mile and captured nine pieces of their artillery. Go in, boys, and give them your best!" The Missourians replied first with a yell, then with their well-intentioned but poorly aimed volley.

After calming down, Gates's Missourians lay down behind Hébert's brigade and listened to the "terrible crash of musketry" that receded from the ridge and then died out.

Lieutenant Colonel Bevier was surprised to receive orders to advance. "It was dark as pitch," he recalled. "I was directed to move my command a certain direction until ordered to halt." Bevier's men marched blindly, not sure they were near the front until they stumbled upon dead and dying soldiers from the Fortieth Mississippi. The Missourians stepped over them and felt their way forward. They splashed through a swamp, and the noise startled Yankee pickets. Bevier was in a fix: "We were getting into place rapidly, when the Federals opened fire, to my great personal inconvenience; for I happened to be between the two lines, and my men, in their eagerness, shot under my horse and over him, and made a rest of both ends of him, and I must say that we—my horse and I—made a pretty good bulwark." Bevier yelled at his men to stop shooting. The Yankees disappeared. Bevier deployed skirmishers and got his men out of the swamp and into a dry glade. "I thought this was getting close enough in the dark; and as no one sent me orders to halt, I halted and waited in-structions, which soon came, informing me that I was right in thus doing wrong." Bevier's men lay down to rest. Through the trees Yankee officers passed hushed commands along their lines.[7]

*   *   *

All night long, scenes similar to the close call of Lieutenant Colonel Bevier were played out across the battlefield, some with climaxes tragic, other grimly comic. The night was cold and the dew fell heavily. Not a breath of air stirred. Voices carried easily. The snap of a twig or the rustle of a bush was enough to set nervous pickets to shooting. Opposing lines were not more than 300 yards apart, and in some places as near as 70 yards.

The careless acts of tired soldiers brought deadly reminders of the enemy's proximity. Said a Missouri Confederate, "One of our company struck a match to light his pipe, when several shots were immediately fired at him without effect. . . . The blaze from the enemy's guns was but a little distance in the brush beyond us." Fires were forbidden, and officers were alert to the slightest infraction. A few soldiers from the Eighth Wisconsin, perhaps assuming they were far enough to the rear to be exempt from the prohibition, lit a small fire to boil coffee. A harsh voice in the dark reproached them. "That's right! Kindle fires, call the fire of the enemy and get your damned heads blown off!" The soldiers smoth-ered the blaze but not before it revealed the angry officer to be General Rosecrans.

Anyone who strayed from his unit did so at his own risk. Some did and were shot; others wandered into opposing lines and were grabbed. A cap-tain of the Fourth Minnesota bumped up against a wounded horse. It fell on him, but he was near enough the regiment that his men heard his cries and pulled him from under the animal. A lieutenant from the Fifth Mis-souri braved death to get water for his thirsty soldiers. "We needed water sorely, and our only chance for a supply was in the little brook half-way between the lines," said Lieutenant Colonel Bevier, who would not think of ordering anyone to hazard his life for it. But Lt. John Lippincott be-lieved he could make it, and Bevier allowed him to try. Lippincott gath-ered an armload of canteens and edged forward. In the moonlight, Bevier traced his progress:

He had nearly reached the creek when he stumbled into an unlucky hole, which drew a scattering shot or two that would have amounted to nothing had he kept still, but he precipitately commenced a disorderly retreat, causing his canteens to make the most infernal clatter ever made by canteens before or since. The enemy supposed our whole army was charging upon them, and opened a most terrific discharge. Lippincott, with his usual judgment, made for the headquarters of the regiment, tumbled over three or four of his superior officers and landed full length on me, nearly knocking the breath out of me and centering all the fire on our devoted heads. We escaped with a few scratches; and, with such pleasant interludes as this, "wore the uneasy night away."[8]

There was nothing remotely humorous in the plight of the wounded, for whom thirst was a hellish torment. Moaning pleas for "water, water; only a drop of water!" struck a chorus of suffering that kept on until daybreak. The doleful refrain was more than some could bear. A "humane soldier, forgetting the foe, would attempt to pass the canteen, but a crack of the musket would warn him to keep close," lamented a Texan.

Others ventured their lives for nothing more noble than ghoulish curiosity. Having missed the fight, Ephraim Anderson of the Second Missouri crept over the battlefield in the night, impelled perhaps by an unspeakable desire to visit death's aftermath. Said Anderson of the experience,

The moon was nearly full, and threw a strong light upon the pale and ghastly faces of the thickly strewn corpses, while it glanced and sparkled upon the polished gun-barrels and bright sword bayonets of the enemy's guns, which lay scattered around. The dead were so thick, that one could very readily have stepped about upon them, and the bushes were so lapped and twisted together — so tangled up and broken down in every conceivable manner, that the desperate nature of the struggle was unmistakable.

The carnage around the battery was terrible. I do not think a single horse escaped, and most of the men must have shared the same fate. One of the caissons was turned upside down, having fallen back upon a couple of the horses, one of which lay wounded and struggling under it; and immediately behind was a pile of not less than fifteen men, who had been killed and wounded while sheltering themselves there. They were all Federals, and most of them were artillery-men. Some of the limbers were standing with one wheel in the air, and strewn thickly around all were the bloody corpses of the dead, while the badly wounded lay weltering in gore.[9]

Around the Ricks house, which Chief Surgeon Archibald Campbell had appropriated as a hospital early in the battle, Federal surgeons sliced and sawed. The wounded came so fast that both the house and the yard were full before dark, and Campbell had operating tables set up in the forest. A new recruit from the Eighth Wisconsin stumbled upon the scene and was forever sorry he had: "This was the most appalling sight I ever beheld, to see men brought in wounded in all manner of ways. About fifteen surgeons were busy dressing and amputating." John Risedorph of the Fourth Minnesota was similarly appalled: "There in great piles were arms and legs, some horribly mutilated while others had simply been penetrated by the wicked minie ball. Some feet were naked while others had their boots or shoes on. Gallons, yes barrels of blood had made the adjacent ground red and muddy." [10]

Not the cries of the wounded, their own thirst, the nearness of the enemy, or the volleys that lit the forest in fitful flashes sufficed to keep some men from sleep or from appreciating the mocking irony of nature. Remembered Major Gilbert of the Thirty-ninth Ohio, "I got a little sleep — the night was very beautiful — stars shone bright and peacefully down upon this scene of strife — they never looked so quiet and good to me before." Most, however, lay awake — listening to the groans and gunfire — and brooded the night away. Five days later a Mississippi soldier penned his thoughts in a letter home: "I know this, that the events of that evening have considerably increased my appetite for peace, and if the Yankees will not shoot at us anymore, I shall be perfectly satisfied to let them alone." [11]

11

Where, in the Name of God, Is Grant?

Capt. Frank von Phul returned to the Coman place, the cottage General Little had taken as his headquarters, after nightfall. Father Bannon was there, keeping watch over the general's body. Phul lingered a moment, then left for army headquarters to attend to one final task as aide-de-camp to Little. At the Moore house Phul was told General Price had taken lodging for the night with his close friend, former Missouri governor Trusten Polk. Phul found the home where Polk had taken quarters and went inside. General Price was there, in his bedclothes, looking tired and distracted.

Phul asked him softly, "General, what shall I do with General Little's body?"

"My Little, my Little; I've lost my Little," muttered Price. Phul quietly contemplated the general: "The lines of sorrow were like furrows on his brow."

After a moment Phul repeated the question. "General, what shall I do with General Little's body?"

"My Little; I've lost my Little, my only Little."

Phul persisted. "General, what shall I do with General Little's body?"

"My Little is gone; I've lost my Little," muttered Price.

Phul turned to leave. "That was the only reply I could get from General Price. He was almost crazed with grief, and I don't believe he knew what I was asking him."

Descending the steps into the dark, Phul met Colonel Snead. With Snead were Col. Wirt Adams, Generals Hébert, Maury, and Armstrong, and a staff officer from Van Dorn's army. Phul explained to Snead his purpose in coming. Snead reassured Phul; he would speak to the general about Little's body.[1]

Colonel Snead had come to Price's lodgings with pressing business of his own. Before retiring for the night Price had told Snead he intended to renew the battle with Rosecrans at daylight, notwithstanding the loss of Little or the presence of Grant northwest of town. The orders he dictated to Snead were vague. Maury was to move his division to the battlefield. General Armstrong would occupy the vacated line with dismounted cavalry. Armstrong was to delay any advance by Grant long enough for Price to defeat Rosecrans. The precise manner of the attack Price left for the morning. Snead wrote out the orders and settled in at headquarters, "determined," he said, "to remain awake all night."

He would not have gotten much sleep in any case. Shortly after midnight General Hébert arrived. The Creole was inconsolable. His brigade had been cut to pieces, Hébert told Snead. The death of Little so disheartened the men that Hébert doubted their will to fight. While Hébert unburdened himself to Snead, General Maury came in. Snead found him as downcast as Hébert. "He was convinced that Grant would attack us in overwhelming force in the morning, brush our cavalry out of his way, destroy our trains, and assail us in rear," said Snead. Wirt Adams and Frank Armstrong spoke next. They shared Maury's views. All four insisted Snead take them to Price. Snead vacillated: "I was still hesitating what to do when one of Van Dorn's staff arrive[d] with important dispatches from Van Dorn, and asked to see the general. I hesitated no longer, but took them to his lodgings."

They found Price asleep. Awakening, Price assumed his adjutant and division commanders had come to call him to battle. "Great was his disappointment when he ascertained the true cause of our coming," said Snead. Maury spoke first. "The old man was hard to move," Maury remembered. "He had taken an active personal part in that battle that evening. . . . The enemy had been so freely driven back, that he could think of nothing but the complete victory he would gain over Rosecrans in the morning. He seemed to take no account of Grant at all."

Price was adamant; retreat was out of the question. "We'll wade through him, sir, in the morning," he told Maury. "General, you ought to have seen how my boys fought this evening; we drove them a mile, sir."

"But," remonstrated Maury, "Grant has come up since then, and since

dark you have drawn me from before him; my brigades are lying in the streets, with their backs to Grant, and the whole wagon train is mixed up with us, so that we can't get into position promptly in the morning. . . . Placed as we are we shall be beaten, and we shall lose every wagon."

Price sat on his bed in his nightshirt. As Maury continued, Price drifted from alert confidence into sleepy dejection. "You can't procure another wagon train like this, not if you drain the State of Mississippi of all its teams," said Maury. "We have won the fight this evening. We decided on going back anyhow in the morning to Baldwyn, and I don't see that anything [that] has happened since we published that decision should detain us here any longer."

Price turned to the others. Snead, Armstrong, and Adams all sustained Maury's views. Reluctantly Price admitted the prudence of returning to Baldwyn to join Van Dorn as planned. He told Snead to have the wagon train depart at 3:00 A.M. on the Fulton road. Maury would detach one brigade to escort the train. Hébert's division would follow. Maury's remaining two brigades were to take up blocking positions on a ridge east of town, a mile and a half from the battlefield. Armstrong's cavalry would act as rear guard.[2]

It was 2:00 A.M. when the gathering broke up. Snead returned to headquarters, drafted the necessary orders, then went out again. One final, painful item of business remained before the army moved out.

Snead walked to the Coman house. "General Little must be buried at once," he told Father Bannon, "for we retreat before dawn."

Snead called on some soldiers to dig a grave in a small garden beside the cottage, while Bannon gathered up the sacral necessities. Three staff officers, a civil engineer, and Little's orderly followed Father Bannon and Snead to the garden. Each carried a candle. The soldiers lowered Little's body into the ground, and Father Bannon spoke a few words. Captain Phul watched as "the last spadeful of earth was placed upon the grave and patted into shape. Our candles still flickered in the darkness, sending out weird shadows." A plain sheet of pine board marked "General Henry Little" was thrust at the head of the grave, and the officers dispersed.[3]

*   *   *

General Rosecrans passed the night pacing the front, speculating on the meaning of the sounds drifting through the forest from the Confederate lines. With no word from Grant, Rosecrans assumed the worst—that Price's entire army confronted him, ready to renew the contest at dawn, and that Grant would do nothing to distract the Rebels.

Rosecrans called together his division commanders, who agreed with him the army should hold its ground. Before midnight Rosecrans quietly replaced Hamilton's exhausted soldiers with Stanley's division, then he sent a reconnoitering party beyond his right flank to search for a route he might use to outflank Price.

While the divisions of Hamilton and Stanley changed places, Rosecrans apprised Grant of the battle and implored him to attack in the morning. Rosecrans sent the dispatch by courier over a long line of vedettes he had taken pains to establish between his army and that of Grant before the battle, then returned to the front.[4]

The racket continued, ominous but inconclusive. The Rebels, thought Rosecrans, seemed to be "cutting, chopping, driving stakes, halting and aligning their men." The noise increased until it was evident the Rebels were on the move, but in what direction was anyone's guess. In his company, said an Ohio private, opinion as to whether the Rebels were "leaving or getting ready to renew the fight was about equally divided."[5]

Rosecrans grew agitated. A new sound, the rumble of wagons, rolled through the forest. "I heard the movement of the train in the distance towards the southeast, and artillery moving apparently along the very heights I desired to occupy, and from which my reconnoitering party had not returned. This gave me no little uneasiness."

Rosecrans was at the front at 4:00 A.M. when "the voices of drivers of artillery or ambulance trains, evidently anxious and in haste," replaced the dull thud of axes and sharp singing of saws. Was Price changing positions to take him in the flank, or was he retreating? Rosecrans returned to his field headquarters and gave orders for the troops to be roused and fed at once; the army would go forward at dawn — in pursuit if the Rebels were retreating, or to the attack if Price was still there.

Rosecrans searched out General Stanley, whose division would lead the advance, to give him his instructions personally. He found the fellow Ohioan sleeping in a fence corner.

"Stanley, Stanley," Rosecrans whispered.

"What do you want, Rosie?"

"You will go in at sunrise on the bayonet; not a shot is to be fired."

"They are five to our one; they have butchered my men like sheep," remonstrated Stanley.

Rosecrans answered with a question: "Where, in the name of God, is Grant?" Then to Stanley: "But go in on the bayonet — don't fire a shot."

"I feel that I shall be killed to-morrow, but your order will be obeyed," Stanley muttered, then folded his blanket about himself and again fell asleep.[6]

*   *   *

At 3:00 A.M. on September 20, Confederate officers shook their troops awake. "The astonishment of the men was indescribable when orders were received to evacuate the place," said Louisianan Willie Tunnard. They had fought hard, driven the enemy, and assumed they would complete the work at daybreak. The logic of Maury's argument to Price that the movement constituted not a withdrawal but merely compliance with previous orders was lost on the men in ranks. "Feds were too stormy for us at Iuka. This is too bad to think about," W. C. Porter of the Sixteenth Arkansas scribbled in his diary. Soldiers of the Third Texas Dismounted Cavalry gathered around the six cannon of Sears's battery they had dragged off the ridge. There were no spare horses to pull them, so the Texans drove steel files into the touchholes. They marched away leaving the spiked guns on the Jacinto road, 100 yards from where they had been captured.[7]

Price was angry, anxious, and fearful Rosecrans would catch his column strung out on the march. Normally gentle with his men, Price vented his rage on unsuspecting teamsters of the army's wagon train, shocking everyone. A slight delay in starting the wagons ignited him. "We never remember to have heard General Price swear, only on this occasion, and he was not choice in his language at this time," said Willie Tunnard. "He ordered the teamsters to drive on, adding, 'If one of you stops, I'll hang you, by God.'"

Price drove the men relentlessly. Tired and hungry — some had eaten nothing but an ear of parched corn in twenty-four hours — soldiers dropped from the ranks by the hundreds. Despite early confusion and heavy straggling, at 8:00 A.M. Price had accomplished what he had refused to consider six hours earlier: he had extricated both his army and his cumbersome trains from between the forces of Rosecrans and Grant. By 2:00 P.M. the Rebel rear guard was eight miles south of Iuka.[8]

*   *   *

Sunrise on September 20 found Stanley's Federals ready to advance. With them, General Rosecrans stared into the morning twilight. The forest was still. There was no movement atop the ridge and no sound but the weak moans of the wounded. Rosecrans told Lt. Col. Edward Noyes to detach a company from his Thirty-ninth Ohio to reconnoiter. Edging their way up the slope, alert for enemy skirmishers, the Ohioans instead discovered a canvas of gore. The dead, said an Ohio veteran, "were found lying in almost every position. One was lying prone upon the ground, his eyes wide open, his gun resting on a log, in the act of firing. A number had guns clutched in their hands in the act of loading, their rammers half drawn. One in dying had grasped a small tree with his teeth, others died while kneeling and taking aim."

No one had seen bodies packed so tightly before or trees so torn by shot. Benjamin Sweet crested the ridge on the Jacinto road. "There was an oak tree right in the center of the road with a wagon track on either side," he said. "This was just as broad as I am across the shoulders, and there were forty-two bullets in it, below a line as high as my head. I counted eleven men and thirteen horses on a place about sixteen by thirty feet." Another Ohioan counted "forty dead bodies on one spot the size of four rods square."

Where the cannon of the Eleventh Ohio Battery had stood, there was a promiscuous tangle of dead horses and men. Said Sergeant Brown of the Fourth Minnesota, who wandered onto the ridge after the Thirty-ninth Ohio had passed,

In the low ground behind the battery twelve horses belonging to two caissons had become tangled together and piled up like a pyramid. Some below were wounded; others, dead, and over and above all, with his hind feet entangled down among the dead and wounded beneath him, stood a noble looking animal with head and ears erect, his right fore leg bent over the neck of a horse beneath him, his eyes wide open and out of his nostrils there extended, like a great white beard, a foam fully a foot long and streaked with purple. He was dead. This scene, and with it that of our dead heroes and those of the enemy lying thickly over the ground and the look of destruction and desolation that abounded in the vicinity, was the grandest and most awful spectacle of war that I viewed during a service of four and a half years.[9]

The ridge looked like the handiwork of a huge firing squad. The dead and dying lay in a long, dense row a half-mile long. Wounded by the hundreds groaned for water and for help, but the soldiers of the Thirty-ninth Ohio had no time to succor them. As soon as he learned the enemy was gone from the ridge, Rosecrans told Stanley to send the regiment to Iuka. Company M of the Second Iowa Cavalry galloped ahead of the Ohioans to reconnoiter the town; Battery M, First Missouri Artillery, brought up the rear to shell any Rebels still in Iuka. Colonel Fuller took charge of the patrol.[10]

Fuller pressed along the Jacinto road until he reached a large sweep of open fields between the road and Indian Creek. There, within sight of town, Fuller deployed into line of battle shortly before 8:00 A.M. The detachment from the Second Iowa Cavalry rode toward town in time to glimpse stragglers from the Rebel rear guard disappear down the Fulton road. The Iowans thundered down the street on the charge, but only Rebel wounded and a few frightened civilians were on hand to witness their martial entrance. Meanwhile, Capt. Albert Powell unlimbered the guns of Battery M, First Missouri, on the west bank of Indian Creek and began shelling the town. One shot crashed through the Iuka Springs Hotel and exploded among the wounded gathered there; another shattered the roof of a private home, setting the rafters ablaze. A deputation of citizens hurried from town to surrender Iuka to the Federals. Carrying a sheet tied to a broom handle, Mr. Samuel De Woody led the group of elderly men through the Federal ranks to Colonel Fuller. They begged him to spare their homes. Price's army had left, they told Fuller, and there were no men of military age in Iuka except the wounded. Fuller ordered Captain Powell to shift his fire away from the town. The Missourian trained his cannon on the last of Dabney Maury's blocking force, which was withdrawing from the heights east of Indian Creek.[11]

Rosecrans was ready to pursue when word of Price's departure reached him at 8:30 A.M. Anticipating a Confederate withdrawal, shortly after dawn Rosecrans had directed Mizner to throw out a mounted dragnet in the direction of the Fulton road. Mizner parceled out his cavalry quickly but keenly, sending eight companies of the Third Michigan Cavalry down the settlement road to slash at the enemy's flank; the remainder of his cavalry he dispatched to the intersection of the Fulton and Jacinto-Tuscumbia roads, six miles south of Iuka, to block the head of Price's column. To chase the Rebel rear guard, Colonel Mizner called on Col. Edward Hatch of the Second Iowa Cavalry.

After 8:30 A.M. Rosecrans added his infantry to the pursuit. He divided his forces so as to snare Price before he could reach Bay Springs, which Rosecrans guessed to be his immediate objective. Rosecrans told General Hamilton to face his division about and return to Barnett's Crossroads. There Hamilton was to pick up the Jacinto-Tuscumbia road and drive eastward. Stanley, meanwhile, would pursue southward over the Fulton road. Hamilton and Stanley were to reunite at the intersection of the Fulton and Jacinto-Tuscumbia roads where, if Mizner's cavalrymen proved able to delay Price, the army would drive the enemy toward the impassable defiles of Big Bear Creek along the Alabama border. It was an ambitious plan, especially the role handed Hamilton. His division had marched hard the day before, fought an unexpected and bloody battle, and had not eaten in twenty-four hours. But Rosecrans was anxious to press his advantage, with or without Grant.

When he ordered the pursuit, Rosecrans had yet to hear from Grant. There had been no reply to his dispatch of 10:30 P.M., in which Rosecrans had reported the results of the battle and begged Grant to attack in the morning. Rosecrans renewed his appeals after daybreak. At 7:00 A.M. he dispatched a message along the vedette line, again summarizing the fight and telling Grant that Stanley was marching on Iuka "as fast as excessive fatigue will admit." At 8:45 A.M., with Iuka secure, he wrote Grant of his planned pursuit. Rosecrans waited an hour, then sent a third message. The enemy, he said, "are retreating with all possible speed — Stanley follows them directly, and Hamilton endeavors to cut them off from the Bay Spring road — The men double quick with great alacrity." His patience exhausted, Rosecrans asked angrily, "Why did you not attack this morning?"[12]

Rosecrans got his answer sometime after 10:00 A.M. Riding with Colonel Fuller west of town, Rosecrans spotted the head of General Ord's column marching smartly up the Burnsville road, "with drums beating and banners flying." Rosecrans accosted Ord. "Why did leave you me in the lurch?" he demanded. Ord pulled a slip of paper from his pocket and handed it silently to Rosecrans. It was Grant's order directing Ord to postpone the attack — an order Colonels Lagow and Dickey were to have apprised Rosecrans of the day before.

Colonel Fuller watched Rosecrans. He saw shock turn to rage. Remembered Fuller, "This miscarriage was the beginning of a misunderstanding which grew into positive dislike between Grant and Rosecrans — a breach that was never healed." [13]

12

A Pursuit Can Amount to Little

Maj. Gen. Edward O. C. Ord was an earnest, aggressive commander who had little patience with incompetent officers, regardless of their rank. As a division commander in the Shenandoah Valley he had shown his contempt for department commander Maj. Gen. Irvin McDowell by feigning illness and relinquishing his command, rather than obey McDowell's orders that he cooperate with John C. Frémont in a movement to trap Stonewall Jackson near Strasburg. Jackson escaped, partly because of Ord's factiousness. But Ord was not blamed. Availing himself of McDowell's low standing with the War Department, he demanded to be relieved from duty under McDowell, whom he castigated as an inept, rude bully. Ord's impudent demand found favor with Secretary Stanton, who granted him a transfer to the West. In a few short weeks Ord ingratiated himself with Grant and won command of the Left Wing of the Army of the Tennessee.[1]

By Ord's standards of duty Grant's performance at Iuka was as inept as any of McDowell's missteps in the Shenandoah Valley. Grant distanced himself from the operation against Price early. Rather than travel with Ord's column, Grant established his headquarters at Burnsville. Perhaps Grant reasoned he could better communicate with Rosecrans from Burnsville, but in exchange for putting himself five miles closer to Rosecrans's vedette line, he sacrificed control over half his force. And Grant made only a minimal effort to keep in touch with Rosecrans. He showed little interest in the progress of his column, preferring to wait passively for dispatches from Rosecrans rather than actively follow his advance. Grant seemed preoccupied with trivialities, such as the erroneous dispatch about Antietam.

When he learned Rosecrans had been delayed and that Stanley's division was twenty miles away, Grant decided to hold Ord six miles short of Iuka. His most aggressive action was to deliver the news of Lee's defeat to Sterling Price.

Ord conveyed the message but made no secret of his disgust with it. Fretting about his headquarters, Ord asked his chief of staff, Col. Arthur Ducat, rhetorically, "Ducat, what do you think of it?" "Why should Price surrender when he can run?" answered Ducat. "If this dispatch goes in, he will get out if he can, because it gives him time," the colonel continued. "The thing to do is to attack him now."[2]

Ducat guessed correctly. Although he received Ord's "insolent demand" in the early morning hours of September 19, Price withheld his answer until mid-afternoon — immobilizing Ord, who was bound by the terms of the truce under which the message was delivered not to move until Price answered.

Grant's seeming indifference to the fate of Rosecrans continued. Early morning reports of suspected Confederate activity near Corinth prompted Grant to send Ord off with part of Davies's division, which was resting at Burnsville, on a reconnaissance toward Corinth. For six hours, from 9:00 A.M. until 3:00 P.M., the commander of Grant's northern column was absent from the front, on a mission that Grant should have entrusted to a subordinate.[3]

Ord's absence evoked no great concern at army headquarters, nor did the lack of precise information on Rosecrans's whereabouts. Apparently assuming Rosecrans could not possibly reach Iuka before dark, Grant dismissed him from his calculations. Not until noon did Grant send Cols. Clark Lagow and Theophilus Dickey to check on the Ohioan's progress and, presumably, to tell him that Ord's attack had been postponed.[4]

Returning from the reconnaissance, Colonel Ducat accompanied Ord to army headquarters. The languor about the place startled him. No one was charting Rosecrans's march on the map or checking road or weather conditions. Although "it was none of my business," Ducat found it "most extraordinary" that Grant failed to follow Rosecrans's column. At 4:30 P.M., as Hamilton's advance collided with Hébert, Grant remarked to Ord that General Rosecrans "was from last accounts from him too far from Iuka for us to attack on our front until further information was received as to his whereabouts." Ord agreed, and Grant told him to move to within four miles to Iuka, there to "await sounds of an engagement between Rosecrans and the enemy before engaging the latter."

But Ord neither moved up as ordered nor claimed to have heard the slightest sound from the direction of Rosecrans's army. He halted his lead division, under Gen. Leonard Ross, seven miles short of Iuka and gave no further thought to Rosecrans. At 6:00 P.M. Ross reported "a dense smoke arising from the direction of Iuka." He concluded that Price was evacuating and destroying his stores, a judgment Ord accepted. Although billowing clouds of smoke might also suggest combat, neither Ord nor Grant—if he was even aware of Ross's report—bothered to confirm Ross's supposition with a reconnaissance toward Iuka.

Ord heard nothing, and so he did nothing. A strong wind, he maintained, "freshly blowing from us in the direction of Iuka during the whole of the nineteenth, prevented our hearing the guns and co-operating with General Rosecrans."[5]

It was 3:30 A.M. on September 20 before Grant learned of Rosecrans's fight with Price. Accepting Rosecrans's admonition that he "attack in the morning and in force," Grant told Ord to engage the enemy "as early as possible."[6]

Ord got started before dawn, but a snarl at the head of his column slowed him. By 7:00 A.M. Ord had moved to within three and a half miles of Iuka. Then, for reasons that remain unclear, he stopped. Ord claimed he was merely complying with Grant's order of the previous afternoon that he await the sound of Rosecrans's guns before closing on Iuka; why he failed to obey Grant's peremptory instructions of 3:30 A.M., Ord neglected to say. He held fast until 8:00 A.M., when the low boom of Capt. Albert Powell's Missouri Battery shelling the town removed any doubt that Rosecrans had met the enemy.[7]

* * *

General Rosecrans was incredulous. The only explanation Ord offered for having left Rosecrans "in the lurch" the day before was that he "did not hear our guns." Rosecrans could not accept that, either then or later. How could Ord, he asked rhetorically, who was only six miles away, fail to hear the noise of his battle with Price, when Col. John Du Bois, who was fifteen miles away "in a straight line over a rolling forest country," heard the noise distinctly?[8]

It was a legitimate question for which there was—and remains—no clear answer. Ord said a sharp wind from the north blew the noise of battle away from him. Most contemporaneous letters from soldiers of his command support him. "We never heard the firing at all," a surgeon from Ross's division wrote his wife three days after the battle. Lt. Col. James Parrott of the Seventh Iowa told his wife the same thing, and in their diaries Sgt. Alexander Downing and Pvt. Henry Clay Adams of Crocker's Iowa Brigade noted an odd silence from the direction of Rosecrans's command. But August Schilling of the First Minnesota Battery wrote that he and his battery mates heard "the thunder of cannon" from the south at 5:00 P.M. and assumed Rosecrans's column "wasn't as lucky as we were."[9]

The testimony of Ord and his men does not square with the recollections of those on the battlefield. Rosecrans's medical director, Archibald Campbell, reported that "the night was calm and without a breath of air stirring, so that, as the battle raged until after nightfall, we were enabled to dress the wounded by candle-light as well as if we had been inside a house." Soldiers in both armies spoke of a heavy, still air that kept the smoke of the battle close to the ground. Assuming everyone's recollections of weather conditions were accurate, the most reasonable conclusion is that the rolling ground close to the fighting dissipated the breeze Ord's men felt, and that the damp air on the battlefield deadened sounds beyond a mile or two.[10]

With no convincing explanation, soldiers speculated on the reasons for Ord's delay. "Everyone thinks we ought to have caught Price at Iuka, and we would have done it if the left had gone in eighteen hours sooner," Pvt. Charles Tompkins wrote from Burnsville. Camp rumors, he added, had Grant drunk at headquarters during the battle. The same rumor circulated among the officer corps. Said Capt. William Stewart of the Eleventh Missouri, "General Grant was dead drunk and couldn't bring up his army. I was so mad when I first learned the facts that I could have shot Grant if I would have hung for it the next minute."[11]

At noon Grant met Rosecrans in Iuka. Grant corroborated Ord's story but seemed more concerned about the developing threat to Corinth than the battle at Iuka or Rosecrans's pursuit of Price. Grant directed Ord to leave one brigade behind to garrison Iuka and with the remainder of his force return at once to Corinth. That done, he turned his attention to matters at hand. Grant approved of Rosecrans's actions, both in bringing on the battle the day before and in giving chase that morning, and told him to "pursue the enemy as far as I thought it likely to result in any benefit to us or injury to them," said Rosecrans. To the War Department Grant telegraphed a short, laudatory report of Rosecrans's performance. He regretted that Rosecrans had not blocked the Fulton road, which would have forced Price to retreat eastward, away from Van Dorn, but a subsequent ride over the ground convinced him that Rosecrans had had too few troops both to hold the Fulton road and to fight Price.[12]

While Grant, Ord, and Rosecrans second-guessed one another, General Stanley drove his division down the Fulton road after Price. The Rebels were an hour ahead of their pursuers. Any real chance at slowing them was lost when Mizner's cavalry failed to reach the intersection of the Jacinto-Tuscumbia and Fulton roads before the Confederates. The Yankee troopers charged but were driven away by a few shots from Hébert's artillery, and the Southerners marched on unmolested.

Colonel Hatch and his eight companies of the Second Iowa Cavalry proved more irksome. Ranging ahead of Stanley's column, they nipped at the Rebel rear guard all morning, and at noon began a running skirmish with a patrol of Armstrong's cavalry. By 2:00 P.M. Armstrong and Maury had grown weary of the Iowans. They had tried once unsuccessfully to lure them into a hastily laid trap. Eight miles south of Iuka they prepared a more deliberate ambush. Maury halted the trailing regiment of the rear guard, the Second Texas Infantry, and placed it in line of battle with Bledsoe's Missouri battery along the fringe of a low, dense stretch of forest through which the road ran. Col. Robert McCulloch's Second Missouri Cavalry formed out of sight, ready to charge. A squadron of Wirt Adams's Mississippi Cavalry Regiment would be dangled out on the Fulton road as bait.

On came the Iowans, yelling and galloping after a Rebel cavalry patrol they had been chasing for four miles. They ran headlong into Adams's Mississippians, who fled as planned. But the abrupt Rebel flight raised Colonel Hatch's suspicions. Just north of the ambush site, atop a hill, he called a halt and ordered his skirmishers to dismount. They edged forward toward a depression filled with blackjack oak, on the far side of which the Second Texas lay in wait. Maury sprang his ambush — prematurely. The Second Texas let go a mighty volley from the timber, Bledsoe's guns filled the road with canister, and McCulloch's Missouri troopers charged. For all the racket, the now rapidly retreating Iowans had lost only six men wounded.[13]

Still, it was enough to disrupt the Federal pursuit. Hatch withdrew to Stanley's sluggish infantry column. Stanley's division had been without rations for three days. The men were tired and hungry, perhaps as hungry as the enemy they chased, and by sunset were played out. Stanley bivouacked his division along the north bank of Crippled Deer Creek, and Hamilton rested his men at Barnett's Crossroads. Both were convinced that the enemy, by marching all night, would reach Bay Springs before daybreak. Further pursuit seemed pointless. Said Rosecrans, "Our rations being exhausted, and the country towards Bay Springs destitute, I was satisfied that further pursuit with our infantry would be utterly unavailing, and directed my command to return the next day to Jacinto."[14]

More worried about his own lines of communication and supply than chasing Price, Grant approved Rosecrans's decision. By midnight on the twentieth he was back in Corinth, remaining there just long enough to see Julia and the children off to St. Louis before boarding a train himself for Jackson, Tennessee.[15]

Grant's sudden shift in priorities was occasioned by unexpected activity on the part of Earl Van Dorn during the three days that Grant and Rosecrans had devoted to Price at Iuka. Never one to wait for the enemy to find him, Van Dorn marched his army first to Pocahontas, where he had suggested Price join him for a joint attack on Corinth. From there Van Dorn struck out for Bolivar, hoping to create a diversion that would allow Price to escape from Iuka. At sunset on September 20 Van Dorn made camp just seven miles from the strategic railroad town. He was too late to prevent the Federals from concentrating against Price, but his feint into Tennessee caused Grant to disperse his forces after Iuka. Rosecrans returned to his old encampment at Jacinto with Stanley's and Hamilton's divisions on September 21. Ord reached Corinth the same day, and Grant hurried reinforcements from there to the garrisons at Bolivar and Jackson.[16]

Unlike Grant, Rosecrans saw Van Dorn's advance more as an opportunity than a threat, and he offered his army to a combined movement against the Mississippian. Having just received word of his promotion to major general, Rosecrans was in good spirits. In the early morning hours of September 21, he wrote Grant, "If you can let me know that there is a good opportunity to march on Holly Springs to cut off the forces of Buck Van Dorn I will be in readiness to take everything. If we could get them across the Hatchie they would be clean up the spout."[17]

There is no record of Grant's response to Rosecrans's exuberant proposal, which Van Dorn's withdrawal into Mississippi to meet Price rendered moot.[18]

That Price would have an army fit to fight with Van Dorn was in grave doubt during the four-day retreat from Iuka to Baldwyn. Morale was bad, discipline lax, and straggling heavy. Soldiers looted homes and robbed crops. Said a correspondent of the *Jackson Mississippian*, "I doubted, on the march up and on the retreat, whether I was in an army of brave men, fighting for their country, or merely following a band of armed marauders, who are as terrible to their friends as foes." David Garrett of the Sixth Texas Cavalry, which had seen no fighting at Iuka, watched men slip away for no apparent reason. Garrett's friend Tom Wood "just got tired and said he would not travel any further, the Feds he said could take him and go to hell with him and that is the last we had of him."

The retreat ended at Baldwyn on September 23. No one bothered to lay out a regular camp, and the men bivouacked as they pleased in the woods outside town. "Our trip accomplished, and we are again at Baldwyn," Major Finley L. Hubbell of the Third Missouri noted in his diary. "But I am totally at a loss, as well as everybody else, to know what we accomplished by it."[19]

*   *   *

What had been lost was as uncertain as what had been won. Confederate casualties undoubtedly exceeded those reported by Price and his lieutenants. Reports from Hébert's brigade conceded 63 men killed, 305 wounded, and 40 missing of 1,774 engaged. Colonel Martin reported 22 killed, 95 wounded, and 117 missing from his brigade, which went into action 1,405 strong. The total reported killed was 85, with 410 wounded and 157 missing.

Not surprisingly, Federal estimates of Southern casualties exceed the 652 given in Confederate reports. Archibald Campbell, the medical director of the Army of the Mississippi, calculated from personal inspection and what he regarded as reliable information that the Rebels had lost more than 520 killed, 1,300 wounded, and 181 captured. General Rosecrans said 162 dead Rebels were found laid out for burial behind the Methodist church and that he counted another 99 on the battlefield. Extrapolating from that, he estimated Price to have lost 385 killed or dead from wounds, 692 wounded, and 361 captured. The Federal provost marshal, Capt. William Wiles, certified that 265 Southerners were found on the field and buried by Union troops.

Federal casualty totals are far more reliable. Rosecrans reported 141 killed, 613 wounded, and 36 missing. Given Union losses of 790, Price's reported casualties of 652 seem low indeed, especially as the Confederates did the attacking.[20]

The dead were buried in long shallow graves where they fell. Green flies descended on the fields and clustered in the bushes atop the ridge. The weather turned hot. Bodies decomposed rapidly. The sickly smell of rotting flesh blew through the streets of town. Ten days after the battle

Cyrus Boyd of the Fifteenth Iowa left his garrison duties in town and followed the odor to the scene of the fighting. Afterward he recorded his impressions: "I have never seen before evidence of such a desperate contest on a small piece of ground. The trees around are almost torn to kindling wood. . . . Twenty-five dead horses lay close together and about forty men belonging to the Fifth Iowa buried in one grave here, besides numerous other graves scattered all through the woods." The ridge was a vast charnel house: "The ground in many places was white as snow with creeping worms. The darkness of the forest and the terrible mortality made it one of the most horrible places I was ever in. Then the silence was oppressive. Not a sound could be heard except once in a while the chirp of some lonely bird in the deep forest. To think of our poor fellows left to sleep in that dark wood. . . ."[21]

13

We Had Better Lay Down Our Arms and Go Home

General Price arrived at Baldwyn somber and reflective. Iuka had convinced him of two things: Rosecrans and Grant could, at Corinth or any other point between Iuka and Memphis, rapidly concentrate 40,000 men—more than enough to crush his and Van Dorn's armies in detail—and only by uniting their commands could the two Southern generals hope to accomplish anything. That dark revelation was enough to cause Price to submit to service under Van Dorn.

From Baldwyn on the morning of September 23, with his "utterly exhausted and demoralized" army resting outside town and stragglers still streaming in, Price wrote Van Dorn, "I will leave here in two days to form a junction with you, and I desire to know at what point we will meet." Price recommended Ripley, a town midway between the two armies that he could reach easily in two days' march. Van Dorn agreed and said he hoped to meet Price there on September 28.[1]

Price's march began inauspiciously. The men were weary and jaundiced. "Without waiting to fix things up and get together our old men we again started on a more foolhardy expedition than the last," said Lt. Wright Schaumburg of Maury's staff.

Friday, September 26, dawned cloudy and cool. Soaked to the skin by a "night spent under the weeping heavens with a chilling blast of a northwest wind, a wet couch of mud and leaves for our reveries," Price's soldiers set out for Ripley.

It was slow going. The road was bad and the country hilly. A heavy rain fell at dusk. "The darkness set in so thick that one could almost feel it; the mud in the road was soon deep, and grew deeper as we advanced," said a Missourian. "I think nearly every man fell down from one to a dozen times that night." The men caught a few hours of fitful sleep before resuming the march at dawn over a road pounded to paste. They made fifteen miles on September 27 and bivouacked five miles short of Ripley. The next morning, "covered with mud and thoroughly drenched with rain," the army traipsed into Ripley. Guides led Price's men through town to the fairgrounds, the site chosen for their camp, and the soldiers pitched their tents under a fine drizzle, "just the kind of day to make a soldier homesick," wrote an Arkansan. At dusk a mild earthquake sent a shudder through camp.[2]

The mood was no better at headquarters. Before reporting to Van Dorn, a duty distasteful in itself, General Price had to contend with a crisis in his own military family. Expressing a sentiment shared by the entire staff, Major Snead told Price he would not serve under Van Dorn. So that there might be no misunderstanding about his feelings, Snead spelled them out in a lengthy letter of resignation. "I had once before . . . quit the army rather than serve under General Van Dorn and his staff. I could not endure the incompetency and rashness of the one, nor the inefficiency of the latter," he began. "As soon as he got control over you," warned Snead, Van Dorn intended to strip Price of his men, arms, and authority. Rather than "endure these wrongs and indignities," Snead asked to be relieved from duty.

Integrity prevented Price from trying to refute what he knew to be true. Instead he appealed to Snead's patriotism. "Every Missourian in the army shared Snead's feelings," Price conceded, but "the military position we occupy toward the enemy, with an engagement where the odds are heavy against us immediately impending, calls imperatively for a sacrifice on my part and that of my army of all that we feel to be due us to secure a victory to our arms in the impending conflict." Snead relented.[3]

Price's interview with Van Dorn on September 28 hardly lessened the uneasiness he and Snead shared. Hunched over two maps of the country around Corinth, Van Dorn explained his plan. Tracing a line northward from Ripley, he announced an "immediate and direct" attack was to be made on Corinth. He reviewed the distribution of Federal forces as he understood it: 6,000 men at Memphis, 8,000 at Bolivar, 3,000 at Jackson, 15,000 at Corinth, and another 10,000 dispersed among minor outposts, including Burnsville, Rienzi, Jacinto, and Iuka. Before the Federals could concentrate their forces to oppose him, Van Dorn intended to move on Corinth with the 22,000 men of his and Price's commands.

Swiftness and surprise were critical, continued Van Dorn. To ensure surprise he would march first to Pocahontas and begin work ostentatiously on a bridge over the Tuscumbia River to give the impression his objective was Bolivar. But rather than cross the river and strike northward into Tennessee, Van Dorn would turn the army abruptly to the east and by forced march descend on Corinth. To impede Federal reinforcements, Armstrong's cavalry was to break up the Mobile and Ohio Railroad north of Corinth.

It was nineteen miles from Pocahontas to Corinth. The road was good, the country gently rolling. One day should suffice to bring the army to the outskirts of Corinth, Van Dorn reasoned. The Confederates would attack the town from the northwest. Van Dorn was well aware he would be assaulting the very earthworks his men had helped construct some four months earlier. But he also knew the earthworks had been most thoroughly developed northeast of Corinth, the direction from which the Federals had approached after Shiloh. In coming from the northwest, Van Dorn would strike the weakest section.[4]

Despite the haste with which he put together his plan, Van Dorn had weighed strategic considerations, even if his conclusions seemed a bit fantastic. The retaking of Corinth, he reasoned, would guarantee the safety of Vicksburg by forcing the Federals to abandon West Tennessee. That in turn would open the way for him to march north and help Bragg to drive the enemy across the Ohio River.[5]

Van Dorn's generals were decidedly divided on the merits and practicality of his plan. Dabney Maury, who despite his loyal service to Price had fallen under the sway of the Mississippian, came out wholeheartedly in favor of the scheme. Maury concurred with both "the great objects he sought to accomplish" and "the means by which he proposed to march to a certain and brilliant victory by which the State of Mississippi would have been freed from invasion and the war would have been transferred beyond the Ohio. Such results justified unusual hazard of battle."

Thirty-nine-year-old Maj. Gen. Mansfield Lovell, a veteran of the Regular Army who had been breveted for gallantry at Chapultepec in the Mexican War, was of two minds. A classmate of Van Dorn at West Point and a general whom Braxton Bragg considered "equal to any officer in our service," Lovell was new to the theater, having arrived only three weeks earlier to take command of the lone division Van Dorn then had with him.

*Maj. Gen. Mansfield Lovell
(Library of Congress)*

Lovell came to the army with a tarnished reputation. Five months earlier he had surrendered New Orleans after a Federal fleet pounded the small Confederate river flotilla there into driftwood. Lovell had had only local militia with which to man a chain of small forts that covered the myriad water approaches to the city, a difficult task that became impossible after the Federal fleet ran past the forts. A military court of inquiry cleared Lovell of blame for the city's fall, saying he had defended New Orleans capably with his small command of green troops. But President Davis held Lovell solely responsible. Consistent with his unforgiving nature, Davis was waiting only for an unequivocal pretext to shelve him.

None of this worried Lovell. Conceited and self-absorbed, Lovell swaggered about Van Dorn's headquarters like a conquering Caesar. Not surprisingly he made himself heard at the commanders' meeting. Lovell suggested an attack on Bolivar. It would isolate the Federal forces at Corinth, cut off their lines of supply, and force them to leave their entrenchments and fight on "an equal and open field." At the same time Lovell conceded that a rapid advance on Corinth might also compel the Federals to evacuate the town. While he preferred to march directly on Bolivar, Lovell granted the merits of Van Dorn's proposal, provided

Corinth could be taken with minimal loss of life. And he, too, fixed his gaze on the Ohio River. "My opinion is that the enemy will evacuate Corinth, fall back to Jackson, Tennessee, and finally (if we get our additional forces from the returned prisoners) we shall be able to drive him to the Ohio River," Lovell wrote his wife. "All this can be done with good generalship without the loss of much life, I think. A few weeks will develop the results of the campaign, and I should not be much astonished if you should hear of us up in Kentucky."

Price categorically opposed Van Dorn's plan. He agreed the taking of Corinth "warranted more than the usual hazard of battle, [yet] was of the opinion that the hazard would have been much less to have delayed the attack a few days" until the 12,000 or 15,000 exchanged prisoners then assembling at Jackson, Mississippi, could be armed and sent forward. Without their added numbers, Price failed to see how the Confederates could hold Corinth, much less exploit its capture. Should Van Dorn attempt to march north from Corinth, the strong Federal forces at Bolivar and Memphis could readily strike at the left flank and rear of his small army. The heavy losses that Van Dorn inevitably would sustain in taking the town rendered doubtful his ability even to hold what he might win. Any gains would be temporary; better to wait for the returned prisoners.[6]

Van Dorn dismissed the objections of both Lovell and Price. An attack must be both immediate and directed against Corinth. The exchanged prisoners at Jackson would not be ready to join the army until the second week of October. "If I waited for their reception all opportunity of striking Corinth with a reasonable prospect of success would be lost," Van Dorn told Price and Lovell; surprise, not added numbers, was the key to success. Van Dorn ordered Price and Lovell to have three days' cooked rations distributed. Lovell's division would lead the advance the next morning, and Price would follow later that day, if possible, or on September 30.

John Tyler, the alcoholic son of the former president and an aide-de-camp to General Price, whom he revered, listened to Van Dorn's self-assured monologue with misgivings. The enemy at Corinth most certainly was as strong as Van Dorn's army, and they enjoyed the protection of elaborate earthworks. Also, Tyler mused, they were led by General Rosecrans, "one of the most skillful and successful officers, alike prudent and sagacious, experienced in strategy, self-poised and courageous." It was becoming clear to Tyler why Rosecrans had graduated fifth in the West Point class of 1842 and Van Dorn fifth from the bottom.[7]

General Lovell left headquarters in a turmoil. Although he had accepted his orders with only mild doubts, Lovell was troubled enough to withhold the plan from his brigade commanders.

Price lingered at headquarters. Anguish contorted his features. "You seem despondent, General Price," said Van Dorn. "No!" Price retorted, brushing aside his doubts with gallant words. "You quite mistake me. I have only given you the facts within my knowledge and the counselling of my judgment. When you reach Corinth you shall find that no portion of the army shall exceed mine either in courage, in conduct, or in achievement."

Price's ire was the greater because, as Colonel Snead had predicted, it was his command that stood to suffer most from failure. Price had brought with him to Ripley 14,363 men: 6,602 in the division of Hébert, 3,896 infantrymen under Maury, 1,428 troopers in Armstrong's cavalry brigade, and 928 artillerymen. Van Dorn had Lovell's division and two cavalry regiments, perhaps 7,000 men in all. Although contributing two-thirds of the force for the expedition, Price had no real voice in its conduct.[8]

*   *   *

Monday, September 29, broke cloudy and warm. It had rained intermittently for three days, but the heat had evaporated the water before it settled. Although commissary officers were slow to load the wagons, delaying Lovell's departure until mid-afternoon, he marched the six miles Van Dorn required of him before nightfall.[9]

Price broke camp at daylight on September 30. Well fitted with provisions, his men marched quickly behind Lovell's column, covering eighteen miles before they stopped for the night outside the tiny hamlet of Jonesborough, nine miles short of Pocahontas. Lovell bivouacked at Metamora, Tennessee, in thickly wooded bottomland west of the Hatchie River.[10]

October 1 was clear and warm. The sandy surface of the Ripley-Pocahontas road scorched the feet of Price's soldiers, and fine particles of dust swirled up and stung their eyes. But they marched hard, covering eight miles before going into camp beside Lovell's division, which had remained at Metamora to await their arrival. The more fortunate of Price's soldiers rested in an open cornfield. Others found themselves sleeping "in a bottom densely and heavily wooded, with a thick undergrowth beneath, and almost a canopy of vines above." Those with strength left to walk wandered over to the Hatchie River, where they enjoyed a cool bath in the gathering twilight.[11]

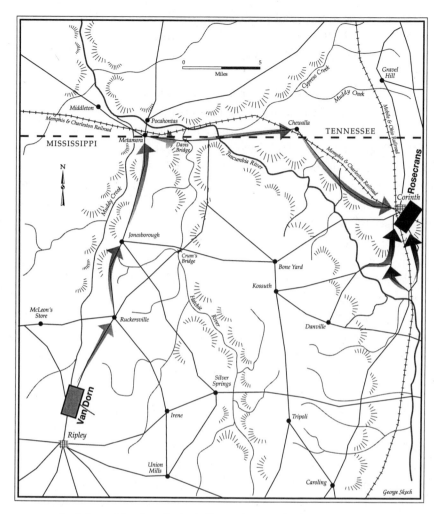

MAP 5. *Confederate March to Corinth, September 29–October 3*

During the day Jackson's cavalry left Lovell and rode to Pocahontas to keep alive the impression that the Confederates intended to cross the Tuscumbia and drive northward. Price scarcely had his command settled for the night when a courier brought an urgent message from Van Dorn, calling on him to assemble lumber-gathering details on the Hatchie River near Pocahontas and begin work on a bridge. Not long after issuing the order, Van Dorn was distracted from his bridge-building ruse by a matter of real urgency. Lovell had discovered that the Federals had partially burned Davis Bridge, a simple wooden structure spanning the deep, sluggish, sixty-foot-wide Hatchie River just east of Metamora. This was the bridge for the road from Pocahontas to Corinth; the next nearest crossing was at Crum's Mill, seven miles south. A detour to Crum's Mill would cost the army at least a day, and with it most certainly the element of surprise.

While Price's men labored through the night to build a bridge they would never cross, Lovell set his men to repairing Davis Bridge. Despite a lack of tools and an encounter with Federal pickets, the Rebels worked quickly, laying the last plank before 4:00 A.M. The division took to the road at once; its destination for the day was Chewalla, Tennessee, a mile and a quarter north of the state line and nine miles northwest of Corinth. Price followed after sunrise. Van Dorn left his unwieldy wagon train two miles east of Davis Bridge, at the intersection of the Chewalla road and a sand trail from Crum's Mill known as the Boneyard road. He detached the First Texas Legion and two batteries of artillery to guard the train and left Wirt Adams's cavalry regiment at Davis Bridge, to warn of any Federals who might approach from the west to relieve Rosecrans.[12]

It was a sullen and apprehensive army that marched toward Corinth. Lovell had waited until the early hours of October 2 to brief his brigade commanders on their destination. Against a moonlit backdrop of cypress trees and muscadine vines, to the accompaniment of hammers and saws, Lovell told the generals of Van Dorn's grand scheme.

The commander of his first brigade, Brig. Gen. Albert Rust, spoke up loudly against it. "It was impossible to succeed in the attack," Rust told Lovell. "The enemy had or could have more men there than we could assault with, such were their facilities for concentrating, and . . . the defenses constructed by General Beauregard were somewhat formidable, and . . . were very much strengthened by the enemy," concluded Rust. "If we [can] not succeed we had better lay down our arms and go home," Lovell replied. Rust, a robust forty-four-year-old former attorney and Arkansas congressman, stuck to his opinion; the plan was pure madness and must be abandoned. Brig. Gen. John Villepigue commanded Lovell's second brigade. What opinion the thirty-two-year-old West Point graduate, whom Beauregard once called a "most energetic young officer," had of Van Dorn's plan is unknown. The commander of Lovell's third brigade, Brig. Gen. John Bowen, a handsome, thirty-one-year-old West Pointer who had resigned from the army to take up architecture in St. Louis, was skeptical but decided to withhold judgment until he learned more.

Lovell and Bowen picked their way through the tangled black woods to call on Van Dorn, who had taken up headquarters at the Davis house.

What Bowen saw and heard there satisfied him that Rust had been right. General Van Dorn was busy sketching a map of the country between Davis Bridge and Corinth. A local guide was helping him draw in the roads. Van Dorn motioned Bowen to his side. He showed him the sketch and pointed out the roads leading to and from the Hatchie River. A superb, outspoken soldier who had risen from the colonelcy of the First Missouri for his combat record and strict standards of discipline, Bowen had only just joined Van Dorn's army. Although he had date of rank on Dabney Maury, he had been relegated to command of a brigade. Not surprisingly, Van Dorn's map failed to impress him: "The map was a crude sketch on a sheet of letter paper, drawn to no particular scale, and such as I deemed utterly unsuitable for the ordinary movements of an army." Apparently no one bothered to show him or Lovell the fine engineer maps that Price had given Van Dorn, and so the two left the Davis house convinced the fate of the march rested on a childlike drawing and the word of a civilian guide.[13]

Regimental officers and the rank and file still knew nothing of the object of their march. Price's Missourians certainly disliked the aspect of things. Their pleasure at crossing the state line into Tennessee the day before had been immense. That they might return to Mississippi was too distasteful to contemplate; "not that we disliked the state," said one, "but we wanted to go north." Texan Newton Keen predicted the worst. Everyone regretted having to serve under Van Dorn again, "for we had been under him at the battle of Elkhorn [Tavern], and had no confidence in him. . . . We went into the campaign whipped."[14]

Such was the mood of the men being shepherded unawares to fight a battle their generals had dismissed as already lost.

*   *   *

General Grant seemed easily distracted after Iuka. His spirits, already low, fell farther when Julia left for St. Louis. He steeled himself long enough to meet Van Dorn's expected attack against Bolivar or Corinth. When none materialized, Grant left Tennessee to visit his wife.

His pretense for going to St. Louis was to confer with Maj. Gen. Samuel Curtis, commander of the Department of the Missouri. The day before the Battle of Iuka, Halleck had wired Grant that the enemy was constructing two ironclad boats on the Yazoo River, in Louisiana. After the damage the Rebel ram *Arkansas* had done to the Federal river flotilla two months earlier at Vicksburg, the mere rumor that more were being built was enough to set the high command to worrying. Halleck wanted

Grant to consult with General Curtis, whose headquarters were at St. Louis, about the feasibility of sending a small force from Memphis or Helena, Arkansas, to destroy the boats.

An exchange of telegrams would have sufficed to coordinate such an operation, but Grant wanted badly to go to St. Louis. To excuse his absence he painted for Halleck a picture of perfect calm in his district, writing him on September 24, "The enemy being driven from his position in front of Bolivar by the rapid return of troops drawn from there to reinforce Corinth, and everything now promising quiet in our front for a short time, I shall go to Saint Louis in person to confer with General Curtis." Perhaps suspecting Halleck would question the need for his trip, Grant added that "another reason for my going is that for several weeks my health has not been good and although improving for the last few days I feel that the trip will be of benefit to me."

What benefit he derived from his untimely departure is open to speculation. Franklin Dick, a well-respected St. Louis attorney and the brother-in-law of Francis Blair Jr., wrote Attorney General Edward Bates on September 28, two days after Grant's arrival, that the general had been seen staggering about town in a drunken stupor. A friend of Dick had spoken to Grant and found him "tight as a brick." A correspondent for the *New York Times* was more charitable and probably more honest. He thought Grant looked "remarkably well, although bearing some marks of the fatigues of his summer campaign."

Before leaving for St. Louis, Grant had had the presence of mind to clarify command relationships in his district by dividing the area into four geographical divisions. He gave Maj. Gen. William T. Sherman command of the First Division, which embraced all territory south of the Hatchie River and west of Bolivar then under Federal control. To Major General Ord went the Second Division, which ranged from the Kentucky state line to Bethel, Tennessee, as well as the additional responsibility of guarding the Memphis and Ohio, Mobile and Ohio, and Memphis and Charleston Railroads—the iron-bounded triangle that was the heart of Grant's district. Rosecrans commanded the Third Division, which consisted of Corinth, Iuka, and a number of small outposts east of the Mobile and Ohio Railroad. McArthur's division of Ord's corps, which was operating around Corinth, was placed under Rosecrans's command. Rosecrans also was to watch the railroad lines south of Bethel and east of Chewalla. The Fourth Division, under Brig. Gen. Isaac F. Quinby, comprised country safely in Federal hands—all of western Tennessee north of the railroads and the posts of Forts Henry and Donelson.[15]

Grant's lieutenants were diligent in his absence. Forgetting his assurances to Halleck that all would remain quiet in the district while he was away, in his *Memoirs* Grant let slip the true nature of things at the time he left: "We were in a country where nearly all the people, except the Negroes, were hostile to us and friendly to the cause we were trying to suppress. It was easy for the enemy to get early information on our every move. We, on the contrary, had to go after our information in force, and then often returned without it."[16]

Ord and Rosecrans did their best to confirm rumors that Price and Van Dorn had joined forces at Ripley to "capture Corinth or break our line of communication and force us to retreat toward Columbus, [Kentucky]." Federal scouts combed the countryside, and strong cavalry patrols visited every town and hamlet in the district.

They turned up nothing. Rather than reassure him, the negative reports only caused Rosecrans to feel more exposed. He was sure the enemy was out there—perhaps west of the wooded valley of the Hatchie River, which his patrols had yet to penetrate.

Other problems loomed. Heavy detachments for outpost duty and patrols left Rosecrans with only the divisions of Davies and McArthur to garrison the town—fewer than 10,000 men in all. Worse yet, McArthur had been compelled by a later date of rank to yield his division to the hopelessly inept Brig. Gen. Thomas McKean, recently arrived from command of the post of St. Louis. A graduate of West Point, McKean had resigned his commission in 1834. He failed to win it back during the Mexican War, serving as an enlisted man through the entire conflict. Afterward McKean wisely returned to his civilian occupation of civil engineer. Only the shortage of officers at the outbreak of the Civil War got him a brigadier general's commission, and he was shuffled off to a series of quiet, backwater garrison commands. At fifty-two McKean was regarded by most as too old for field duty. Somehow he had wangled a combat command, and Rosecrans was stuck with him.[17]

Rosecrans put the soldiers in Corinth to work strengthening the fortifications. When he set up headquarters in town on September 26, Rosecrans was dismayed to find the interior defensive works still consisted only of Batteries Robinett, Williams, Phillips, Tanrath, and Lothrop, which Captain Prime had built on the College Hill line earlier in the summer. Rosecrans ordered the lunettes connected by breastworks and the ground to the west and north covered by abatis. He organized all able-bodied Black contrabands into twenty-five-man squads and put them to work beside his soldiers. Finally, Rosecrans directed that the line of lunettes be extended to cover the approaches from the north. Construction began on Battery Powell, a six-gun artillery redoubt situated a half-mile north of town and a mile east of Battery Robinett that would command the Purdy road. The men labored at a frantic pace.

For four days the work went on. By September 30 a line of breastworks had been raised between College Hill and Battery Robinett. Enough trees had been cut to erect a thin abatis from Battery Robinett to the Mobile and Ohio Railroad. But from the track to Battery Powell the ground lay open.[18]

The closing days of September brought reports of enemy movements west of the Hatchie River, but where or why the Rebels were marching remained a mystery. Captured prisoners themselves did not know their destination. Like most Rebel prisoners they dissembled, telling their captors that Van Dorn and Price had marshaled an army of 40,000. Rosecrans doubted but could not entirely dismiss these numbers. Instead he rationalized away his fears. Whatever their strength, the Confederates would "have the impression that our defensive works at Corinth would be pretty formidable. I doubted if they would venture to bring their force against our command behind defense works. I therefore said: The enemy may threaten us and strike our line entirely, get on the road between us and Jackson and advance upon that place, the capture of which would compel us to get out of our lines and fight him in the open country." (This was the course of action Lovell had urged upon Van Dorn.)[19] However, Rosecrans did not give himself over completely to wishful thinking. He ordered Davis Bridge burned on the night of September 30, but the cavalrymen charged with the task were sloppy and succeeded in burning only the floor planking. The next morning he called in Crocker's brigade from Iuka and ordered Hamilton's division, which was strung out as far south as Rienzi, to draw closer as well. To slow any Rebel advance against Corinth from Pocahontas, Rosecrans told McKean to send two more regiments from Col. John Oliver's brigade to Chewalla, then held by the Fifteenth Michigan Infantry and a company of cavalry.

Oliver's Fourteenth and Eighteenth Wisconsin regiments broke camp at 2:00 P.M. on October 1 and, accompanied by the First Minnesota Light Artillery, hurried out the Chewalla road. To better defend against an attack from the west, Rosecrans told Stanley to move his division by way of Kossuth to the Hatchie River.[20]

Stanley had a miserable march. The soldiers of Fuller's brigade rose at 7:00 A.M.; Mower's men, with farther to march, had been awakened four hours earlier. All day long the two columns marched, "through a region of few settlements, poor log cabins, rolling oak ridges and sluggish streams," most of which had dried up under the summer sun. The temperature climbed to ninety-four degrees before noon. Fuller halted his men for a meal and short rest in Kossuth at noon, and the brigade reached the Hatchie River at dark. It was midnight before Mower reached Kossuth. He entered the town almost alone. "The heat and dust and swift pace were too much for the men," recalled a member of the Forty-seventh Illinois. "One by one the boys dropped out, unable to continue further. Those who reached their destination threw themselves down in the furrows of an old corn field, too weary to build fires or seek refreshment."[21]

Ten miles to the north the Fourteenth and Eighteenth Wisconsin regiments trudged toward Chewalla. They halted at sunset, a half-mile short of the village. Blocking their further progress was the Fifteenth Michigan, intent on using the road to march in the other direction. Having been fired on by enemy scouts, the Michiganders told tales of 40,000 Rebels waiting on the far bank of the Hatchie River to sweep down on them. Sgt. William Tucker of the Fourteenth Wisconsin found their plight rather amusing. "On our arrival near Chewalla we found the Michigan boys badly broken up and with their camp equipage making for a more friendly country," Tucker wrote. "At first we were inclined to be somewhat surprised at them for being in so much of a hurry to move from that locality. It was supposed by our command that it was nothing more than a band of bushwhackers, which were frequently prowling around our outposts."

The Michiganders gave as good as they got, confessed Tucker. They "insisted that they were ready to remain with us, and that we would have all the fun we wanted in the morning, if not before." Colonel Oliver re-formed the regiment and placed it in line of battle with the Fourteenth and Eighteenth Wisconsin on a long low ridge south of Chewalla commanding the road from Davis Bridge. He placed two twelve-pounder howitzers from the First Minnesota Light Artillery on the crest astride the road and sent the rest of the battery back to Corinth. Skirmishers edged off the ridge and melted into the dense timber.

The spirits of Tucker and his Wisconsin comrades sank in the gathering darkness. "Our pickets could very plainly hear what seemed to be artillery and wagons moving, and in the distance see what appeared to be lights from camp fires," he remembered. "We could not estimate the enemy's strength. . . . Still all indications very clearly showed that there was an army in force in our front."[22]

*   *   *

General Grant was back in Jackson. Whatever tonic his trip to St. Louis might have provided evaporated in the heated confusion at headquarters. "I returned here from St. Louis just in time to find my presence very much required," he wrote Julia with almost comic understatement. No more able to read Van Dorn's intentions than were Generals Rosecrans, Ord, or Hurlbut, Grant felt obliged to leave his forces "where they were until the enemy fully exhibited his plans."

Grant's instructions, then, necessarily were vague and tentative. To Hurlbut he wrote on October 1, "If Van Dorn is in west of you it will not do to detach too much of your force to look after the Rebels about Pocahontas—I have instructed Rosecrans to follow them if they move towards Bolivar." To Rosecrans he wrote, "Inform yourself as well as possible of the strength and position of the enemy and if practicable—move on them as you propose. Inform me if you determine to start and I will give you all the aid possible from Bolivar." To Halleck Grant penned what amounted to a confession that his trip to St. Louis had been ill considered and his confident declaration that all would remain quiet had been premature: "My position is precarious but hope to get out of it all right."[23]

14

More Trouble Than We Could Care For

October 2 dawned cloudy. A light rain fell, just enough to cut the heat and settle the dust. At Corinth General Rosecrans waited hopefully for a report that might clarify the enemy's intentions. He expected it to come from Oliver's brigade, and to hasten contact he told Oliver to send mounted parties from Ford's Independent Illinois Cavalry Company, on detached service with him, to reconnoiter toward Kossuth and Pocahontas.

Colonel Oliver had patrols on the road at daybreak. Early reports were inconclusive but disturbing. A twenty-man patrol sent to Kossuth saw nothing but heard drums beating in the forest. Their report caused Oliver to retire his command, less two companies from the Fifteenth Michigan left as skirmishers, to the junction of the Chewalla and old Kossuth roads.

Contact occurred at 10:00 A.M., when Company F of the Fifteenth Michigan, deployed in a loose skirmish line on a hill west of Oliver's new position, was charged by the entire Seventh Tennessee Cavalry. The Michiganders fired a few wild shots, then ran headlong down the hill onto their reserves, a squad of pickets from Company E. Together they formed in an open field beside the Busby house — six and a half miles northwest of Corinth — and charged the Seventh Tennessee Cavalry, which dismounted to meet them. The Rebel troopers gave ground readily, losing one man — Pvt. John Young of Memphis, the first Confederate casualty in the advance on Corinth — in the encounter. When they saw infantry and artillery replace the cavalry, the Michiganders stopped, fired a volley, then slowly withdrew.[1]

The infantry that had stopped the Michiganders' brave little charge was Rust's brigade of Lovell's division, the vanguard of Van Dorn's advance. Shuffling along under a fine, warm drizzle, the Southern soldiers were as downcast as their generals. Only Van Dorn, and perhaps Maury, were optimistic of the outcome. Said Willie Tunnard, "The men pushed toward Corinth and . . . remembered the fortifications around this entrenched position, strengthened under the energetic labors of the enemy . . . and their hearts misgave them as to the final result when it was known where they were going."

Both they and their commanders were jittery, particularly after meeting Oliver's outpost. Riding with his staff at the head of Bowen's brigade, General Lovell caught sight of several wheels standing on the horizon, partly obscured in tall grass between the Chewalla road and the track of the Memphis and Charleston Railroad, which ran on a high dirt fill. Lovell ordered Bowen's lead element, Caruthers's Mississippi Battalion, to face to the left. Drawing his saber, he pointed to the wheels, which looked to him like a battery of enemy artillery, and ordered a charge. "In an instant we were scattered over all the ground we could cover at ten paces apart and advancing at a full run to drive in the enemy's skirmishers, if there were any," recalled Lt. William C. Holmes. "I kept my eye on those wheels, in their solemn stillness, till I got so near that there could be no mistake as to their reality. . . . [They were] the wheels of an old abandoned sawmill." Holmes looked behind him; the entire brigade was advancing in line of battle. "I did not know at what moment insanity might possess the brigade and they might fire on us, as there was nothing else to fire at." Holmes kept his head. He tied his pocket handkerchief to the end of a cornstalk and, waving it madly, ran back toward the brigade. Holmes blocked the path of the Twenty-second Mississippi, which stopped when its officers heard his explanation of the oddly silent enemy. The rest of the brigade stopped as well, and Holmes could congratulate himself on having "saved further injury to that innocent old mill."[2]

There was little damage done to either man or machine the remainder of the day on the Chewalla road. Late in the morning Rosecrans ordered Oliver to withdraw to Alexander's Crossroads, the junction of an old country lane with the Chewalla road, four miles northwest of Corinth. Oliver's Federals fell back firing throughout the afternoon but never held a piece of ground long enough to inflict or receive serious casualties.

Oliver performed his delaying action well, forcing General Rust to move his brigade in line and the Seventh Tennessee Cavalry to screen their advance dismounted until sunset when, weary of the chase, Rust halted his brigade a few hundred yards east of the Busby house. Oliver withdrew the final two miles to Alexander's Crossroads unmolested. There he met General McArthur, who had ridden forward with the Sixteenth Wisconsin to reinforce him. McArthur's presence cheered Oliver's exhausted soldiers. "A noble man every inch of him, and a common every day sort of person too," said a Wisconsin soldier of the thirty-five-year-old Scottish immigrant who was temporarily without a command. "He will speak to anyone, no matter what his rank." McArthur made himself useful. He helped Oliver place his pickets in the dark and stayed with him through the night, alert to any sounds of enemy activity.[3]

*　　*　　*

General Rosecrans had passed a fitful day at Corinth. While the skirmishing on Oliver's front convinced him the enemy indeed was in great force near Pocahontas, it failed to demonstrate conclusively whether Van Dorn intended to march on Corinth or, as the Ohioan conjectured, "cross the Memphis and Charleston Railroad, go north of us, strike the Mobile and Ohio road and maneuver us out of our position," in which case the push against Oliver would merely be a feint.

Regardless of Van Dorn's intentions, it was imperative Rosecrans concentrate his army at Corinth. Should Van Dorn choose to attack the town, Rosecrans would need every man he could muster in its defense. If Van Dorn marched northward to force Rosecrans into battle on open ground, having his whole force on hand would assume even greater importance.

Rosecrans called in his scattered forces during the day. Outposts at Rienzi and Danville were abandoned. Crocker came in from Iuka and made camp two miles northwest of Corinth. Hamilton's division was reunited at Camp Big Spring, two miles south of Corinth. Davies's division rested at Camp Montgomery, a mile farther south. Stanley passed the day in the neighborhood of Kossuth, watchful for a Confederate crossing of the Hatchie River. As the sun set on October 2, the only troops in town were two brigades of McKean's division.[4]

For reasons known only to himself, Rosecrans hesitated to summon his troops into Corinth. Not until 1:00 A.M. did he dictate orders that at last would bring his army together. Even then he was unsure whether Van Dorn would move against Corinth or, if he did, from what direction he would strike. Two good roads ran from Chewalla to Corinth. The Chewalla road entered the old Confederate entrenchments from the northwest. A second road joined the Purdy and Monterey roads northeast of town. Van Dorn might approach over either, or both.

Although Rosecrans assumed Van Dorn would come from the northeast, where the ground was open, he had to protect all possible approaches. Said Rosecrans of his evolving tactical plan, "The controlling idea was to prevent surprise, to test by adequate resistance any attacking force, and, finding it formidable, to receive it behind the inner line that we had been preparing from College Hill around by Robinett." In other words, Rosecrans would place part of his army in the old Confederate earthworks two and a half miles from Corinth. From there they were to fight an intentional retrograde, forcing the enemy to reveal the direction of his main attack while drawing him toward the Federal lunettes and rifle pits on the edge of town.

Rosecrans never committed his plan to paper. Rather, he briefed his division commanders orally as they arrived with their commands. Issued at 1:30 A.M., the only written order that night merely told Rosecrans's generals where, in broad terms, they were to form their divisions. McKean was to remain in place. Davies was to occupy a line between the Memphis and Charleston and the Memphis and Ohio Railroads. Hamilton was to form behind the Rebel earthworks from the Purdy to the Hamburg road, and Stanley was to remain in reserve south of town.[5]

Simple though these instructions were, their execution proved problematic. The orders came as no surprise; before midnight, division commanders were put on notice to have their commands ready to "move at a moment's notice." Delays were inevitable whenever officers tried to move their units over narrow country lanes in the dark of night, but lapses by the division commanders and their staffs compounded the problem.

General Hamilton had neglected to have tents struck and rations cooked, and it was 3:00 A.M. before the soldiers of Sullivan's brigade were on the road to Corinth. Sullivan was slow in clearing camp, so that Brig. Gen. Napoleon Buford (commanding the brigade Colonel Sanborn had led at Iuka) was unable to get his men started before 6:00 A.M. Ahead lay a five-mile march to their assigned positions north of Corinth.

That Buford should be slow in moving out surprised no one. The fifty-five-year-old half brother of cavalryman John Buford, he had graduated from West Point in 1827. Buford resigned his commission in 1835 to take up engineering and banking in Rock Island, Illinois. He prospered until 1861, when Southern states holding bonds of his bank repudiated their debt. Assigning his property to creditors, Buford accepted the colonelcy of the Twenty-seventh Illinois. Although well meaning and kind, Buford was a failure as a combat officer. The men of his brigade dismissed him as a silly old man in his dotage, an opinion strongly seconded by John Rawlins. Buford's antirepublican sentiments appalled him; the graying brigadier general hoped and fervently believed the "final result of this war will be the overthrow of our present system . . . [giving] us Dukes and Lords and titled castes and that his family will be among the nobility." General Oglesby ignored Buford's ranting, but to Rawlins it "evidenced a diseased brain, a weak and foolish old man."[6]

General Hamilton had his division on the road by daybreak. General Davies, on the other hand, had yet to break camp. Somehow he misunderstood the intent of the 1:30 A.M. instructions and thought he was to await explicit marching orders before starting. Consequently, although rations were cooked, haversacks packed, and "arms stacked on the color-line" before 2:00 A.M., Davies did nothing. Roused and formed on the color line at 3:00 A.M., the men were left to stand sleepily as precious hours slipped by. At dawn Davies learned that Hamilton's division had left. He telegraphed Rosecrans for instructions and, after getting a reply that never made it into the records—perhaps because the language was unprintable, Davies hastened his division onto the road at 7:00 A.M. Six miles of hard marching lay ahead.

In his haste Davies forgot to call in three of his regiments: the Eighth, Twelfth, and Fourteenth Iowa, part of the old "Union Brigade" that had been bled white at Shiloh and attached to Davies's first brigade during the summer. In consideration of their service at Shiloh the Iowans — so few in number they were led by a lieutenant colonel — had been allowed to retain their old sobriquet of Union Brigade. Davies had ordered them out on a reconnaissance beyond the Tuscumbia River two days earlier, and they were bivouacked near Danville on the night of October 2. Not until 8:00 A.M. on the third did they receive orders to rendezvous with the division in Corinth. Theirs would be a ten-mile march under a sun bright and hot.[7]

General Stanley gave a better account of himself and was well served by his brigade commanders, but with farther to march than Hamilton or Davies he could not possibly reach Corinth before mid-morning. Mower had his brigade under arms an hour before the order to march came and on the road by 3:00 A.M. Fuller got started at dawn. His men heard a low boom off in the distance, and they debated its meaning. Recalled Capt. Oscar Jackson of the Sixty-third Ohio, "Some said it was a gun, others said not and that it would rain. A few minutes afterward, boom, boom, boom, a whole battery opened, stopping the argument, as the evidence was now on one side. It was curious to observe the effect it had on the men, who, tired and strong, were dragging themselves along. Instantly every head was raised, the step quickened and all forgot they were tired."[8]

Thomas McKean had little to do. His second brigade (Oliver's) was at Alexander's Crossroads waiting for the enemy. Despite the confusion at army headquarters, someone remembered General McArthur was without a command. In the course of the night a special order was issued giving him field command of the First Brigade. It was not much of a brigade, just three regiments strong, but McArthur intended to make the best of it. First, however, he had to reassemble his new command. At 6:00 A.M. McArthur left Alexander's Crossroads with the Sixteenth Wisconsin and established brigade headquarters beside the Memphis and Charleston Railroad, a few hundred yards southeast of the old Confederate earthworks. There McArthur was joined by his second regiment, the Twenty-first Missouri, which had returned from outpost duty at Kossuth shortly before midnight. The Scotsman retrieved his remaining regiment, the Seventeenth Wisconsin, which had started off at daylight to reinforce Oliver. At 5:00 A.M. General McKean's Third Brigade, commanded by

Marcellus Crocker, formed near Battery F, one of a chain of six artillery lunettes built a mile and a half outside Corinth to guard against Confederate attacks from the west and south.

At dawn, then, McKean's was the only Federal command standing between the enemy and Corinth. Davies was just breaking camp six miles south; Hamilton was on the road, five miles from Corinth; and Stanley's division was six hours away.

Rosecrans could do no more than hope his army would reach Corinth before the Rebels. He, like Crocker's Iowans "in front of those wooded western approaches, on the morning of October 3, waited for what might happen, wholly ignorant of what Van Dorn was doing at Chewalla, ten miles away through thick forests."[9]

*   *   *

The Confederates were closer than Rosecrans realized. They had, as Rosecrans guessed, passed the night near Chewalla but were on the road at 4:00 A.M. Lovell's division continued to lead, with Villepigue relieving Rust as advance guard.

Impelled by a fear that reinforcements from Grant would reach Corinth first or strike his rear, Van Dorn drove the army hard. He launched the infantry on a nine-mile forced march, intending to attack the moment he reached Corinth.[10]

In spite of the mad pace of the march and prospect of impending battle, morale seemed to improve as the Army of West Tennessee neared Corinth. John Tyler thought "the spirit of the men was excellent. All were alike emulous of courageous bearing and gallant achievement." Or perhaps their resolute bearing masked fear. Efforts at humor seemed forced. The Second Missouri Infantry was hurrying past Bledsoe's battery, parked by the roadside waiting its turn on the road, when Ephraim Anderson heard a voice ring out, "Hello, Anderson!" It came from an artillerymen who had been his neighbor at home. "We are in for some fun today," responded Anderson. "We will sleep in Corinth tonight," his friend rejoined before the regiment disappeared in a swirl of orange dust. Whatever their states of mind, there is no doubt Villepigue's Mississippians were exhausted when they ran up against their first obstacle: Oliver's command drawn up at Alexander's Crossroads.[11]

Col. John Oliver seemed like the right man to direct the first phase of Rosecrans's planned delaying action. The thirty-two-year-old graduate of St. John's College, Long Island, had an intensity incongruous with his soft, pudgy frame. Oliver's gaze was piercing and purposeful, his devotion to duty almost pathological. Oliver enlisted as a private and within a year was colonel of the Fifteenth Michigan. At Shiloh he had taken the regiment, which could have passed the battle safely in the rear as part of the headquarters guard, into the thick of the fight. Besides costing several dozen Michigan boys their lives, that act of audacity earned Oliver a citation for "conspicuous gallantry" and command of a brigade. But at the moment Oliver had only two regiments and two cannon. During the night the Eighteenth Wisconsin had been ordered away, leaving Oliver with 500 men to contest the Confederate approach.

Scattered shots from the skirmish line announced the enemy at 7:00 A.M. The two howitzers of the First Minnesota Light Artillery joined in, shelling the road beyond the wooded horizon. But Villepigue's Confederates barely paused, so Oliver sent the Fourteenth Wisconsin forward to try to force them to deploy. Col. John Hancock pushed his regiment 500 yards before waving the men into line astride the road, along the slope of a gentle, timbered rise. "Pretty soon we heard the rebels down in front of us, the officers giving orders, but the men did not seem to obey them very willingly," said Pvt. John Newton. "At any rate they did not come on very fast. As soon as they came in sight we fired and ran. The rebels ran too, but they ran away from us, as well as we ran away from them."[12]

A timely order from General McKean spared Colonel Oliver further embarrassment. Acting on his own volition McKean directed Oliver to fall back across Cane Creek and destroy the bridge, if time permitted. Cane Creek meandered from north to south through a long, patchwork expanse of pastures and cultivated fields paralleling the trace of old Confederate works, which ran a mile to the east. The Chewalla road bridged Cane Creek a mile behind Oliver's advanced position at Alexander's Crossroads. Just six feet wide, Cane Creek was more of an obstacle than it appeared. Its banks were twelve feet high and steep, and the ground nearby was spongy.

Oliver did not stand upon the ceremony of his going but pulled his brigade back from the crossroads at once. The soldiers of the scattered Fourteenth Wisconsin already were nearing the creek, so that Oliver only needed to extricate the Fifteenth Michigan and his two howitzers. But even that proved daunting. After the Fourteenth Wisconsin disintegrated, the lack of any serious resistance emboldened Villepigue's Mississippians, and they rapidly closed on Oliver's withdrawing regiments. Bullets peppered the pastures and roadbed and around the Federals, and they shoved across the bridge with growing alarm. During the brief cannonade at Alexander's Crossroads an axle of one of Oliver's howitzers had splintered. The crew dragged it as far as the creek before the rope snapped. The Minnesotans spiked the gun, shoved it into the creek, where it wedged between the banks and the dry bed, then hurried across the bridge.[13]

Safe on the east bank of the creek, Oliver's men rallied. Skirmishers eased into place beside the road and opened on Villepigue's Mississippians. It was 7:30 A.M. The sun hung at the level of the treetops. Its rays burnt a warm sheen on the dew lingering in the meadows and fields. Dismounting on the bridge, Yankee troopers tore up the flooring, felled a tree across the frame, and then sprinted past the skirmishers to rejoin their mounts.

Colonel Oliver chose as his next position the first defensible high ground beyond Cane Creek, a timbered hill that abutted the track of the Memphis and Charleston Railroad. Rising a half-mile east of the creek bridge, the hill was one of a two-mile-long, semicircular chain of hills, low knolls, and short ridges that followed the trace of Beauregard's earthworks. The railroad ran through a deep cut between the western face of Oliver's hill and a hill equally high on the south side of the track. The track itself lay on a bed of fine blue and white marl, a sparkling contrast to the orange dirt of the hills.[14]

Oliver was pleased with the position. It commanded both the Chewalla road and the railroad and "was a strong one and easy to hold against anything but an overwhelming force." But he was troubled by his standing order, which was to fall back and avoid pitched battle. At 8:00 A.M., as Oliver was placing his two regiments and one remaining howitzer on the crest, General McArthur arrived and eased his qualms. McArthur also liked the ground and told Oliver bluntly to "hold the position at all hazards." Well aware Oliver's command was too small to defend even the most ideal terrain without being flanked, McArthur rode off to bring up his brigade.[15]

Whether McArthur exceeded his authority in ordering Oliver to stand firm is uncertain. In his report McArthur said he "determined to make a stand on Cane Creek Bluff." In a telegram written shortly after Oliver fell back from Alexander's Crossroads, Rosecrans told Grant that he had ordered McArthur to take charge of matters along Cane Creek and to "push forward to make strong reconnaissances." When McArthur reported the hill to which Oliver had withdrawn seemed "of great value to test the advancing force," Rosecrans authorized him to "hold it pretty firmly with that view" — that is to say, only long enough to compel the enemy to reveal his strength, not "at all hazards," as Oliver understood his orders.

Rosecrans later attributed McArthur's zeal to his "Scotch blood," which boiled at the prospect of battle.[16]

McArthur's decision to fight on Oliver's hill was undoubtedly wise. Rosecrans was in a tight spot and needed time. At 8:00 A.M. he had no troops besides McArthur's with which to resist a Confederate advance down the Chewalla road. Far from being neatly aligned in a continuous arc north of town, as Rosecrans had intended, his army was scattered about the countryside. The command nearest Oliver's hill was Crocker's Iowa Brigade at Battery F, a mile away. General Hamilton had his division deployed a mile and a half northeast of Corinth, watching the Monterey and Purdy roads. The head of Stanley's column was still at least two hours away, and Davies's division was only then entering town.

When Davies reported to army headquarters for instructions, he found Rosecrans doubtful and a bit dazed, not at all sure where the main blow might fall or if the Rebels intended to attack Corinth at all. Rosecrans still entertained the notion that the movement against Oliver might be no more than a diversion to cover a move against Grant. Rosecrans told Davies to take his division out a mile and a half and form line of battle astride the Memphis and Charleston Railroad as quickly as possible. Written instructions followed: "Rush forward your skirmishers on your front and feel what you have got to handle, if anything." What Davies was to do after that was unclear. "We may assume the offensive very soon; it depends upon the pressure that may come on the right," the order concluded.

The muddle outside headquarters mirrored Rosecrans's addled mind. In the commotion Capt. W. H. Chamberlin of the Eighty-first Ohio felt the tension of coming battle. "It was evident that something was going to happen. Troops were moving in every direction, teams were driving at break-neck speed, and all the usual business appearance of the town was giving way to inextricable confusion. At the same time, the sound of artillery grew more distinct and nearer, and orderlies and staff officers were dashing by on hurried hoof."[17]

15

Well, Boys, You Did That Handsomely

Earl Van Dorn was as free of anxiety as Rosecrans was consumed by it. At daybreak Van Dorn's army was concentrated precisely where he intended it to be, not dispersed beyond reach as was Rosecrans's command. And unlike his West Point classmate, Van Dorn harbored no doubts about the enemy's plans. He knew of the chain of lunettes the Federals had built near town—some said that a lady spy smuggled sketches of the fortifications and periodic reports of Yankee troop strength to him—and, said Maj. John Tyler, Van Dorn had deduced Rosecrans's design of fighting a delaying action from the old Confederate earthworks. To Tyler's surprise, Van Dorn arrayed the entire army in full view of the earthworks, retaining no reserve. One mile north of Alexander's Crossroads General Price detoured to the left, deploying his two divisions 400 yards short of the earthworks, between the Memphis and Charleston and Mobile and Ohio Railroads. Price formed Hébert's division on the left and Maury's on the right. Brig. Gen. Martin Green's Second Missouri Brigade held the extreme left, its flank resting on the Columbus road. Frank Armstrong brought his troopers over to cover the three-quarters of a mile of ground between Green's left and the Mobile and Ohio Railroad. Next came the Mississippians and Alabamians of Col. John Martin. Gates's First Missouri Brigade fell in on Martin's right. Perhaps out of consideration for its losses at Iuka, Hébert retained his own brigade, now led by Col. W. Bruce Colbert, in reserve. Dabney Maury deployed two brigades forward. Brig. Gen. Charles W. Phifer took the left with his brigade of dismounted cavalry, and Brig. Gen. John C. Moore's brigade lined up with its right touching the track of the Memphis and Charleston Railroad. Brig. Gen. William Cabell formed his Arkansas brigade in supporting distance of Phifer and Moore.[1]

Price played his role well. He hid his misgivings from his men behind all the martial pomp he could muster. Dressed in the multicolored plaid hunting shirt the soldiers had come to call his "war coat," Price rode the length of the line. With him was the mounted headquarters band, which paused now and then to encourage the troops with popular tunes. The musicians unwittingly provided Gates's Missourians with a good laugh as well. Midway through a rendition of "Listen to the Mockingbird," a Federal shell struck an oak tree, splitting it in two and showering the musicians with limbs and leaves. The band darted for the rear, trailed by hoots of derision. More shells followed. Bunched together in closed ranks, the men surrendered themselves to the whim of a chance hit.

Neither Price's presence nor the brief serenade could distract the soldiers from another source of growing agony. The temperature, which was climbing with the sun, hovered about the ninety-degree mark. Men by the hundreds had fallen out during the march, and those left standing swayed from heat and hunger. The 400-yard-wide abatis of fallen timber between them and the earthworks, into which Yankee infantry was pouring, did not help their humor. William McCurdy of the Thirty-seventh Mississippi later wrote that only thirteen men stood with him from his company; "the balance were broke down and left by the road side. Marched to death and suffering with hunger, we were soon drawn up in line of battle near the breastworks of the enemy. . . . The timber was cut down between us and fell[ed] across each other."[2]

On the south side of the Memphis and Charleston Railroad Mansfield Lovell struggled to bring up his division. For nearly an hour he was forced to wait on the west bank of Cane Creek while a detail from Villepigue's brigade repaired the bridge. Colonel Oliver did what he could to slow their work. General Davies lent McArthur a section of Battery I, First Missouri Light Artillery, which the Scotsman passed on to Oliver. Colonel Oliver placed one of its pieces, a James six-pounder rifled gun, beside the remaining howitzer of the First Minnesota Battery.

The two cannon opened a long-range fire on the bridge. At the same time Capt. Levi Vaughan edged his Company E, Fourteenth Wisconsin, down the forward slope to skirmish with Villepigue's men. Few if any Mississippians were hit, as the spent Yankee bullets splashed harmlessly in the water or thudded against the bank. Oliver's cannoneers also missed their mark, but the chance of a shell bursting on the bridge kept the Mississippians from giving their full attention to the work.[3]

The bridge was hammered back together by 9:00 A.M. Two of Villepigue's regiments hurried across and into line in a rolling pasture below Oliver's hill. John Bowen got his brigade rapidly into position on Villepigue's right and deployed skirmishers to help subdue the harassing fire. Albert Rust came over and extended Lovell's front nearly a half-mile beyond the left flank of Oliver's tiny command. A squadron of Col. William H. Jackson's cavalry was inserted between each brigade of infantry. With his eight remaining companies Jackson rode to the extreme right, intending to reconnoiter west and south of town as soon as the attack began.[4]

At 9:30 A.M.—ninety minutes after Oliver's cavalry had wrecked the Cane Creek Bridge—the Confederate army was ready to advance against the outer works of Corinth. General Van Dorn took up headquarters in the Murphy house on the Columbus road, behind Price's center. There he called a final conference before launching the assault, which he meant to open on the right with Lovell's division. Certain of success, Van Dorn waited for his generals and let the minutes slip by.[5]

*　　*　　*

General McArthur moved with a will, galloping to his brigade headquarters, a half-mile down the Chewalla road, to rouse the Sixteenth Wisconsin and the Twenty-first Missouri. After marching most of the night, said a Wisconsin volunteer, both regiments "had just settled in their tents when the bugle call to arms summoned the men to rush out and fall into line of battle." There was hardly time "to eat something hurriedly and get ready to meet the enemy." From the northwest came the low rumble of cannon firing, then a staff officer appeared to show them the way forward. McArthur left the Seventeenth Wisconsin behind to guard the brigade camp and provide the nucleus of a fall-back position.

By the time McArthur returned to the front, it was evident Oliver would need more than the two regiments the Scotsman brought with him. He had repelled Rebel skirmishers who had come down from the far ridge to challenge the Fourteenth Wisconsin and the Fifteenth Michigan for possession of the railroad cut, but that small victory merely portended a larger danger, should the enemy occupy the southern hill in force. To delay them Oliver sent the Twenty-first Missouri across the track and onto the hilltop.

The Confederates also were probing heavily up the Chewalla road. The Fifteenth Michigan commanded that approach and had succeeded in keeping Villepigue's skirmishers at bay. In expectation of an all-out attack by the Mississippians, Oliver retained the Sixteenth Wisconsin in reserve.

For McArthur the greatest threat lay to the northeast, beyond Oliver's right flank, where the enemy was clearly deploying "so as to gain the old rebel breastworks." Leaving the defense of the hill again to Oliver, McArthur went off in search of reinforcements.

He found them quickly, thanks to a confused but concerned General Davies. Rosecrans's oral orders that Davies stop short of the outer breastworks were contrary to his thinking of the night before, when Rosecrans seemingly had settled on fighting a delaying action from there. Not having been privy to Rosecrans's thinking, Davies accepted his instructions without question. Indeed, Davies may have been relieved that Rosecrans did not expect him to form at the outer breastworks, where the distance between the railroads—the sector he was to defend—was nearly two miles. With six regiments absent on various details, Davies had with him only 2,924 officers and men: Brig. Gen. Pleasant Hackleman mustered just 1,097 men in his First Brigade, Brig. Gen. Richard Oglesby counted a mere 720 in the Second Brigade, and Col. Silas Baldwin had 1,117 troops.[6]

Davies led his small command out the Memphis road. After marching a mile and a half as ordered, he halted at 9:00 A.M. at the fork of the Chewalla and Columbus roads and formed a compact line of battle in the timber. Whatever relief Davies felt at having his division drawn together gave way to misgivings over the larger fate of the army. Artillery fire from the northwest had grown louder and more regular, leading Davies to conclude Oliver had withdrawn to the breastworks.

Davies's better instincts prevailed: "Thinking a movement forward on the Columbus road would support Colonel Oliver and prevent the enemy flooding down too rapidly upon us I sent to General Rosecrans for permission to move forward and occupy the rebel breastworks on the Columbus road. He replied that I could do as I thought best." With his commanding general's vague blessing, Davies marched northward.[7]

Rosecrans was behaving oddly. The Ohioan faced his first test in a real battle, and that realization apparently left him dazed. The campaigns in western Virginia had been seriocomic, brief encounters between half-trained mobs. The fight at Iuka had happened suddenly and developed a logic of its own before Rosecrans had time to respond. Now a veteran army confronted him—a foe, however, who offered Rosecrans ample

Brig. Gen. Thomas A. Davies (Massachusetts Commandery of the Military Order of the Loyal Legion, and the U.S. Army Military History Institute)

time to put his plan into play. Yet Rosecrans hesitated. Receiving only halfhearted or inscrutable instructions, it is no surprise McArthur and Davies chose to respond as the situation on the ground seemed to dictate.

A fifteen-minute hike brought Davies's division to the edge of a large, open field, in the center of which the Columbus road forked. The main road continued due north; a well-worn bridle path diverged toward the northwest. While contemplating how, or if, he should divide his force between the two trails, Davies sent the Seventh Iowa and a section of Battery K, First Missouri Light Artillery, up the Columbus road to reconnoiter to the breastworks. His calculations were interrupted by a winded courier from McArthur, who told him that Colonel Oliver would be overrun unless he sent him two regiments at once.

As the whole purpose of his forward movement had been to assist Oliver, Davies decided to oblige. He immediately detached two regiments from Baldwin's brigade, the Seventh and Fifty-seventh Illinois, and a section of Battery D, First Missouri Light Artillery, to reinforce the Michigander.[8]

Before Davies was able to consider his next move, Lt. Col. Arthur Ducat, who had been transferred to Rosecrans's staff before the battle, rode up with new — and more definite — orders from headquarters. Ducat reiterated Rosecrans's approval of Davies's request to advance to the Rebel breastworks and told him "not to let the enemy penetrate beyond" them, an admonition implying far more strenuous resistance than the developing action Rosecrans had earlier intended. Ducat reminded Davies that "in no event must he cease to touch his left on McArthur's right" and cautioned him to keep watch over the bridle path between the Chewalla and Columbus roads.[9]

Davies parceled out what remained of his depleted division along the breastworks. The departure of Baldwin's two regiments left him slightly more than 1,500 men with which to hold a front two miles long. The men in ranks, if not Rosecrans or Davies, knew the task was impossible. "The regiments were stretched to their utmost capacity, in a thin line, but yet there were immense gaps which could not be filled," Captain Chamberlin lamented. Pvt. Erastus Curtis agreed: "We formed in a single line of battle, with intervals between each man, of from three to four feet, the like of which I had never seen before, [and] never after."

The position was not a particularly strong one, either. Said Captain Chamberlin, "The old abatis, formed by felling the timber for the space of three hundred yards in front of the works, had lost much of its strength by time. . . . Beyond this was thick woods, whose abundant foliage, yet unhurt by the frost, formed an impenetrable cover for the movement of the rebel troops." The works themselves, observed Cpl. Charles Wright, were "a slight affair, and would serve only as protection against musketry."

Chamberlin and Wright might have added that their regiment had been reduced by detachments to five companies, a mere 238 men. With this small force, augmented by a twenty-four-pounder howitzer and a ten-pounder Parrott gun from Battery H, First Missouri Light Artillery, Col. Thomas Morton started down the bridle path to defend the breastworks where the trail crossed. Davies inserted the rest of Oglesby's brigade to the right of Morton's Eighty-first Ohio. Col. Augustus Chetlain, who had just reported from Burnsville with six companies of the Twelfth Illinois, formed 300 yards beyond the flank of the Ohioans. A gap nearly as wide yawned between the Twelfth Illinois and the Ninth Illinois on the brigade right.[10]

Pleasant Hackleman's brigade manned the breastworks between the Columbus road and the Memphis and Charleston Railroad. Before committing the brigade, Hackleman sent the Seventh Iowa and Battery K, First Missouri Light Artillery, scouting up the Columbus road. They found the breastworks empty and the enemy forming in the woods beyond. Hackleman deployed his remaining two regiments, the Second Iowa and the Fifty-second Illinois, to the right of the battery. Col. Thomas W. Sweeny of the Fifty-second despaired of his predicament; 500 yards of undefended breastworks lay between the right flank of his regiment and the railroad. Sweeny detached two companies to watch the gap, called for artillery support, and settled in to defend the extreme right of the division as best he could. The time was nearly 10:00 A.M. Not a shot had been fired, but several dozen of Davies's Federals already had fallen, victims of sunstroke. The number climbed with the sun.[11]

*   *   *

By 9:00 A.M. Rosecrans's mind had cleared. With his blessing, General Hamilton occupied the breastworks on Davies's right. Again Federal flanks failed to touch: 400 yards of breastworks lay empty between Davies and Hamilton.

Of far greater moment was the mile-wide gap between Davies's left and the small force McArthur had marshaled atop Oliver's hill. Davies had not forgotten Rosecrans's admonition that he "in no event cease to touch his left on McArthur's right." But he either miscalculated the distance involved or assumed McArthur would use the two regiments he had sent him to cover the ground. The Scotsman, however, had his own problems. With Lovell's entire division arrayed to attack the hill, and J. C. Moore's brigade deploying just beyond Lovell's left, McArthur needed every available man close at hand. He used the reinforcements from Davies accordingly, hurrying the Fifty-seventh Illinois to the left of the Fourteenth Wisconsin and inserting the cannon from Capt. Henry Richardson's Battery D, First Missouri Battery, between the regiments.[12]

A rattle of musketry from the rail bed announced the return of Rebel skirmishers. In the same instant Col. Andrew Babcock's Seventh Illinois arrived at the double-quick, and McArthur sent it to the hill north of the Chewalla road. It was a weak position. A deep ravine just beyond the line of abatis offered shelter to an enemy attacking in front. Babcock threw a company into the ravine but was helpless to close a 200-yard interval between his left and the right flank of the Fifteenth Michigan.

To fill the gap, McArthur appealed to Davies for more troops. The request startled Davies, but he nonetheless released the Fiftieth Illinois, his only reserve. "While moving to the front there suddenly fell upon our ears, low cannonading and the low sharp roll of musket firing," said the regimental historian. It was 10:00 A.M. After nearly two hours of deploying, skirmishing, and waiting while Van Dorn briefed his generals, the Confederates had moved to the attack.[13]

*   *   *

When it came, Mansfield Lovell's attack was irresistible. The Confederates approached from three directions, subjecting McArthur's five regiments to a cross fire that fast made their positions untenable. Although the outcome was never in doubt, McArthur's small force inflicted sharp losses on Lovell's Confederates. What Albert Rust called the "extremely animated" resistance of the Twenty-first Missouri stopped his brigade briefly below the brow of the ridge. When it got going again, thick timber on the slope split the brigade in twain. The Third and Seventh Kentucky settled into a close-range duel with the Twenty-first Missouri. The three regiments on the brigade left, screened by the Fourth Alabama Battalion, drifted past the Missourians' right flank, over the ridge and down its east slope toward the railroad cut.

That detour doomed the Twenty-first. The regiment had had some bad moments but was holding its ground against the Kentuckians. The color-bearer had sneaked away at the first volley, trailed by several dozen soldiers perhaps hopeful the standing order to "follow the colors" would exculpate them. It did not. Col. David Moore drew his saber and waded into them. To Moore's relief Cpl. Jess Roberts redeemed matters. "He gallantly seized the colors," Moore recalled, "and advanced on the line of battle . . . causing great enthusiasm among the men."

No sooner had Corporal Roberts's band returned than Moore's horse was killed, pinning the colonel to the ground. Moore was pulled free, badly dazed. The regimental surgeon could not help him—he too had been shaken up when his mount threw him—so Moore stumbled to the rear, no longer fit to command.

Maj. Edwin Moore took charge. Soldiers spilled past him by the dozen, and gradually the regiment yielded the crest to the Confederates. Moore and his line officers restored order in the railroad cut. Joined by the Sixteenth Wisconsin the Missourians charged back up the ridge, holding it until Rust outflanked them. Faint from the heat and lack of water, at 10:30 A.M. the Missourians fell back, having lost only a dozen men in sixty minutes of sharp fighting. The Sixteenth Wisconsin went with them.[14]

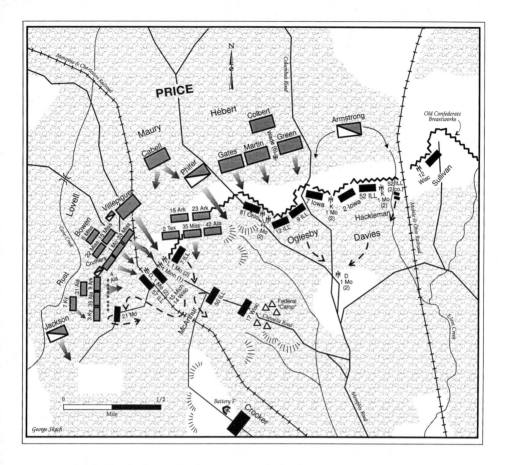

MAP 6. *The Battle Opens, October 3, 10:00 a.m.*

Preceded by skirmishers from the Fourth Alabama Battalion, Rust's left regiments glided past the Missourians' right flank to the steep eastern edge of the ridge. As the Alabamians stepped from the protection of the forest and started down the railroad embankment, a bitter fire erupted from across the track. The Alabamians broke and "fled in wild confusion," exposing the main line to a deafening chorus of rifle volleys and shelling. The Ninth Arkansas, the Thirty-first Alabama, and the Thirty-fifth Alabama charged off the slope toward an enemy entrenched not sixty yards away. Solid shot replaced shell, remembered Pvt. W. G. Whitfield of the Thirty-fifth, and one "struck a large tree, just a few feet from my head, and tore it to pieces. One of my company, who was deaf, turned his head to one side and looked up as though he heard it."

When the Rebels crossed the track, the Yankee gunners turned to canister. The carnage was terrific, and Rust's men retired up the slope.

Encouraged by several craven officers, the Thirty-first Alabama fell apart, and neither its colonel nor Rust could rally the men. After Rust left to re-form the Ninth Arkansas and the Thirty-fifth Alabama, the regiment simply marched off the field.

Rust had come up against the Fifty-seventh Illinois and Lt. George Cutler's two-gun section of Richardson's Battery D, First Missouri Light Artillery. The Illinoisans held a strong line on the hill but lacked the cover afforded by the old Confederate entrenchments, which ended 100 or 200 yards to their right.[15]

"Reforming, they come again with that cold-blooded yell which has to be heard to be appreciated," said one Illinoisan. Again the Rebels were repelled. A third time they charged. Averred Capt. James Zearing of the Fifty-seventh, "Our men stood the shock nobly, delivering the most steady and effective fire that I have seen during the war." But this time the Confederates kept coming. A hundred men fell in a few seconds, reported Rust; he concluded, however, that "to have halted would have brought certain destruction upon my command." Rust directed the Ninth Arkansas to make a detour to the left to avoid the deepest part of the railroad cut, ordered bayonets fixed, and sent his line forward. The fire from the Fifty-seventh intensified, said Captain Zearing, "but it had no effect in checking their march. They advanced on the double quick in the utmost disregard of human life."

The Illinoisans were stunned. The Ninth Arkansas and the Thirty-fifth Alabama swept on, intent on taking the two guns of Cutler's section. Cutler limbered them just a few dozen yards ahead of the Rebels. One cannon bounced off the hill to safety, but the limber team of the second gun ran wild when a wheel driver panicked and jumped from his horse. The limber pole broke, and the cannon slammed into the dirt. A few members of the battery tried to push it off, giving up only when the Rebels were near enough to bayonet them. Lt. Albert Goodloe of the Thirty-fifth Alabama watched Private Whitfield, who earlier had almost been beheaded by a shot from that very cannon, chase after the last Yankee to leave. He nearly caught the cannoneer, said Goodloe, "but the Yankee was too fleet for him." The rest of the men were not far behind Whitfield. They left behind the cannon — on the breech of which the name "Lady Richardson" was boldly painted in white — and followed the fleeing Federals.[16]

To the right of Cutler's section the Fourteenth Wisconsin put up an even stronger stand against half of John Bowen's brigade. Companies E and K had been thrown forward to the railroad embankment; the remainder of the regiment was tucked behind the old Confederate earthworks along the brow of the hill. From the hilltop the two cannon of Battery I, First Missouri Light Artillery, and remaining gun of the First Minnesota joined in to shell the Rebels.

The target of their fire was Capt. C. K. Caruthers's Mississippi Sharpshooters Battalion, an untested unit more anxious than most to prove itself. Explained Lt. William C. Holmes, "The battalion had assumed a new role in the great science of warfare. We had been drilled in the 'skirmish drill' almost exclusively, and our place in the army was in the front, leading the entire division in the advance, but always in the rear in the retreat. Individually, to a man, we were anxious to show our hand in the line of warfare set apart to us, for our small size made us the jest of the other regiments."

At 10:00 A.M. Caruthers's battalion was ordered out in front of the Twenty-second Mississippi. "We felt our way to the front," continued Holmes, "when all at once an entire battery turned loose solid shot whose unearthly screams through the air set up the most gigantic dodging I have ever seen. . . . On we went, and musket shots were added to the horribly sublime music, till suddenly an exhilarating frenzy seemed to take possession of every man, and we swept away the enemy's skirmishers, despite the huge limbs of trees cut off by the solid shot."[17]

Young James Newton was among the Yankee skirmishers driven in by Caruthers's Mississippians. "I could not run so fast as I had ought on account of a sore foot, and for one spell I was right between two fires," Newton remembered. "I had sense enough left to know that such a thing would not do for me so I went off to the left and the secesh went past me."

Newton was at least momentarily safe. Several of his comrades, however, were gathered in by the pursuing Mississippians, who spilled into the railroad cut. There they paused. Not fifty yards away — "belching forth fiery destruction, simply grand in its sublimity," said Lieutenant Holmes — stood the two guns of Cutler's section. Although it made them an easier target, the men of Caruthers's battalion instinctively huddled together. Unable to see the enemy through the smoke, they fired blindly uphill, losing men and hitting nothing until Capt. Tom Atkinson of Company C mounted the embankment, waved his tall white hat, and shouted, "Come on, boys, let us take it!" Lieutenant Holmes watched for a response. A sergeant from his own company lifted his musket and yelled, "Come on boys, let us follow him up." Men started forward, a few at a time, until "all seemed infused with the sublime horror of the occasion, and to a man we rushed over that embankment, up the hill to the very mouth of the cannon."

They got no farther. Concealed by the smoke and their breastworks, the Fourteenth Wisconsin let loose a volley that Holmes said sent the Mississippians stumbling "down that hill much faster than we ran up it." Halfway down they met the Twenty-second Mississippi coming up. No orders were given; remembering their skirmish drill, the men shouted to one another to "form on its left" and with the Twenty-second Mississippi went up the hill again.

This time the Rebel line made it to within a few yards of the breastworks before stopping. Instead of giving back, the Mississippians took cover and returned the fire. Lieutenant Holmes and his company "took refuge in an old gully, and for a few minutes a torrent of Minie balls passed over us, cutting the leaves and twigs from the trees. We were almost covered, but we were out of danger."[18]

Not so the Fourteenth Wisconsin, thanks to the pathetic resistance of the Fifteenth Michigan, which broke before Bowen's First Missouri and the Thirty-third Mississippi of Villepigue's brigade coming down the Chewalla road. The Rebels spilled over the breastworks to the right of the Fourteenth Wisconsin, catching that regiment in the most concentrated cross fire of the morning. Men fell at a dizzying rate. Captain Vaughan of Company E, the most popular officer in the regiment, was shot at near point-blank range. Pvt. Andrew Sloggy caught him as he fell. "Tell my wife, and all my friends that I did my duty, and died for my country," he muttered to Sloggy, who tried to coax two soldiers into carrying Vaughan off. "It was of no use," lamented Sloggy. "The captain urged me all the while to leave him before the rebels would get me. I stayed with him until there was no choice for me but to leave." Another well-liked member of the same company, Pvt. J. M. Vandoozer, was shot through the head while taking a chew of tobacco. Two days later, friends found Vandoozer's body, the plug of tobacco still grasped in his fingers. More seriously for the morale of the regiment, the entire color guard was killed or wounded. The bearer of the regimental colors was bayoneted, and the standard fell between the lines. A soldier from the Fifteenth Michigan, in a small act of redemption for his regiment's poor conduct, dashed into the fray, grabbed the colors, and returned them to the regiment that evening. The bearer of the national colors, Sgt. Dennis Murphy, came away with his flag. When the regiment entered the fight, Murphy had remarked, "I'll come out a dead sergeant or a live lieutenant." He emerged with a field commission, a cripple for life. Although repeatedly hit, he had hung onto the flag until it was in tatters and soaked with his blood.[19]

Bowen's Confederates came over the breastworks with clubbed muskets. The slaughter of the color guard had paralyzed those nearby. They were too startled to fire or run, and the Rebels were among them before they regained their composure. Twenty-one Wisconsin men surrendered. In the confusion the regimental sergeant major, returning from the brigade train, walked into the Rebel ranks with a supply of ammunition. Also captured was Private Newton. His dodge away from Caruthers's skirmishers brought him into the breastworks simultaneously with the Twenty-second Mississippi: "I could see the Butternuts on both sides of the ditch; when they came up . . . I surrendered."[20]

Lieutenant Holmes got himself a Yankee, though not in the manner he may have wished. Holmes had hidden in the gully until after the Twenty-second Mississippi cleared the breastworks. When the firing died down, he got up and wandered forward to find it or his own battalion. "I found no Twenty-second. It had gone clear out of sight after the fleeing enemy, but left behind a sure and terrible token of its work." With his own troops gone as well, Holmes became "an interested spectator." He walked along the breastworks just abandoned by the Fourteenth Wisconsin. There before him "was presented a sight that could but appeal to the heart of the hardest soldier—men dead, dying, wounded, all in line just as the regiment stood." (Seventy-eight to be precise—27 dead and 51 wounded—of the 225 men who had answered roll that morning.) Among them was one unhurt soldier. In his arms he cradled the body of a man just dead, shot through the head. "'My friend, are you hurt?'" Holmes asked, thinking to himself what a "travesty" he had done the word "friend." "No, but this is my poor dead brother, and I could not leave him," the Federal answered. "I would not mind it for myself so much, but his poor wife and little baby!"

As the only unwounded Rebel present, Holmes might claim the man as his prisoner, but "I could stand no more of that, and I went to the next man sitting down and gritting his teeth with pain." Holmes repeated his question: "Are you hurt, my friend?"

"Oh, yes," the man answered, showing Holmes a bubbling bayonet wound in the groin.

"Well, here is a grain dose of morphine I had for myself should I be wounded. Will you take it?"

"Oh yes, with pleasure and thanks."

Holmes handed it to him, reflecting on the absurdity of the scene. "I thought what ought to ring down the ages of humanity to great and small: 'Well, what foolishness! Here we have been trying to kill each other, and now we are trying to do all we can to repair the damage done.'"

Holmes passed to the next live man, whom he found clutching a tree, shot through the lungs and groaning with every breath. Holmes gave him a dose of morphine. He handed his last dose to a young boy who sat crying and cradling his mangled arm, then returned to the bayoneted man. The morphine had eased his pain. Holmes sat down beside him — "an intelligent, talkative Yankee" — and the two fell into a friendly discussion, punctuated by the moans of the dying.[21]

The fight was not quite over. Still on the field was the Seventh Illinois. The random retreat from Oliver's hill, which neither Oliver nor McArthur could control, left the Illinoisans alone and apparently forgotten. The weatherworn abatis and the ravine in front of their position helped them stop a frontal attack by Villepigue's Thirty-ninth Mississippi. A few minutes later they spotted a thick line of Rebel skirmishers emerging from the timber 200 yards northeast of their right flank. Cautiously the Rebels moved across abatis-laced open ground toward the empty breastworks beyond the flank of the Seventh. Col. Andrew Babcock had his right companies train their rifle-muskets on the flank of the slowly advancing line.

The skirmishers belonged to J. C. Moore's brigade of Maury's division. When the Thirty-ninth Mississippi moved to the attack, Moore was directed to advance as far as the timber line at the edge of the abatis, there to await orders. Moore sent part of the Second Texas Infantry and one company of the Thirty-fifth Mississippi into the open, where they ran into skirmishers from the Seventh Illinois. The Illinoisans ran back to their breastworks, and for a moment the field fell silent. Then the Thirty-ninth Mississippi broke apart, and Moore charged.

The Rebels swept from the woods five regiments strong. The Second Texas was on the right of Moore's first line, and the Forty-second Alabama was on the left. Both were fine commands. Sandwiched between them, where it could do the least harm, was the Thirty-fifth Mississippi, a new regiment of dubious reliability that had a camel to carry the officers' baggage and a colonel who drank too much. The Fifteenth and Twenty-third Arkansas trailed in support.

Colonel Babcock's flank companies opened fire too early, most of their bullets passing harmlessly among the skirmishers of the Second Texas. At the same instant a shell fired from a ten-pounder Parrott gun at a range of

500 yards exploded in front of the color guard of the Forty-second Alabama. Eleven men collapsed in a bloody pile, including the color-bearer. A few spent bullets thudded against soldiers of the left-flank company, slightly wounding several. Whipped into a frenzy by the unexpected fire and the momentary fall of its colors, the Forty-second Alabama broke into a running charge. In their anger they mistook the skirmishers of the Second Texas for the enemy, sending a volley into their backs that killed a lieutenant and wounded six privates.[22]

The chance fire that stirred the Alabamians came from the Eighty-first Ohio and the section of artillery posted with it on the extreme left of General Davies's attenuated line of battle. In front of the charging Confederates were vacant earthworks.

Colonel Babcock saw that his predicament was hopeless. He sent to General McArthur for instructions, which by Babcock's telling never came. McArthur claimed to have told him to "change front to the right and charge the enemy with fixed bayonets"—a foolish order that Babcock could not have obeyed. By then Babcock and his men knew they were alone. Dense timber blocked Oliver's hill from view, but the Illinoisans had been following the fight by sound. Said Sgt. Leib Ambrose of the Seventh, "Up to this the battery and the force on our left have been making the woods ring with their terrible thunder, but they are silent now; their cannons are still; their musketry is hushed." From the woods instead came shots from skirmishers of the Thirty-third Mississippi, who had left the Chewalla road and were probing north after the retreat of the Fifteenth Michigan. The Thirty-ninth Mississippi renewed its attack across the abatis in front of the Seventh. Having crossed the breastworks to the Illinoisans' right, Moore's Confederates were re-forming their lines to front south. Sergeant Ambrose summed up the situation: "This is our position; rebels in our front; rebels on our right and rear; rebels on our left and rear; soon their right and left columns will meet; soon we will be surrounded if we remain here." Reaching the same conclusion, Colonel Babcock gave the command, "By the right of companies to the rear," and the Seventh slipped away before Moore could close off its line of retreat.[23]

Mansfield Lovell was pleased with the work his division had done. To the men of the Thirty-fifth Alabama, who had helped capture the Lady Richardson, he remarked jovially, "Well, boys, you did that handsomely." But the struggle for Oliver's hill took the fight out of Lovell. He recalled Rust, whose brigade had pushed 300 yards beyond Oliver's hill in pursuit of the fleeing enemy, and told him to return to the south side of the railroad. To Bowen and Villepigue, who had lost control of all their regiments except the Thirty-third Mississippi, he gave orders to reassemble on the hill.

Lieutenant Holmes paused in his conversation with the color-bearer of the Fourteenth Wisconsin to watch the return of his brigade. Twenty minutes had elapsed with no one but dead and wounded Yankees, a few stragglers, and himself on the hilltop. Remembered Holmes, "The Twenty-second returned in due time, and we formed in line as best and as quickly as we could to await orders; but a battle is to a company, regiment, brigade, or the whole army what a blow is to a beautiful rose — it is broken into innumerable pieces, and it looks as though it can never be replaced, but still it must be done."

Mansfield Lovell, to employ Holmes's metaphor, was not about to lose any more petals. Rust, Villepigue, and Bowen — the latter in growing disgust — waited for orders to advance as minutes slipped into hours. Finally, at 3:00 P.M., Bowen asked Lovell for permission to lead his brigade to the support of the First Missouri, which had strayed during the fight for Oliver's hill and attached itself to J. C. Moore's brigade. Lovell instead told Bowen to recall the Missourians. Bowen complied. Seeing the pointlessness of further entreaties to Lovell, he allowed his men to disperse to collect and bury the dead, carry off the wounded, and begin repairs on the damaged Lady Richardson. Lieutenant Holmes heard distant firing and wondered at their own inactivity: "We had captured the outlet line, and the battle raged with redoubled fury. We remained there; not a regiment of the brigade or the division [was] engaged."[24]

16

They Ran Like Hens

The soldiers of Richard Oglesby's woefully understrength brigade had plenty of time to contemplate their fate. Six Confederate brigades assembled deliberately in the timber opposite them, with no thought to disguising their purpose. They did everything with a theatrical, almost disdainful flair that suggested they knew how few Federals were crouched behind the breastworks they were about to assault. At 10:00 A.M. their skirmishers emerged from the woods and spread out among the abatis with a calmness infuriating to the Yankees. "We were looking over the works and scanning everything in front with all the eagerness displayed by troops in the beginning of a fight," Cpl. Charles Wright said. Nerves frayed quickly, and men squeezed off shots haphazardly. A sergeant next to Wright screamed, "Boys, there's a sharpshooter in that bushy oak down there in the ravine!" and let go a shot. "I think he tumbled and fell behind the tree," affirmed another Ohioan. Wright had seen nothing; he was preoccupied with what was going on behind the Rebel skirmishers. "Through an opening in the woods I could see the rebels forming their lines for the assault on our works. The woods seemed to be full of the men in gray. A great number passed this opening; tramp, tramp, tramp, they kept coming." Wright grew angry. "Knowing our number and speculating on theirs, I was convinced that back a mile and more in our rear, under the guns of Robinett, was the place to make a fight."

The Rebel skirmish line thickened until it contained more men than did all of Oglesby's brigade. Only then did the Southerners return the Federal fire. Their first shots fell short, kicking up little puffs of orange dust in front of the breastworks. One of Wright's comrades broke under the strain of waiting. He stuck his head above the breastworks and yelled, "Shoot higher, you damned rebels, you're doin' no good!"[1]

As if to taunt the Rebels, Col. Thomas Morton stepped atop the breastworks and jammed the regimental colors into the dirt. Nobody hit him. Climbing down, he paced behind his Ohioans, gazing steadily at the woods beyond. A rider on a white horse caught his eye. Galloping from a clump of timber, the man paused to scan the Federal breastworks with a field glass, evidently to get the range of Morton's supporting artillery.

Morton stopped and yelled to the first sergeant of Company E: "Say, can't you stir that fellow·up a little?" He could not but knew a man who could. The sergeant called over nineteen-year-old Pvt. Erastus Curtis, a crack shot. Colonel Morton bade Curtis to fire. "What is the distance, Colonel?" asked Curtis. Morton raised his field glass and guessed 500 yards. Curtis cocked his rifle-musket and stuck the barrel over the breastworks. Before he could fire, six cannonballs came crashing down on the works near the regimental colors. The Confederates had shown their contempt of Morton's gesture, and of the Union defenses, by running Wade's Missouri Battery onto a knoll in front of the Eighty-first Ohio.

Capt. John Welker's Battery H, First Missouri Light Artillery, had the final say in the affair. Welker had split his battery, placing one section on the bridle path alongside Colonel Morton's Eighty-first Ohio and the other 200 yards to the right, beside the Twelfth Illinois. Welker's two sections opened a converging fire that killed most of the Rebel limber horses and drove the guns back into the woods. The battery commander, Lt. Samuel Farrington, was cut in half by a solid shot. General Price nevertheless wanted to bring his remaining batteries into the open to support the impending assault, but his chief of artillery resisted, pointing out that Wade's "horses were stricken down too fast for a gun to be planted." As though to drive home the point, a shell burst among Price's party, killing the horse of an aide and just missing Maj. John Tyler.[2]

Apart from the good aim of their artillerymen, who were the only ones to enjoy the preliminary proceedings, all the 720 Federals had going for them was the quality of their leaders: in Oglesby a fine brigade commander, and in Cols. Augustus Chetlain of the Twelfth Illinois and August Mersy of the Ninth Illinois two talented regimental commanders.

"Uncle Dick" Oglesby, as his men called him, was among the most

Brig. Gen. Richard J. Oglesby
(Illinois State Historical Library)

colorful and prominent officers in blue at Corinth, an intimate friend and political ally of Lincoln whose early life closely resembled that of the president. Born in Kentucky on July 25, 1824, Oglesby was orphaned at age nine and taken in by an uncle in Decatur, Illinois. He received a rudimentary education before quitting school to work variously as a farmhand, a rope maker, and a carpenter. With the money he saved, Oglesby studied law in Springfield and was admitted to the bar. Quitting his fledgling practice to enlist in an Illinois regiment when the Mexican War broke out, he won election as commander of his company by "outwrestling, out-running, and out-talking" his fellow volunteers. Oglesby strove hard to keep the respect of his men. He led them on a 450-mile march from the border deep into Mexico, "walking every step of the way in twenty days," he told his sister. "Tis true my feet often wore into blood blisters, and the skin came off in pieces as large as half-dollars, but I had to go it. I knew it would not do to despair whilst there were so many of the men looking for my example in that respect."

Two years later Oglesby again journeyed overland, this time to California to take part in the gold rush. He returned to Decatur with $5,000 in gold dust, speculated in land, and became one of the richest men in the county. He also threw himself into Republican politics, becoming one of the dozen party leaders, known as the "original Lincoln men," most responsible for the future president's nomination. At thirty-seven, with a friend in the White House and a natural gift for leadership, Oglesby looked like a man with a bright future.

So, too, did Augustus Chetlain. Also thirty-seven years old, Chetlain had been born in Switzerland and migrated with his parents to Galena, Illinois, as an infant. Starting as a clerk in a local store, Chetlain went into business for himself in 1852. Seven years later he was rich. Chetlain had accepted a commission at Grant's suggestion, and his service to date confirmed the general's estimation of him.

August Mersy had neither wealth nor political influence. He came to America from Prussia in 1849, a penniless former general of the revolutionary army. He found employment as a bank clerk in Belleville, Illinois, but was unable to rise higher than cashier. When war came, however, Belleville's German community remembered Mersy's accomplishments during the revolution, and they turned to him for leadership.

Mersy and his largely German Ninth Illinois had fought well and suffered much. At Fort Donelson the regiment held off a series of furious attacks during the abortive Rebel effort to break out of the Federal cordon. The Ninth paid dearly for its tenacity, losing 210 men in two and a half hours. At Shiloh the Ninth was again brutalized, losing an incredible 366 men. Whether the survivors could stand up to another test was an open question, but one that would not be answered here. Fifteen minutes after General Price waved his 9,000 men forward at 11:00 A.M., the Confederates were in the breastworks.[3]

Cpl. Charles Wright of the Eighty-first Ohio saw his earlier fears confirmed. All around him was "conclusive proof that the battle had better be fought under the guns of Robinett." A long Butternut line—J. C. Moore's brigade—climbed the empty breastworks far beyond the regiment's left. Closer at hand Phifer's entire brigade bore down on the five small companies from front, while Martin's brigade made for the gap in the line between the Ohioans and the Twelfth Illinois. "The musketry fire was now terrific," said Wright. "I could hear the steady voice of Lieutenant Chamberlin, 'Load fast, boys; load fast!' which was the one thing needful at this particular time." Wright and his comrades got off between fifteen and twenty rounds before Phifer's Rebels swarmed over the head-log of the earthworks, pinning several slow-footed Ohioans to the bottom of the ditch with their bayonets. Wright had obeyed the command to "spring out of the ditch" and got away unhurt. So, too, did most of the men of the Eighty-first, leaving Lt. John Conant's section of Welker's battery unprotected. With nearly all his horses shot down, Conant and his gunners abandoned their cannon to the Ninth Texas Dismounted Cavalry and ran for the rear.

Welker's right section gave a good account of itself, tearing huge gaps in the Rebel ranks. A single discharge of canister killed twelve members of the Forty-third Mississippi; another mortally wounded Col. John Martin, and a third blast tore apart General Green's horse. But Colonel Chetlain, knowing a hopeless fight when he saw it, ordered his men to fall back "as soon as possible." Left without infantry support the remaining two guns of Welker's battery escaped just ahead of the Rebels.[4] The Ninth Illinois was the last to go. Stunned by the ferocity of the enemy attack and the rapid collapse of the Twelfth Illinois on their left, three officers and thirty men were captured before they could escape. The rest of the regiment, less seven dead or wounded, fled into a thick underbrush of briars and mulberries behind the breastworks. "They ran like hens running from a hawk, hiding behind every log and in every place they could find," recalled James Fauntleroy of the First Missouri. Astonished prisoners told their captors they thought them drunk, for the recklessness of their charge.

Delayed by the abatis, mounted officers galloped up a few minutes later. Ready to lead their men in pursuit, they instead found most already chasing the Yankees on their own. Lieutenant Colonel Bevier of the Fifth Missouri took it philosophically. He dismounted, hitched his horse to a log, and walked over to the abandoned cannon of Welker's left section. There Bevier found a soldier of his regiment sitting on a gun tube and wiping the sweat from his face. Bevier himself was "panting and blowing" from having negotiated his horse through the obstacle-laden field under a sun now midway across the sky. The temperature had climbed to 100 degrees and was still rising.

"Well, Colonel, you mounted fellows are tolerably useful in camp, and serve a good purpose on the drill ground, but we don't need you much in a fight."

"No, I'll swear you don't," gasped Bevier, "and you boys can out-run the devil when you are after a Fed."

"You bet we can!"[5]

Price's advance had been confined to the west side of the Chewalla road, which placed the men of Hackleman's brigade in the safe but exasperating position of watching the spectacle without being able to contribute a shot on Oglesby's behalf. When the Rebels crossed the breastworks to his left, Hackleman was outflanked, and he ordered his three regiments to fall back.[6]

To Charles Colwell, the moment looked desperate. On detached service from the Ninth Illinois with the division trains, Colwell had advanced his wagons to the first field behind the breastworks just as Oglesby's line gave way. "I met soldiers coming running, infantry, cavalry, and artillery. They said the lines were broken and nearly all taken prisoners." General Davies rode up a moment later. His staff was as nonplussed as the soldiers. "General, the lines are broken, the rebels came in behind us, [have] taken us by surprise, and [are] pouring a deadly volley into us from behind us as well as before," stammered an aide.

Davies kept his head. To Colwell he said calmly, "Charley, you can take your train back to town." Colwell remonstrated, "General, if I move the train it will raise a stampede and wouldn't it be better if I got the train in such shape that I could drive out at a minute's notice and then wait till you form your line?" Davies agreed. Turning to the task of consolidating his division and restoring order to Oglesby's brigade, Davies chose the open field as a rallying point for Hackleman and Oglesby. Davies sent a courier to Baldwin — wherever he might be — directing him to retire to the intersection of the Columbus and Chewalla roads. Hackleman's brigade marched through the field in perfect order and assembled in the timber along its southern fringe, on the east side of the Chewalla road. Re-forming Oglesby's brigade took longer, but ultimately the presence of Oglesby steadied the men. Obviously embarrassed by the retreat, the burly Illinoisan drew rein in the center of the field and bellowed, "Men, we are going to fight them on this ground. If there's dying to be done, men, I pledge to you my word I'll stay with you and take my share of it."[7]

The sincerity of Oglesby's vow went untested. Davies had no intention of making a fight there. "The second line of my two remaining brigades was only intended to attract the attention of the enemy and cause them to form line of battle in my front, which they did," he explained. No sooner had Oglesby finished his speech than a messenger from Davies rode up and handed him a paper. Corporal Wright watched Oglesby: "He read, placed the paper in his pocket, and gave the order to 'about-face,' and in battle-line we moved through the woods toward Corinth."

Davies had two other good reasons for not engaging the enemy. From a returning messenger he learned that McArthur, who had promised to re-form on his left, had been forced to withdraw south of the Chewalla road, beyond the fortified camps of the Seventeenth Wisconsin and the

Twenty-first Missouri. A strong Rebel column, the messenger said (a reference to J. C. Moore's brigade), was driving down the road between their two commands. The courier dispatched to Baldwin also returned, bleeding badly. He had been shot trying to get through to Baldwin, who had been cut off by Moore and Phifer and forced to withdraw with McArthur.

It was between 1:00 and 1:30 P.M. when Davies ordered Oglesby and Hackleman to retire. They were to form a line of battle at the junction of the Chewalla and Columbus roads, "with the same view and same effect as their previous movement." [8]

But Price refused to play Davies's game a second time. Indeed, the Missourian would have been pleased to stop for the day. His men were tired and thirsty. Most had emptied their canteens, and scores fainted while chasing the Yankees from the breastworks. When the enemy disappeared from the far side of the field, Price elected to rest his command. [9]

*   *   *

Brig. Gen. John Creed Moore was a solid troop commander—"an officer of fine ability and courage," said Dabney Maury—well capable of acting on his own. His men found him to be a strict disciplinarian with a grim and haughty demeanor, but they recognized his ability. Here at Corinth Moore did not disappoint them. Unlike Lovell, who stopped after capturing Oliver's hill, Moore pursued the Federals along the Chewalla road, forcing them to seek refuge behind the entrenched camps of the Seventeenth Wisconsin and the Twenty-first Missouri, a mile southeast of Oliver's hill. There, independently of each other, Oliver and McArthur began to piece together a battle line. [10]

Oliver was resting his shattered brigade between the camps and Battery F, an earthen redoubt on the crest of a long, gentle ridge a quarter of a mile to the south, when the Seventeenth Wisconsin, which had passed the morning comfortably in reserve, and Battery F, Second Illinois Light Artillery, reported for duty. Col. John Doran told Oliver that he had orders to report either to him or to McArthur. Oliver had no idea where McArthur had gone but suggested Doran form the Seventeenth on high ground across the railroad tracks, fronting northward. Oliver promised to support him, but in truth his small command was in no condition to fight. The Fifteenth Michigan was nearly out of ammunition, and the survivors of the Fourteenth Wisconsin were broken in spirit. Even Colonel Hancock, revered by the soldiers of the Fourteenth as one of the "coolest and bravest" officers in the army, lost his composure. Recalled a member of Company E, "After we fell back . . . the colonel came around and asked how many men each company had. When he came to our company, I told him we had seven, but could tell him nothing of the balance; told him Captain Vaughan was killed and two or three more I knew of. I never saw such a look on a man's face before. He said nothing, but sat down on a log and cried."[11]

General McArthur rode up a few minutes later. After taking one look at Oliver's troops, he relieved them of further duty and told Oliver to report to General McKean, who had made his headquarters behind Crocker's brigade at Battery F. McKean already was showing signs he was too old to cope with the stress of combat. The gray-haired Pennsylvanian had done nothing during the morning, seemingly content to allow McArthur to assume, de facto, the mantle of division command.

The Scotsman discharged his expanding duties gracefully. He sent skirmishers from the Seventeenth Wisconsin to occupy their former camp. As other units reported in, McArthur fed them into line on the left of the Seventeenth, parallel with the railroad and facing northeast. Col. Frederick Hurlbut showed up first with his Fifty-seventh Illinois. Hurlbut had redistributed ammunition and had combed the nearby forest and fields for stragglers; he was able to muster a respectable force, but the men were on the brink of collapse. "The heat is intense," wrote the regimental historian. "There is no water, and the men are famishing. Some of the Fifty-seventh fall in their tracks, fainting and exhausted under the rays of the scorching sun. Teams had been sent to the rear for the purpose of hauling water, but as yet none reached us." McArthur told Hurlbut to wait beside the Seventeenth Wisconsin for further orders. Colonel Baldwin tried to intervene and reassert his authority as brigade commander, but Hurlbut ignored him. So, too, did Col. Andrew Babcock, who complied with McArthur's command that he place the Seventh Illinois next to the Fifty-seventh Illinois. Lt. Col. William Swarthout likewise reported directly to the ubiquitous Scotsman, who directed his Fiftieth Illinois into line on the left of the Seventh. Anchoring the left flank on Battery F, McArthur rounded out his hastily assembled, half-mile-long line of battle with the Sixteenth Wisconsin and the Twenty-first Missouri of his own brigade.

No sooner had McArthur completed his dispositions than the Wisconsin skirmishers returned to report their camp swarming with Confederates. Already scattered shots from Rebel skirmishers were striking men in McArthur's main line. As the Southerners drew nearer, their fire increased in precision. Cannoneers and horses from Battery F, Second Illinois Light Artillery, which had unlimbered a section in front of the Seventeenth Wisconsin, became the principal targets.[12]

At 2:00 p.m. McArthur acted to save his artillery and rid the camps of Confederates: "I determined to drive them out of it, and ordered the line to charge with the bayonet en echelon of battalion from the right." The Scotsman rode to Col. John Doran and asked him if his Seventeenth Wisconsin "could charge successfully on the brigade doing such execution on our battery." Doran said he could, and he did. Baldwin's brigade took up the charge on Doran's left, and the Sixteenth Wisconsin and the Twenty-first Missouri swung north from Battery F.

The Federals drove Moore's brigade through the camps and beyond, until McArthur stopped them along a grassy ridge above the Chewalla road. Moore regrouped in nearby woods. Acquitting himself as well as McArthur had, Moore restored order to the Second Texas, the Thirty-fifth Mississippi, and the Forty-second Alabama, which had borne the brunt of the Federal charge, and then called up his reserve regiments, the Fifteenth and Twenty-third Arkansas, to extend his line to the left. With a front reaching beyond McArthur's right, Moore counterattacked. At the same instant, the errant First Missouri of Bowen's brigade, which had continued on after the fight on Oliver's hill, emerged from the timber behind McArthur's left flank. Threatened with a double envelopment, McArthur called on McKean for reinforcements and slowly backed his men off the ridge.[13]

McKean's first combat order was a catastrophe. Although he had at his disposal the 2,189 soldiers of Col. Marcellus Crocker's Iowa brigade—the only truly fresh troops in Rosecrans's army—he chose not to call on them. Instead McKean told Colonel Oliver to gather the parched and exhausted survivors of the Fourteenth Wisconsin and the Fifteenth Michigan and march to McArthur's relief. Crocker's Iowans remained beside Battery F, where they had stood since dawn.

The consequences of McKean's decision were predictable. Oliver's men marched off the naked ridge, away from Battery F and into a narrow strip of wood that hid the track of the Memphis and Charleston Railroad. Emerging from the timber, they came upon McArthur's command falling back. The noise and confusion sent a tremor of panic through their ranks. McArthur was compassionate; he simply reported the "reinforcements were unable to comprehend the situation." Colonel Doran understandably was more direct. After trailing his men "tardily, at a long interval of both time and distance," Oliver's two regiments "discharged their muskets into the Seventeenth, turned, and ran."[14]

Moore's Confederates were too broken up to pursue closely, allowing McArthur to pull back to the foot of the ridge. There, protected by a

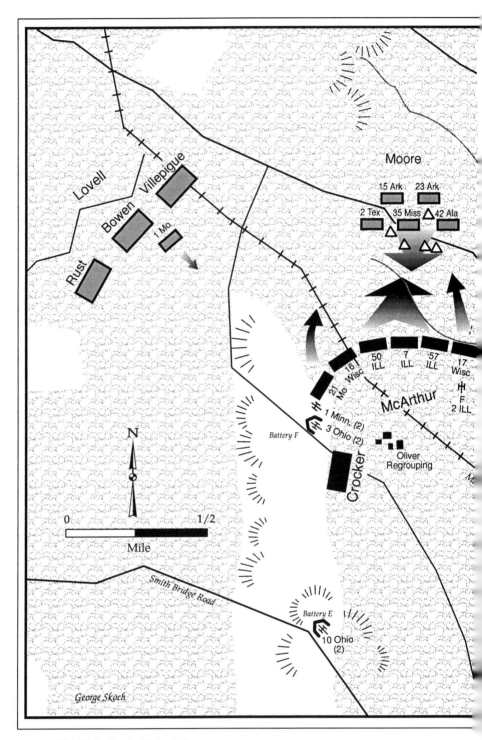

MAP 7. *The Fight for the Federal Camps, 2:00 p.m.*

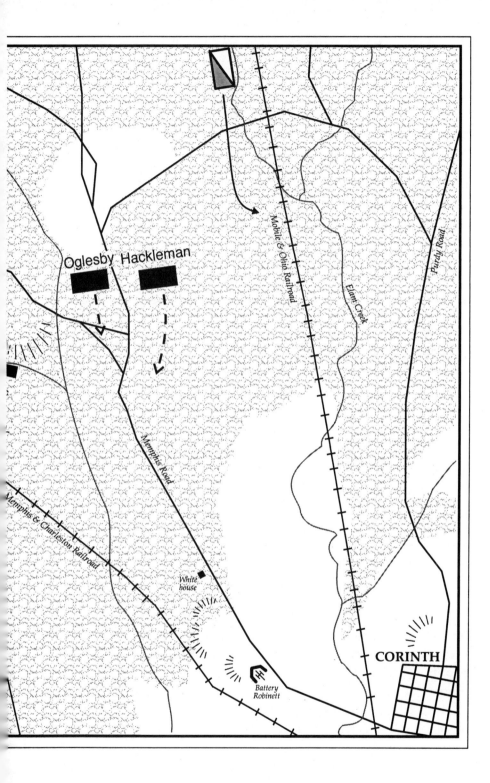

Oglesby Hackleman

Mobile & Ohio Railroad

Elam Creek

Purdy Road

Memphis Road

Memphis & Charleston Railroad

White house

Battery Robinett

CORINTH

blanket of shot and shell, he re-formed his command a third time. McArthur's presence was electric. The Fiftieth Illinois had nearly lost its regimental colors when the color-bearer was shot down. While a corporal of the color guard rushed out to retrieve them, the bearer of the national standard stepped forward to rally the men, but they shoved past him, heedless of the colors. McArthur rode up as the men paused on the friendly side of the timber to catch their breath and, said an Illinoisan, "observing the colors with only its guard and a few men, called out as he lifted his Scotch cap, 'What regiment is this?' Being informed that it was the Fiftieth he at once, with the assistance of the officers of the regiment, placed the colors in position, and in a few moments the regiment was ready for work." [15]

The bold front the Fiftieth showed him fooled neither McArthur nor the regimental commander, Lieutenant Colonel Swarthout. The men were played out. Said Swarthout, "Quite a number had to be taken from the field, some suffering from sunstroke and others from utter exhaustion." Matters were the same in the other regiments. McArthur gave no thought to their fighting again, and McKean had the good sense not to demand it. He permitted McArthur to retire a quarter-mile to the south and ordered Crocker to form a new line of battle to block the inevitable advance of Moore's Southerners.

It was 3:00 P.M. when McArthur's men passed through Crocker's line. The day was over for the Scotsman's two regiments and the five he had appropriated. They marched to Corona College, drew water, and dispersed to rest. Colonel Hurlbut of the Fifty-seventh Illinois deemed it his duty to report his regiment to Colonel Baldwin. He found him at the Tishomingo Hotel, nursing a slightly wounded hand. When Baldwin chose to leave the field is unknown, but he was not missed. The day, and his brigade, had belonged to McArthur. [16]

*　*　*

Marcellus Crocker was another of the talented Union brigade commanders at Corinth. Only thirty-two years old, Crocker had sad, sunken eyes and a "slender, nervous" frame, both the consequences of chronic poor health. He had been forced to leave West Point after two years because of tuberculosis, a disease that went into remission upon his return to Iowa but never completely left him. A lifetime of illness made him edgy as well. Attested a sympathetic captain of the Seventeenth Iowa, "He has a passionate temper, and is plain-spoken, often saying things which, in his calmer moments, he would leave unsaid." Crocker compensated for both his infirmity and his peppery disposition by being a natural leader. He studied law after leaving the military academy and at the outbreak of the war was among the best lawyers in Iowa. Leaving his Des Moines practice, he was mustered into the service as a captain in the Second Iowa Infantry in May 1861. Six months later he was promoted to colonel of the Thirteenth Iowa. At Shiloh Crocker fought brilliantly in some of the heaviest combat of the battle. Afterward he was given command of the newly created "Iowa Brigade," which, in addition to his own regiment, consisted of the Eleventh, Fifteenth, and Sixteenth Iowa. As a brigade commander Crocker displayed a "mode of discipline severe and uncompromising, and a careless blunder he would never excuse." But he was neither cruel nor shrunk from danger, and if they did not adore him, his troops admired him. They undoubtedly were proud to be part of what already was conceded to be one of the finest brigades in the western armies.[17]

The skill of Crocker's Iowa Brigade was evident in the ease with which the men changed front, despite having to contend with what one regimental commander called "an almost unmanageable thick underbrush in the rear of its former line." Lining up at right angles to its earlier line, the brigade deployed with the Fifteenth Iowa — its left touching the open, eastern flank of Battery F — and the Sixteenth Iowa in the front line. The Eleventh and Thirteenth Iowa regiments formed in close column fifty yards to the rear. The brigade stood on "a naked high ridge with nothing to protect us from the fire of the enemy" except underbrush and a smattering of skinny oak trees.[18]

Under the burning sunlight the Iowans squinted toward the forest along the railroad, from which McArthur's retreating regiments had yet to emerge, and listened as the fighting drew nearer. Clint Parkhurst of the Sixteenth Iowa remembered the agony of uncertainty: "We were close enough to hear the combat well, but not close enough to see it. A few heavy explosions of musketry broke on the air, in quick succession. Then . . . the Union fugitives poured into view like scattered sheep, and reaching our line rushed on to the rear, scores of them being bloody from wounds." Beyond rose the clear, wild cheers of the Southerners.

The Rebels broke out of the timber, were visible for an instant, then disappeared behind a roll in the ground, 200 yards short of the Iowans. The wait was brief but excruciating. Remembered Clint Parkhurst, "We knew that preparations must be going on to attack us, and to stand idly there awaiting the onset was a trying ordeal — a test of manhood keener than fighting." The officers were no less affected. Riding along the line, reminding the men not to fire until ordered, were Crocker and Col. William Belknap, commander of the Fifteenth Iowa. Belknap was also thirty-two years old and a former lawyer, and between the two was a warm, open friendship built on mutual respect. The two paused a moment behind some small trees. A bullet struck a sapling that stood between them. "Do you know, old fellow, what I am thinking about?" said Crocker. "No, colonel," Belknap answered. "I wish I was back in Des Moines." So, too, did Belknap.

A silence heavy as death fell over the field; Moore's Rebels seemed slow in coming. Then everyone's attention was drawn to a new sound. "While we waited with intense interest and much anxiety the next move in what was to us a momentous drama, an appalling burst of martial thunder came from a locality a mile or more to the right of us," wrote Parkhurst. "Musketry and artillery mingled in one awful and prolonged peal. It was not an affair of a regiment or two, but seemed like the collision of two heavy lines of battle, and the roar was incessant as long as I was conscious of listening to it. Our thoughts, however, were almost immediately concentrated on events in front of us."

Bullets cut the air around the Iowans. Rebel skirmishers appeared and then vanished in the tall weeds. Remarkably, at that instant General McKean was handed an order from army headquarters to withdraw Crocker's brigade to the inner fortifications of Corinth. The Pennsylvanian elected to obey, but in a manner that showed his judgment was improving. He would pull back in three stages. First, Crocker's second-line regiments, the Eleventh and Thirteenth Iowa, would retire 400 yards to cover the withdrawal of the artillery from Battery F. In the meantime Crocker's front-line regiments, the Fifteenth and Sixteenth Iowa, would fight a covering action to allow both the artillery and the Eleventh and Thirteenth Iowa regiments to get away and, if necessary, launch a limited counterattack to confuse the enemy long enough to make good their own escape.[19]

Crocker remained with the Fifteenth and Sixteenth Iowa regiments. The Confederates came on deliberately, with bayonets fixed. They were confident and, under the circumstances, well fed and rested. The delay between their first appearance and attack occurred because so many men had fallen out to plunder the Yankee camps that Moore had been obliged to pause. The troops were delighted. Recalled a Texan, "In this camp we found bread, butter, cheese, crackers, and other food in abundance, and while enjoying a short rest, partook of the enemy's unwilling hospitality during his enforced absence — the first food we had tasted that day." Morale improved further when two regiments from Cabell's reserve brigade came up. Moore put them on the left of his line, which gave him seven regiments, or nearly 2,500 men, with which to confront Crocker.

Clint Parkhurst despaired of ever hearing the order to fire. Captains and lieutenants walked up and down in front of their companies, swords in hand, slapping up rifle-muskets that were leveled prematurely by nervous and impatient soldiers. The Confederates closed to sixty paces, stopped, and cocked their rifle-muskets to fire. Those of the Sixteenth Iowa already were cocked. Company officers jumped behind the ranks, the Iowans took deliberate aim, "and with a crash we fired." The Rebels returned the fire. A wall of dust and smoke arose, and the opponents were lost to sight from another. The soldiers of the Sixteenth Iowa stood their ground. At Shiloh they had ducked behind trees or rocks at the first fire. Crocker's incessant drilling and strict discipline since that battle had had an impact. Said Parkhurst, "A few men fought on one knee, but not a man lay down, and the great majority stood erect on the color line, and loaded and fired in drill-ground fashion." Not all the credit was Crocker's. The men had learned at Shiloh that those hit while fighting erect, with only their left side facing the enemy, were less liable to receive fatal wounds than if struck while on one knee or lying down. Parkhurst's own experience here at Corinth was proof of that: "Once, while standing erect, I turned my left side to the enemy, to drive down a musket ball. The next instant a big bullet passed through my left pantaloons pocket, where I carried a package of ten rounds of ammunition. It tore the paper cartridges to pieces, but I was unhurt. Had I been facing squarely to the front I would have had a mortal wound."

Matters were considerably more unsettled in the Fifteenth Iowa. The entire front line got down on one knee. The men had their rifle-muskets cocked and ready; but Colonel Belknap waited an instant too long, and the enemy fired first. The blue line shivered. "We heard the Rebel shout and yell," remembered Cyrus Boyd. "Then somebody commenced firing and we shot away in the smoke and not knowing exactly where to aim as the enemy were in lower ground than we. But their first volley laid out many a man for us." (Getting in the first volley often made a dramatic difference in a regiment's casualties: the Fifteenth would end the day with 11 dead and 67 wounded; the Sixteenth would count only 1 man dead and 20 wounded.) The men fell back a few yards and broke ranks to take cover behind the scattered trees and logs, despite Colonel Belknap's efforts at keeping order. Nevertheless, they maintained a constant fire for forty-five

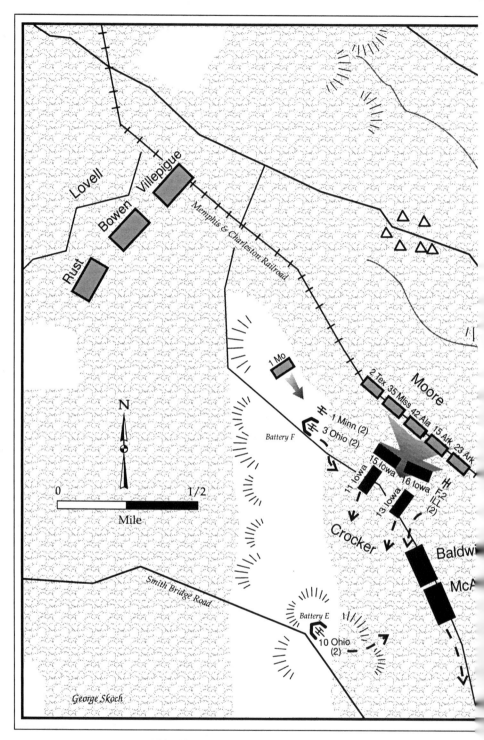

MAP 8. *Battery F and the White House, 3:30 p.m.*

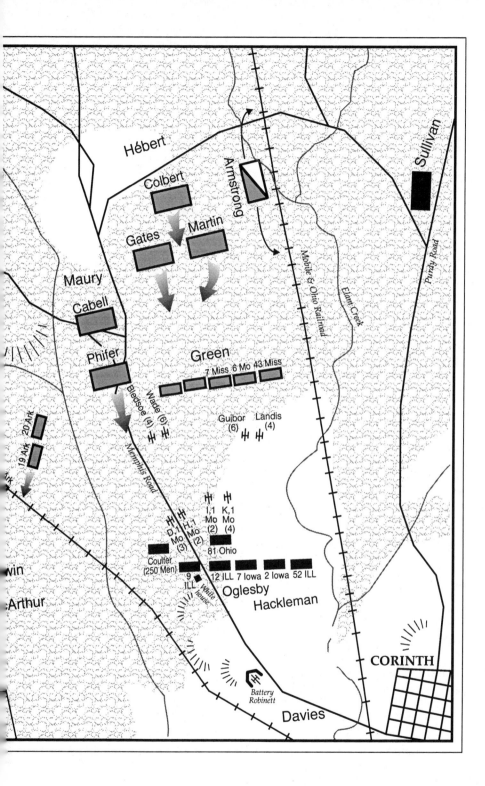

Hébert

Colbert

Armstrong

Sullivan

Gates   Martin

Maury

Cabell

Phifer

Green

7 Miss 6 Mo 43 Miss

Wade (6)

Bledsoe (4)

Guibor
(6)

Landis
(4)

Memphis Road

Mobile & Ohio Railroad

Elam Creek

Purdy Road

20 Ark

19 Ark

I,1   K,1
Mo   Mo
(2)   (4)

D,1 H,1
Mo   Mo
(3)   (2)

81 Ohio

Coulter
(250 Men)

9
ILL   12 ILL 7 Iowa 2 Iowa 52 ILL

White house

Oglesby

Hackleman

win

Arthur

CORINTH

Battery
Robinett

Davies

minutes and with the Sixteenth Iowa succeeded in holding the enemy to their front in check.[20]

The sudden appearance of the First Missouri on their left flank broke the will of the Iowans. Cyrus Boyd was a member of the flank company of the Fifteenth, which bore the brunt of the Missourians' attack. "They swarmed around on our left and fired from behind trees and logs and kept pressing forward. Our ranks became much confused . . . and men fell in great numbers."

Boyd's company was cut to shreds. A corporal standing to Boyd's left was shot in the stomach and fell with a dull groan; the man to his right collapsed with a bullet in the head. Their lieutenant was killed in the first volley from the Missourians. Boyd's company disintegrated, and the entire regiment began to break up. To forestall a rout, Colonel Belknap ordered the Fifteenth off the field by the right flank. The Sixteenth, which had fought the Rebels to its front to a standstill, marched off in good order 200 yards ahead of the Missourians. Once again the Federals were spared greater damage by the rapacity of the Rebels, who stopped to plunder Crocker's camp. After running through the rows of lowered tents himself, Boyd cast a backward glance. "I could see the Rebs tearing the sutlers' tents away and going for the goods. . . . By this time everything was on the retreat toward Corinth and the firing had almost ceased in the woods."

General Moore never contemplated pursuit. In capturing two Federal camps and what he proudly considered to be a "strong, well constructed work," the Tennessean felt he had done enough — particularly as Lovell's division had stood idle a half-mile away while he took on the equivalent of two Federal brigades.[21]

That Lovell could have accomplished much is undeniable. When Crocker changed front to face Moore, he left open the northwestern approach to Battery F. McKean had no other force available to resist an attack from that direction. Had Lovell moved aggressively, he could have routed Crocker and the remnants of McArthur's and Oliver's brigades, which had yet to withdraw to town, and perhaps capture much of McKean's artillery. Undoubtedly he would have shattered the division too badly for it to have reassembled that day. Confronted with the collapse of his left at the same time Davies was being pressed hard in the center, Rosecrans may have been forced to concede Corinth to the Confederates.

General Bowen saw the opportunity and had begged Lovell to allow him to support the wayward First Missouri with the remainder of his brigade. Instead Lovell had insisted that he recall the First Missouri. Fortunately for Moore, Bowen apparently chose to disregard that order. Albert Rust shared Bowen's disgust. He considered his own men to be "in first-rate fighting condition" after their "comparatively brief" engagement on Oliver's hill and by late afternoon regarded the enemy as "whipped." Aggressive by nature, John Villepigue probably shared a similar opinion.

Not until 5:00 P.M., nearly an hour after Moore had captured Battery F, did Lovell consent to an advance, and then only after Van Dorn had told him to move forward. Lovell contented himself with making contact with Moore's right. With the Tennessean, Lovell halted near sunset on a ridge a half-mile short of the Federal inner fortifications, into which the brigades of Crocker and McArthur were only then filing. Even after dark, Moore still hoped to attack. He was waiting for word from Lovell that he, too, would advance when General Maury recalled him to his own division.

An opportunity as golden as the hot October sun had been lost to the Confederates. But Lovell never explained his inactivity. His support for the operation had been predicated on a minimal loss of Southern lives. When it became obvious that Corinth would not be taken without a bloody fight, Lovell may have chosen to register his objections — and reduce the losses in his division — simply by doing the minimum required. What he would accomplish on the morrow remained to be seen.[22]

17

I Bid Them All Good Bye

Rosecrans was less charitable with General Davies than Davies had been with his brigade commanders after the clash at the breastworks. Updating Grant at 1:00 P.M. on the course of a battle he had done little to direct, Rosecrans complained Davies's men "did not act or fight well." Nevertheless Rosecrans thought the army would "handle" the enemy, in part because he believed the attack on Corinth to be a feint. While the Rebels swarmed over Oliver's hill and Davies reassembled Oglesby's brigade south of the outer breastworks, Rosecrans wrote Grant with a confident air that "one unusually reliable [scout] gives their entire force not exceeding 30,000 and is satisfied that they intend to make their main move on Bolivar. . . . If we find the force not meant for Corinth or we are in position to do it, shall move on them steadily with everything we can spare."[1]

Rosecrans's missive reflected both wishful thinking and a gross misreading of affairs on his own front. Already Davies had begged him for reinforcements, a good sign Rosecrans would have few troops to spare to move on the enemy. Minutes after Rosecrans sent his update, Davies again pleaded for more troops. His two brigades were regrouping at the junction of the Chewalla and Columbus roads, their second delaying position since quitting the breastworks. Even as they formed a battle line, Davies made plans to withdraw to what he determined would be his final position: two huge, partially cultivated fields 950 yards southeast of the fork of the Columbus and Chewalla roads and 725 yards in front of Battery

Robinett. Explained Davies, "This position was selected as the only one where the small force under my command had any hope of meeting the enemy with success. At all other points it could have been flanked and surrounded. . . . Here, in the edge of the woods, the men could lay partially concealed, with an open view in their front."

The New Yorker's reasoning was sound. The Confederates had made clear (to Davies's satisfaction at least) that they intended to approach Corinth primarily over the ground between the two railroads. The tracks converged as they neared town; that reduced the amount of available space for maneuver — an advantage for Davies, who could better cover a contracted front with the few men left to him. As they drew close, the Confederates would have to cross either the two fields on the east side of the Memphis road where Davies had determined to deploy most of his force, or a field west of the road nearly as large. On the west side of the road, between the open expanses, was a white frame dwelling that Davies's Federals called the White House, and for which they named their final line.[2]

Davies enjoyed other natural advantages of position. A swamp adjacent to the Mobile and Ohio Railroad would protect his right flank; the guns of Batteries Robinett and Williams, which could easily range the west White House field, his left flank. Regardless of which route they chose, the Confederates thus would be exposed to protracted artillery and small arms fire while moving over nearly 800 yards of open, gently rolling ground.

Davies deployed his command well. He placed his First Missouri field artillery batteries so as to rake the east White House fields, command the Memphis road, and deny the enemy the wooded approach between the road and the fields. After losses at the old Confederate breastworks and subsequent detachments, Davies had on hand eleven cannon. He sent the three twenty-pounder Parrotts of Richardson's Battery D into the timber west of the Memphis road. The one ten-pounder Parrott and one twenty-four-pounder howitzer left from Welker's Battery H he placed in the woods east of the road. Brunner's two-gun section of Battery I returned from its morning duty with McArthur and with Green's four ten-pounder Parrotts went into battery in the east White House fields.

The infantry fell in behind the artillery, along a ridge of dead timber at the southern edge of the field, in the same order they had formed behind the breastworks that morning. Hackleman held the right. The Fifty-second Illinois settled in with its right flank touching the swamp near the railroad. The Second and Seventh Iowa regiments completed the line to the left. Oglesby covered the Memphis road on the division left, with the Twelfth Illinois deployed in the southwestern fringe of the field, the Eighty-first Ohio thrown forward 200 yards to support the artillery, and the Ninth Illinois on the west side of the road.

Although General Rosecrans had yet to send reinforcements, Davies got some help from the timely arrival of the Union Brigade. Its members, however, were in no better shape than their comrades in the White House fields. They had marched eight miles to reach the battlefield. It was their first serious trek of the summer — the brigade had been on outpost duty since June — and men dropped out by the score. No more than half of the brigade — perhaps 250 men — were with Lt. Col. John Coulter when he reported to General Davies shortly before 3:00 P.M. The New Yorker bade Coulter hasten to the extreme left to protect his two batteries posted near the Memphis road. Coulter led his men into an open wood behind Richardson's battery, where, on a slight rise, they lay down in line.[3]

Davies would need every advantage the terrain offered. His division was on the brink of collapse. Hundreds of men were prostrated from the heat or sheer exhaustion. A hasty roll call revealed 1,211 officers and men present in Hackleman's brigade (including those of the Union Brigade), while Oglesby had a mere 576 of all ranks. The case of the Eighty-first Ohio is instructive. It began the day 218 strong. Thirty men were lost in the fight at the old Confederate breastworks, but only 112 answered roll in the White House field. Over the three-hour withdrawal the sun had struck down 71 soldiers.

The leadership was also failing rapidly. After shepherding his men into line, Col. Thomas Morton of the Eighty-first Ohio slunk away. Joseph Nelson of Company C caught a glimpse of his colonel's solitary retreat. He was "going to the rear afoot, leading his horse, resting his hand on the horse's neck in front of the saddle." Perhaps, Nelson speculated, Colonel Morton "considered it safer to retire quietly alone than to be encumbered with the regiment." Unable to face the prospect of another slaughter like Shiloh, Col. August Mersy drank himself into a stupor. When Charles Colwell came forward from town with several wagonloads of water, he found Mersy "so drunk he could hardly sit on his horse. I could hardly keep from laughing at him, he looked so comical." Mersy offered Colwell an unsteady hand. "Charley, I am so glad to see you alive," he slurred. Colwell thanked him and then looked to the distribution of his precious freight of water barrels. He passed among the men of the Ninth. "I had a fine talk with the boys and bid them all good bye or what few were left, for I did not know whether I would ever see them again."[4]

The soldiers on the line held out little hope as well. Everyone understood they were expected to withstand an attack by an enemy whose superiority had been made clear at the breastworks that morning. Up and down the half-mile-long line the Federals watched and waited. Some tried to make light of matters. A solitary Rebel shell hissed, screeched, and tore up the ground in front of Corporal Wright's company of the Eighty-first Ohio, "causing our prostrate line an immense amount of anxiety." Wright cast a glance at Sgt. John Mader, who lay directly behind their company commander. Mader moved his head behind the captain's foot. He looked at Wright, pointed to the foot, and said "It's got to go through *that* before it hits me."

In the minutes before the first shells fell, Cpl. John Bell of the Second Iowa contemplated the obscene contrasts in his surroundings. Said Bell,

We [lay] flat on our faces. Details are made to hurry off and fill the canteens of the various companies. The sun beats down with terrible force. A battery dashes up and takes position on our left. Just in front of us is an old log cabin, evidently occupied, and afterwards there is a legend in camp to the effect that in the cellar of that old cabin is a poor woman, alone. . . . As we wait, the silence grows oppressive. Little birds twitter in the trees about us and mark the only break in the terrible stillness, save the occasional low whispers of the soldiers as they lay quiet, with pale, determined faces, grasping their weapons and oppressed with the conviction that in all probability they will never again see the sunrise.[5]

*   *   *

Not until 2:00 P.M., when Davies was abandoning his second delaying position, did Rosecrans become convinced that Corinth — and not Bolivar or some other point in western Tennessee — was the real object of Van Dorn's offensive. With McKean's front quiescent and no Rebels evident east of the Mobile and Ohio Railroad, "it was pretty clear that we were to expect the weight of the attack to fall on our center, where hopes had been given by our falling back." Rosecrans neglected to convey his sudden certainty to his division commanders. With the exception of Davies's loan of Baldwin's brigade to McArthur, the division commanders operated independently of one another, as though the railroad tracks dividing their sectors were inviolate. Nor had Rosecrans made his presence felt. He remained at his headquarters until noon and transmitted no orders of importance. Since reporting to Rosecrans on his way to the Confederate breastworks that morning, Davies had heard nothing from the commanding general. Col. John Du Bois, an aide-de-camp to Rosecrans, appeared briefly before Price's first attack to tell Davies that reinforcements would be forthcoming, a promise that proved empty. Davies withdrew to the White House line on his own initiative and not as part of Rosecrans's alleged plan to fight a delaying action back to the inner fortifications.

Davies was genuinely perplexed, then, when shortly before 3:00 P.M. a headquarters courier handed him an order from Rosecrans that read, "For fear of a misunderstanding in relation to my orders, I wish it distinctly understood that the extreme position is not to be taken till driven to it." The dispatch did not say what that position might be. Davies might infer it to be the inner works of Batteries Williams and Robinett (which is what Rosecrans apparently intended); but not being privy to Rosecrans's thinking, Davies was left to guess at the meaning of his ill-tempered instructions. He pocketed the order and prepared to fight at the White House.[6]

The 2:00 P.M. circular order confused General Hamilton even more than it had Davies. He had heard from Rosecrans only once since morning, and what the commanding general had said hardly inspired confidence. When Rosecrans learned that Hamilton had occupied the breastworks, he assaulted him verbally: "I had no intention to have you occupy the whole front of the rebel entrenchments, but to straddle the Purdy road, covering it effectually, resting your right flank upon the works wherever it may happen to come and placing your left within supporting distance of Davies. . . . I would be glad to come out and examine that ground, but do not think you need it." Perhaps Rosecrans should have examined the ground. Had he done so, he would have discovered that Hamilton could not both straddle the Purdy road and support Davies.

Ignoring Rosecrans's tirade, Hamilton remained at the breastworks with his two brigades while, a half-mile to the west, the Confederates routed Davies. By the time Hamilton received his copy of the circular order that so perplexed Davies, the enemy was nearly a mile in his rear, west of the Mobile and Ohio Railroad. On his own authority Hamilton withdrew Sullivan's brigade a few hundred yards from the breastworks and changed its front to face west, but a mile-wide gap remained between Sullivan's left and the right flank of Davies's White House line. While Sullivan's skirmishers engaged in desultory firing with Rebel cavalrymen guarding Price's left, Hamilton mused over the meaning of the circular. "The extreme position was not understood by either Davies or myself, but probably meant an advanced position," he surmised incorrectly. "How we could be driven to it by an enemy in our front is difficult to understand."[7]

Hamilton's ruminations were interrupted by a second order, which served only to muddy matters more. It read, "The general commanding desires me to say to you not to be in a hurry to show yourself. Keep well covered and conceal your strength. The enemy will doubtless feel your position, but do not allow this to hasten your movements."

Hamilton had no idea what Rosecrans wanted him to do, nor could the bearer of the dispatch, Lt. Col. Arthur Ducat, shed any light on its meaning. Ducat was an able but arrogant officer who had been transferred from Ord's staff to that of Rosecrans before the battle. He had Rosecrans's confidence but no well-defined responsibilities; when he wrote orders on the commanding general's behalf, it was with the ornate but empty title of "Chief of Grand Guards and Outposts."

Rosecrans had sent Ducat to carry the enigmatic order in part to find out, as Ducat phrased it, "what General Hamilton was doing, or could do." Rosecrans's curiosity nearly cost Ducat his life. Riding up the Purdy road Ducat stumbled upon a Rebel skirmish line. The conclusion was apparent: Hamilton had lost touch with Davies and allowed the enemy to throw "a heavy flanking, well supported skirmish line" between his left and Davies's right. Ducat spurred his horse through the Rebel skirmishers.

When he reached Hamilton, Ducat had no interest in explaining the meaning of Rosecrans's order, if he even knew it. To Ducat Hamilton's duty was clear: he should "close in and attack the enemy's left flank" to relieve pressure on Davies. Hamilton declined to act on the advice of a mere staff officer. He wanted clear and precise instructions from Rosecrans himself.[8]

Ducat negotiated the Rebel skirmish line a second time. It was 3:00 P.M. The air resounded with the opening salvos of an artillery duel: Davies was engaged.

Ducat found Rosecrans in the rear of the White House. Rosecrans had ventured from his headquarters to issue instructions that would at last set in motion a coordinated response to Van Dorn's attack. He ordered McKean to fall back from Battery F to a ridge three-quarters of a mile to the southeast, where he was to redeploy so as to connect with the left of Davies's White House line.[9]

Oddly, Rosecrans did not communicate directly with Davies. He had not, however, ignored Davies's requests for reinforcements. Rather, until Joseph Mower's brigade arrived from Kossuth at noon, he lacked a reserve from which to draw troops. Rosecrans called for Mower's brigade at 1:00 P.M. An hour and a half passed before the aide entrusted with the order, Col. John Du Bois, handed it to Stanley; he had been unable to find Stanley or anyone who knew the general's whereabouts. The order was brief but emphatic: "The General Commanding directs you to send a brigade across on the Chewalla [Memphis] road through the woods by shortest cut. Reinforce Davies from your left. Close in conformity with that movement. You had better send Mower." Mower's men were near collapse, having marched thirty-four miles in twenty-four hours, but the order obviously brooked no delay. Stanley told Mower to move out at the double-quick. Col. William Thrush, commander of the Forty-seventh Illinois, pleaded restraint: "The men are already almost utterly exhausted; to move them faster will render them unfit for action." The roar of artillery drowned out Thrush's words of caution, and Stanley urged the men to march faster.[10]

Having attended to Davies's immediate needs and rectified McKean's front, Rosecrans was disposed to hear Lieutenant Colonel Ducat out. The sorry state of affairs east of the Mobile and Ohio Railroad startled Rosecrans, but he immediately embraced Ducat's suggestion that Hamilton change front and fall on the Confederate left rear. Until that moment Rosecrans had alternated between bewildered inaction and feckless improvisation while the tide of battle rushed toward town. Now an opportunity had been presented him to affect events decisively. His enthusiasm for Ducat's proposal was boundless, but his sense of obligation was minimal. In his report of the battle and subsequent testimony before the Joint Congressional Committee on the Conduct of the War, he not only claimed the idea for his own but also pretended it to have been part of a carefully considered plan for trapping the Confederates. Said Rosecrans to committee members in the august calm of a congressional chamber,

> The enemy pushed steadily in, and by 11:00 o'clock it became apparent that, instead of a feint, the enemy was in full force. About 3 o'clock in the afternoon he had been pressed into the wooded angle between the Memphis and Charleston and the Corinth and Jackson [Mobile and Ohio] railroads, and had advanced within range of the defensive line under construction. The opportune moment appeared now at hand, and I directed General Hamilton, whose division was on our right,

beyond the range of the enemy's operations, to face to the westward, and move on to the enemy's flank and rear.[11]

Of course this was self-serving fiction. Nothing had been apparent at 11:00 A.M., and the only one being pressed at 3:00 P.M. was Rosecrans.

The dark side of Rosecrans's genius—a tendency toward intolerance and impulsiveness—was showing itself under the pressure of battle. One of his most loyal corps commanders later suggested that Rosecrans issued far too many orders during combat. Critics took this observation one step further. Contemporary accounts suggested that Rosecrans was hampered by stuttering—that he stammered and faltered even while addressing soldiers during a review. In combat, if his defamers may be believed, his stuttering rendered him incoherent. Said New York *Herald* correspondent William Shanks, "I have known him, when merely directing an orderly to carry a dispatch from one point to another, grow so excited, vehement, and incoherent as to utterly confound the messenger. In great danger as in small things, this nervousness incapacitated him from the intelligible direction of his officers or effective execution of his plans."

Although William Shanks seldom had anything good to say about anyone, the order Rosecrans dictated to Colonel John Kennett to put in play his plan for Hamilton suggested there was more than a little truth to Shanks's allegation. It read, "Davies has fallen behind the works, his left being pressed in. If this movement continues until he gets well drawn in you will make a flank movement if your front is not attacked, falling to the left of Davies when the enemy gets well in, so as to have full sweep, holding a couple of regiments looking well to the Purdy road. Examine and reconnoiter the ground for making this movement."[12]

Ducat took the order, apparently without reading it, and set off again for Hamilton. He reached him at 3:30 P.M. For thirty minutes Davies's artillery had been trading salvos with Rebel batteries hidden in the woods north of the White House fields. Confederate infantry could be expected to pour from the timber at any moment. It was imperative that Hamilton attack the enemy's flank and rear before they drove in Davies. Yet Hamilton hesitated to obey. Rosecrans's order made no sense to him. "Falling to the *left* of Davies?" Explained Hamilton, "Now bearing in mind that Davies's division was to the left and in front of mine, if this order meant anything it was that my division should abandon its position on the right of the army entirely, and pass either to the rear or front of Davies in order to reach the place indicated, and would therefore have destroyed every possible chance of attacking the enemy in the flank, and would also have left the right of Davies exposed, and the way into Corinth open to the enemy." Hamilton flipped the sheet over and scribbled, "Respectfully returned. I cannot understand it." Ducat remonstrated in vain. He tried to explain the intent, but Hamilton declined to obey without written clarification.

A disgusted Ducat wheeled his horse and set off on the two-mile return ride. Written clarification of an error that, though unquestionably material, was so egregious as to be obvious: Rosecrans had said, or Kennett had written, "left" when he meant "right." Years later the mere recollection of Hamilton's inertia sent Ducat into a tirade: "The writer doesn't believe there is in the record of the war such a confession of weakness, and such a wretched attempt to bolster up a failure to obey instructions, and not to perform the clear duty of a general . . . to march with the instincts of a soldier to the music of battle as this one. Who, pretending to be a general, would not have found out for himself the situation, but this one?"[13]

18

No Time to Cool Off Now

Earl Van Dorn and his retinue came upon General Price at the junction of the bridle path and Columbus road a few minutes before 2:00 P.M. There, surrounded by parched and panting soldiers, they conferred. Price was for stopping. Van Dorn, determined to take Corinth before Federal reinforcements arrived, insisted the attack be pressed. They reached a compromise of sorts. The troops were allowed to rest half an hour more, but the attack would go on.

In the mile-wide expanse between the railroads, Price had six brigades concentrated. He made no change in alignment. Hébert's division held the left, with the brigades of Green, Martin, and Gates in the front line and Colbert's in reserve. Maury was on the right. Phifer and Moore were still in the first line, but Moore had drifted so far west as to be lost to the assault. Maury had reinforced him with two regiments from Cabell's brigade, the remainder of which trailed Phifer.[1]

The ground played havoc with the Confederate advance. In trying to negotiate the forest, Martin's men marched obliquely to the right instead of marching directly forward. They masked the front of Gates's brigade, which had lagged behind a few dozen yards, bringing the Missourians to an abrupt halt. While the troops on the front line sorted themselves out, Colbert, too, was forced to stop. Meanwhile Martin Green's brigade continued on the correct course, alone and unsupported.[2]

It took Green thirty minutes to shepherd his men across the 800 yards of forest between his brigade and the White House fields. Through briars, around trees and boulders, and over logs they marched at the double-quick until, catching a glimpse of open ground through the trees, Green paused to re-form his broken battle line. At 3:00 P.M. a single shot from a Yankee twenty-pounder Parrott whistled overhead, then the forest came alive with the crash of bursting shells and falling solid shot. "All of our eleven guns were soon at work, and poured a steady stream, staggering their advancing column. . . . The infantry essayed time and again to advance and . . . the most murderous fire on their column . . . kept the whole Confederate force at bay," reported General Davies proudly. Said his chief of artillery, Maj. George Stone, "Here . . . one of the most fierce artillery duels on record raged with the fury of desperation, the enemy being repulsed by the double-shotted guns of our batteries."

Davies and Stone badly exaggerated the effect of the cannonade. Green had no intention of exposing his men to canister in an open field; had he charged repeatedly as Davies claimed, there would have been scarcely a man left in the brigade to tell the story. Rather, Green held his infantry in the woods and brought up his own artillery, the two twenty-four pounders and two twelve-pounders of Capt. John Landis's Missouri Battery and the six guns of Capt. Henry Guibor. Landis unlimbered his cannon along the forest's edge, where Green intended his artillery to deploy. Guibor began to do the same, until a staff officer appeared with an order for him to advance his guns. To the astonishment of Landis and the dread of Guibor's men, Captain Guibor limbered up guns and led his Missouri battery into the White House fields. He halted on an open rise fewer than 400 yards from the Federal main line. Yankee skirmishers were only 100 yards off. They peppered the battery with a few poorly aimed rounds that knocked down just two horses, then scurried away before the Rebel cannoneers loaded their guns. Federal artillery threatened to do far more damage. As soon as their front was clear of friendly skirmishers, Batteries I and K of the First Missouri Light Artillery opened fire on Guibor.[3]

The first Federal shells fell short, bursting near a small cabin standing between the lines. Two women, waving a bed quilt and screaming with fear, rushed out of the cabin. Guibor's attention was called to their presence, but the safety of his battery demanded an immediate reply to the Yankee guns. "Fire away, there!" he shouted to his section leaders. The women ducked back into the cabin, just ahead of the blast.

Help came. Captain Landis went into battery beside Guibor, and from knolls at opposite ends of the east White House field, Missouri Federal and Confederate gunners traded salvos. In the timber near the Memphis road were five more Federal cannon, also manned by Missourians. To combat them General Price committed two Missouri batteries of his own: Bledsoe's four-gun battery and Wade's six-gun battery. By 3:15 P.M. the ground around the White House reverberated with the fire of thirty-one cannon.

For thirty minutes the shelling went on without either side yielding. Major Stone had sent his caissons well to the rear; six guns standing with their limbers on a knoll in an open field he judged a sufficiently grave risk. Twice the limbers were emptied of ammunition, and twice Major Stone supervised their reloading with ordnance carried to the knoll by two six-mule-team wagons from Corinth. When the ammunition failed a third time and the wagons did not appear, Major Stone reluctantly removed the cannon from the knoll.

General Davies watched them pass. "The artillerists filed slowly to the rear, men looking more like coal-heavers than soldiers, with perspiration streaming down their faces blackened with gunpowder, and the wounded horses leaving a stream of blood in the road." Davies had reached the limits of his patience. "The artillery had fired, of all calibers, over 1,500 rounds of artillery ammunition, and still no reinforcements had arrived and no attack made on the right and left flanks and rear of the enemy to support me," he complained in his report. "I again sent to General Rosecrans asking for reinforcements, telling him I feared I could not hold my position unless they were sent."[4]

Davies had little time to contemplate his fate. No sooner had the last of Stone's cannon cleared the White House fields than the Missourians and Mississippians of Green's brigade raised a cheer and swept into the open, past the batteries of Guibor and Landis. The Confederates edged to within fifty yards of Davies's supine Federals, who had yet to fire a shot. Succumbing to the tension, the Rebels let go the first volley. It passed overhead. Grateful for the chance to fight under nearly even odds, the Yankees rose to their feet and returned the fire. The effect was terrible. "We opened on them and their lines melted away into death and confusion as quickly almost as I can write it," recalled Lt. Col. John S. Wilcox of the Fifty-second Illinois. "Still they pushed ahead and returned our fire warm and heavy." Said Sgt. James Payne of the Sixth Missouri, on the receiving end of the Yankee fire, "They gave us a volley that left a line of gray where it struck." The Sixth ran into the cabin in the middle of the field, its yard unkept and overgrown with briars and brambles. The right companies glided around the yard, but those of the center and the left fell into confusion trying to negotiate the undergrowth. Sergeant Payne's Company A was caught in the yard when the Federals opened up. The company commander was killed and the senior lieutenant wounded in the first fire. The company began to unravel, as did those on either side. Officers darted to the front to show the way forward and were cut down at a dizzying rate. Four company commanders were shot. The regimental commander, Col. Eugene Erwin, had his foot sliced apart but stayed in the saddle. His lieutenant colonel was carried off with a mortal wound, and the major was shot through the head. The Sixth Missouri held on for perhaps fifteen minutes before giving back with the rest of the brigade.[5]

No one could live long, it seemed, in the east White House fields. But General Green ordered the brigade into the open again. This time the men edged forward, seeking cover behind scattered logs and boulders. The fight degenerated into a static exchange of volleys neither side seemed anxious to break.

Survivors struggled to find words to convey the frenzy of the scene. Wrote Sergeant Payne, "The rattle of musketry became a roar like the plunging of mighty waters. The combatants were not more than thirty yards apart and the smoke made a blue haze about them that rendered outlines indistinct. . . . All the animal in man was aroused. No one seemed to think of death; the ruling impulse was to destroy." Said Corporal Wright of the Eighty-first Ohio,

> The afternoon sun shone in the powder-stained faces of the men, the polished rammers glint and glisten amid the wreaths of battle-smoke. . . . I touch elbows on the left with a young comrade whom I had learned to love and respect for his many good qualities; a rebel bullet had torn its way along the side of his head, and I requested him to go to the rear; he responded, "I am not hurt bad enough yet!" A few feet away a comrade staggers from the ranks, drops his gun, sinks to the ground, and he is dead on the field of honor![6]

That blistering autumn afternoon yielded scores of acts as gallant as that of Corporal Wright's friend. The misery of the men can only be imagined—bathed in sweat, choking on acrid gun smoke, and breathing a swirl of blood, urine, and burning flesh. But in succumbing to the collective frenzy of the moment, they surmounted their agony, if only for a short time. Gun barrels became so hot the hands of the men blistered as they held their rifles to reload, and cartridges burst before they could be rammed home. The men swayed and stumbled in the ranks but fought on. When told that cartridges were exploding in the faces of his men and that their rifles might burst, Colonel Sweeny of the Fifty-second Illinois retorted, "Let them burst; there is no time to cool off now."[7]

Col. James Baker, commander of the Second Iowa, lacked Sweeny's phlegmatic streak. He certainly looked unflappable, sitting silently upon his horse, waiting for orders, but the stalemate tried his patience. Said a fellow Iowa officer of the thirty-nine-year-old lawyer turned colonel, "He had a stocky and vigorous form, a dark, or olive complexion, black hair, and dark, lustrous eyes. In personal appearance he was extremely prepossessing." Yet he had one peculiarity in court, which his friends attributed to modesty, that he could never overcome. Said an admirer, "He never attempted to address a jury or a public assembly without at first showing signs of fear. It could be seen in his pale face and in the nervous tremor of his hand." Perhaps it was that same hint of terror that led a lieutenant to approach Baker and suggest, "Colonel, let us charge the enemy." The lieutenant's appeal touched another of Baker's traits — an independence of character bordering on insubordination, and the Iowan called to his men to fix bayonets and charge.

Into the field they spilled at 4:00 P.M. In their path lay the Sixth Missouri. Fortunately for Baker the Missourians were in no condition to resist. All but four commissioned officers and six noncommissioned officers were dead or wounded. The color-bearer had quit the field after taking nine wounds, and Sergeant Payne followed him, his hand pierced by a bullet. With Payne's departure, his company, which had mustered thirty-two that morning, was reduced to six enlisted men. The Sixth wavered, as did the Forty-third Mississippi on its left and the Seventh Mississippi Battalion on its right. General Davies ordered his entire division to follow Baker's example, and Green's Confederates fled a second time.[8]

Baker missed the denouement of the charge; a bullet had penetrated his bowels. Also down was Oglesby. At the conclusion of the charge, with his aides absent on other errands, the portly Illinoisan had ridden forward to speak with Colonel Chetlain. A bullet stopped him in mid-sentence. It struck him below the left armpit, bore through his lungs, and came to rest against a dorsal vertebra. Chetlain summoned an ambulance and took charge.

Sadly, the triumph for which Baker and Oglesby seemingly had given their lives proved short lived. The enemy showed signs of renewed life across the Memphis road, where the Union Brigade had been fighting alone and badly outnumbered, and in the forest northeast of Davies's right flank. Fearing an envelopment, the New Yorker gave the command to retire.[9]

The Union Brigade had been having a hard time. Its 250 members, together with the left companies of the Ninth Illinois Infantry, bore the brunt of Brig. Gen. Charles W. Phifer's attack down the Memphis road. The five Federal cannon posted near the road slowed the enemy advance, which had been none too aggressive. When the artillery withdrew, the Union Brigade fell back with it, re-forming along the southern edge of the west White House field. Phifer's dismounted Arkansas and Texas cavalrymen left the road and lined up in the timber north of the field. They stepped into the clearing, said Corporal Soper of the Twelfth Iowa, in "two unbroken and continuous lines of battle, extending to the right and left as far as the eye could reach and far beyond our lines." The Union Brigade fought feebly. When it became clear Phifer's line extended beyond their left flank, the brigade fled across the Memphis road. But Phifer failed to press his advantage, permitting Davies to counterattack Green.[10]

It was 4:30 P.M., and the soldiers of Davies's division were played out. Those fortunate enough to have refilled their canteens before Green's attack had long since drained them. Few had more than a handful of rounds left in their cartridge boxes. Davies's sense of relief can well be imagined when he saw the first companies of Joseph Mower's lead regiment, the Twenty-sixth Illinois, break into the southwest corner of the east White House fields at the double-quick. Their timing could not have been better; at the very instant the Illinoisans appeared, Green's Confederates resumed firing from the woods. Only a few Rebel bullets carried the length of the field, but they were enough to scatter the Twenty-sixth. Davies was dumbstruck; for eight hours he had been calling for reinforcements, and now those that finally came broke at the first fire.[11]

General Hackleman, too, had witnessed the abrupt collapse of the Illinoisans, and in his fury he galloped across the field to rally them. He got no farther than battle line of the Eighty-first Ohio before a bullet found him. Corporal Wright had been watching Hackleman: "I saw him quivering in his saddle and slowly fall from his horse. [Pvt.] Calvin McClelland laid down his gun and stepping back took the mortally wounded general in his arms and gently laid him on the ground." Brigade command passed to Colonel Sweeny of the Fifty-second Illinois. The same ambulance that had retrieved Oglesby paused to pick up Hackleman, and the two horribly wounded Illinoisans — both personal friends of President Lincoln — were carried off the field together.[12]

Mower saw to it the remaining regiments of his brigade gave a better account of themselves. The Eleventh Missouri, which had comported itself so well at Iuka, settled into the woods behind the rise toward which

Hackleman's brigade was withdrawing. The Fifty-second Illinois stood in the field to the Missourians' front. Thanks to its mascot, the next of Mower's units to appear was the most celebrated regiment in the Army of the Mississippi. That afternoon, as it had on every occasion before, the Eighth Wisconsin went into action with a bald eagle roosting on a perch beside the regimental colors. The eagle was named "Old Abe," and in the Eighth to carry his perch was an honor greater than bearing the colors.

Abe had come to the regiment in August 1861, a few days after the men were recruited. An Indian had brought the young eagle to Chippewa Falls to barter. Two of the townspeople gave him a bushel of corn for the bird, then offered to sell it to Company C of the Eighth Wisconsin, which was recruiting at Eau Claire. The members chipped in a dime apiece to buy him, but before they could complete the transaction, a citizen of Eau Claire bought the eagle and made a gift of him to the company. Capt. John Perkins at first hesitated to accept such an odd member into his company but finally agreed to take him to the front. Caparisoned with red, white, and blue ribbons around his neck and a rosette of the same colors on his breast, Old Abe accompanied the regiment southward.

David McLain, one of the members of Company C to have offered up a dime for the bird, bore him into battle at Corinth. McLain was proud to have taught Old Abe to drink water from a canteen — "thus saving him much suffering on long, hot marches in a dry country" — but otherwise found the duty a mixed blessing. Army life had agreed with Old Abe, recalled McLain, and "by this time he was getting rather heavy. I think he weighed about twelve or fourteen pounds, and on hot days was very troublesome. When he got tired on his perch, he would want to fly to the ground to rest."[13]

Before Mower could bring up his last regiment, the Forty-seventh Illinois, the Rebels reappeared. Davies and Mower conferred and agreed to a passage of lines. The Eleventh Missouri would replace the Fifty-second Illinois, and the Eighth Wisconsin would substitute for the Seventh and Second Iowa regiments.

Mower's line proved too short to cover Chetlain's front, but Davies withdrew the brigade nonetheless; the men were simply too exhausted to fight any longer. The Union Brigade, it seems, was left in the west White House field to cover Cpt. Nelson T. Spoor's battery. Davies marched the remainder of the division back to the protection of Battery Robinett. Nearly crazed with thirst, the men attacked wagons laden with water sent from Corinth.[14]

* * *

The men of Mower's brigade had no time to contemplate their surroundings. The Eleventh Missouri and the Eighth Wisconsin were under fire the instant they marched over Davies's recumbent soldiers. Although they had been fighting for two hours, Green's Confederates drove
through the fields with a vigor that startled Mower's men, who were
nearly spent from twenty-four hours of constant marching.[15]

The Missourians and Mississippians had at last been reinforced. By
4:00 P.M. General Gates had extricated his brigade from the tangle with
Martin's errant command and moved up behind Green. He rested the
brigade 400 yards behind the fighting, in weeds four feet high, "under
cover of a ridge, where the minie balls whizzed over our heads in showers," recalled Lieutenant Colonel Hubbell of the Third Missouri. Landis's battery of twenty-four pounder howitzers played upon the enemy
from atop the ridge. That Green needed help was obvious to every man
in Gates's brigade, including Gates. But absent orders, he felt he must remain where he was. It apparently occurred neither to General Hébert,
whose whereabouts during the fight at the White House are unknown,
nor to Price to commit Gates to the attack. Not until Green applied directly to the Missourian to reinforce his left, where the Forty-third Mississippi was being cut to pieces by an oblique fire from the Fifty-second
Illinois, did Gates move. He set off through the woods at about 4:45 P.M.
intending to bring his entire brigade in on Green's left, not merely to
steady the Forty-third Mississippi but perhaps also to outflank the enemy
in the bargain.

Gates apparently never communicated his intention to Green, who
waited only long enough for Gates's lead regiment, the Second Missouri,
to deploy on the left of the Forty-third Mississippi before resuming his
assault. For a third time his brigade sallied forth into the east White
House fields.[16]

The addition of the Second Missouri to Green's command and the departure of Oglesby's brigade gave the Confederates a decided advantage.
Phifer renewed his attack west of the Memphis road simultaneously with
Green's third assault. He easily brushed aside the ragged remnants of the
Union Brigade, then made for Spoor's Second Iowa Battery, posted on
the road. The Confederate Second Missouri opened an enfilading fire on
the Federal Eleventh Missouri, which gave back slightly, and Mower
found himself threatened with a double envelopment.

He left the Eleventh Missouri to fend for itself and gave his attention to the threat Phifer posed. As fugitives from the Union Brigade streamed across the Memphis road, Mower called upon the Forty-seventh Illinois to meet the Rebel onslaught. Coming up late, the Forty-seventh had gotten entangled with the left companies of the Eighth Wisconsin, which were themselves trying to change front to meet Phifer. As soon as Lieutenant Colonel Thrush extricated his men, he ordered them into the west White House field, where, said an eyewitness, "the fire from the Confederate lines became so fierce that it seemed as though a magazine had exploded in their very faces."[17]

The Forty-seventh Illinois presented too narrow a front to challenge the whole of Phifer's brigade. But before the Confederates could avail themselves of their superiority in numbers, Lieutenant Colonel Thrush ordered a bayonet charge. Spoor's Battery contributed a steady fire of canister from the road. Thoroughly surprised, the Confederates gave way. Thrush fell, shot through the heart. Capt. Harman Andrews took command, and the Illinoisans kept up the pursuit for 250 yards before Phifer regained control of his brigade. Then the Illinoisans halted, and the Confederates counterattacked. "The battle took a turn and for a while seemed to hang in even scale," said Clyde Bryner of the Forty-seventh. "The men were falling like autumn leaves." Captain Andrews was killed, three company commanders were hit, and nearly 100 enlisted men fell, dead or wounded. The Forty-seventh was leaderless. Rebels swept around their left flank, yet the Illinoisans held their ground. Not until the enemy began to work its way behind the regiment did the surviving company officers lead their men from the field. "Terribly worn from their severe march and lack of water," said Bryner, the Illinoisans were happy to leave, stopping only when under the protective guns of Battery Williams.

On the brigade right the Eleventh Missouri fought for thirty minutes before retiring toward Battery Robinett, where the Missourians were reunited with the truant Twenty-sixth Illinois and Spoor's battery, which had gotten away intact.[18]

In the brigade center the soldiers of the Eighth Wisconsin gave only a marginally better account of themselves. Their leaders certainly set a poor example. The usual commander of the regiment, Robert Murphy, was still under arrest for evacuating Iuka. His successor, Lt. Col. Josiah Robbins Jr., proved a coward. Early in the action a spent ball bruised him in the stomach. Robbins allowed himself to be carried from the field, and regimental command passed to Maj. John Jefferson. He was a conscientious officer but had been sick with typhoid fever for several days and never should have left his cot. Gaunt and burning with fever, Jefferson was knocked senseless by a flesh wound in the shoulder. Capt. William Britton took charge and promptly had his horse shot out from under him.

Even Old Abe had more of the chicken than the eagle in him. At the first Rebel volley a bullet cut the cord holding Old Abe to his perch. Another clipped his wing, carrying away three quill feathers. Old Abe screamed, spread his wings, and flew a few feet upward, then along the line of battle. David McLain rammed the perch in the ground and chased after him. Overtaking Old Abe some fifty feet from his perch, McLain scooped him up and carried him back to his perch. But Old Abe would have no more of that. Said Capt. James Greene of Company F, "He hopped off his perch to the ground and ducked his head between his carrier's legs. All attempts to make him stay on his perch were useless. He was thoroughly demoralized."

So were the men of the Eighth. Hit while still deploying from column into line by Missourians who yelled "like so many screech owls or devils," the soldiers had to be coaxed into place by their captains. That they proved pliant was due to the efforts of the regimental quartermaster. Riding ahead of the regiment on its march to the White House, the quartermaster had found and knocked open a barrel of whiskey. He was ready when the men passed. "They had been on the run for several hours and were in a state of exhaustion," said Captain Greene. "As the boys marched by every man who wanted to dipped his cup or canteen in and took a drink." The effect of liquor on hot, tired, and thirsty men can be imagined.

The Wisconsin men were lucid enough to see they could not fight after the troops on both their flanks had fallen back. Confessed Captain Greene, "They broke and ran before the advancing Rebel charge—the carrier of the eagle picking him up and carrying him under his arm as fast as he could run. It was a new experience for us, for heretofore we had always been the victors. The regiment and brigade dissolved so quickly that it was impossible to see what had become of them." Captain Greene ran into the captain of the color company—Abe's protectors. He had with him no more than a dozen men. As they ran, the color-bearer was shot dead. The next man to grab the colors fell wounded. Captains Greene and Wolf picked them up and ran "with the enemy at our heels." The cannon of Battery Robinett drove off the Rebels, and the two captains brought in the colors.[19]

Lagging a bit behind his regiment, Cpl. James Bradley of the Third Missouri Dismounted Cavalry caught a glimpse of General Green a few moments after Mower's brigade collapsed: "This writer will never forget the appearance of General Green as he rode, smiling, along the rear of the brigade that evening just after a desperate charge of his men had driven the enemy flying to their entrenchments. He had a revolver in each hand, and his hands were black with powder, showing that he had, indeed, been in battle. He actually led the charge of his brigade."[20]

Green had reason to feel pleased. At 5:00 P.M., while his soldiers paused to catch their breath at the southern edge of the forest below the White House fields, the Virginian surveyed the ground before them. Four hundred yards to the southwest was Battery Robinett. Only one of its three twenty-pounder Parrott guns was trained to cover the field Green would have to cross, should he press on toward town. The entire complement of guns could be withdrawn from the redan and trained on the field, but a simultaneous attack by Phifer down the Memphis road against Robinett would prevent such a measure. With the cannon of Battery Robinett directed elsewhere, Green would have only to negotiate a quarter-mile of cut timber to reach the outskirts of Corinth. The trees had been chopped down hastily and left where they fell. The Federals called the clutter of felled trees and stubble of stumps an abatis, but it was a poor obstacle that would scarcely serve to slow a determined charge.[21]

On the Federal side of the field, a mere 300 yards from the junction of the Memphis and Charleston and Mobile and Ohio Railroads, all was confusion. Davies's troops were beyond regrouping for the night. General Stanley had ridden into the field to meet Mower's men as they spilled rearward, but they also were too tired and thirsty to heed his call to rally. Captain Spoor had his battery resting behind Robinett, but the Iowan's caissons were almost empty. Because Rosecrans had chosen to wait until Mower broke to summon Fuller's brigade, which rested a mile and a half away near Corona College, there was no help immediately in sight. Green, on the other hand, had Phifer's brigade — a bit unsteady but still capable of fighting — on his right and, by 5:00 P.M., Gates's comparatively fresh brigade coming into line on his left. Also, William Cabell had reported in with the two regiments and two battalions left him after Maury had detached the Nineteenth and Twentieth Arkansas regiments to reinforce Moore earlier in the afternoon. Cabell's infantry, tired from a day of marching but unbloodied, fell in behind Phifer, ready to support an attack. Green's own command had lost heavily, especially in field-grade officers, but the Virginian never doubted for a moment that his brigade was up to one final assault. "So far as my brigade was concerned I could have gone into the town. There was nothing in the way."

With Fuller only then receiving orders to move, the Confederates had been handed thirty minutes of daylight with which to take Corinth.[22]

*   *   *

Mingling with the soldiers who had captured the White House fields, General Price saw matters quite differently than did General Green. The ubiquitous Major Tyler was at Price's side and witnessed the same spectacle. Exhaustion was everywhere evident. Wrote Tyler,

> The day was now far spent and, the excitement of the contest subsiding, scores of our men sank down exhausted by heat and desperate for water. General Price rode among them, sympathizing with their suffering. He himself had been hit in the left arm by a shell fragment; though not serious, the wound was ugly and painful. Wherever Price went, cheers followed him. The wounded threw up their caps and even the dying waved to him their hands. He dispatched his entire body guard with canteens for water, and sent off courier after courier to hasten up surgeons, ambulances, and restoratives.[23]

It was not the fatigue of his men that led Price to caution against further attacks, but the unaccountable failure of Lovell to do more. For two hours a strange — and, to Price and his heavily engaged subordinates, infuriating — silence had reigned west of the Memphis and Charleston Railroad. Price was convinced that "with a cordial support from General Lovell's command we would have carried their works and held them." Not knowing what had become of Lovell, Price concluded "it prudent to delay the attack on the town until the succeeding morning." He told Van Dorn as much when the Mississippian met him on the front line a few minutes after 5:00 P.M. Maury joined the gathering. He was for pushing on: "After a hot day of incessant action and constant victory, we felt that our prize was just before us, and one more vigorous effort would crown our arms with complete success." Van Dorn agreed and wanted to storm the town at once. Price dug in his heels. Without the support of Lovell, he doubted that his men could do the job. "I think we have done enough for today, General, and the men should rest," Price told Van Dorn. Maury agreed that nothing would be lost waiting until the next morning. Van Dorn acquiesced, and Price's frontline brigades stood down. Commanders perfected their lines and looked to their wounded, and the soldiers sank to the ground to sleep.[24]

Discouragement ran high. Some already sensed that their sacrifices might have been in vain. Ephraim Anderson of the Second Missouri expressed what seems to have been a common sentiment, at least in Gates's brigade: "If the works had been immediately charged, we could easily have carried them. Their present force had been defeated during the day's engagement, must have been considerably demoralized, and a rush upon them promptly would have left but little time to make proper dispositions."[25]

On the balance Price was probably prudent to counsel a halt. The exuberant hopes of Maury and Green notwithstanding, it is a military axiom that a hard-fought victory can leave a command as disorganized as a defeat, and the brigades of Phifer and Green certainly had done their share of fighting.

Matters were far different on Lovell's front. During the two hours that Moore had battled McArthur and Crocker, Lovell had stood idle. When he at last moved, it was only to make contact with Moore's command. Lovell stopped on the ridge near Battery F and told Moore to wait with him there until he said otherwise. Thoroughly disgusted with Lovell's indolence, Moore looked for an excuse to return to his own division. He found it at dark. "While waiting for a notification from General Lovell to advance, which he said he would give when ready, we received orders from General Maury to rejoin the division and take position on Phifer's right." Moore obeyed with alacrity.

Lovell's listlessness angered his own brigade commanders as well. Albert Rust, who had strenuously objected to the attack on Corinth, now was all for continuing the assault: "We had come much nearer achieving success than I had hoped for. I believed at the end of the first day's fight that the place was nearly taken." He did not see any reason why Lovell should not move on the enemy's inner works at once; Rust's men, at least, were in "first-rate fighting condition."

General Bowen agreed that Corinth could have been carried. Lovell had missed two opportunities to do it. The first, thought Bowen, came when Crocker withdrew: "The enemy's center was broken near the railroad. I saw it retiring in confusion, pursued simply by a line of skirmishers. If Lovell's division had moved directly forward we could have entered pell-mell with them into town." The second came after Mower's retreat. "If the line had formed within an hour and the advance made directly upon the center I think the place would have fallen."

Bowen's men chafed with him. Recalled Lieutenant Holmes of Caruthers's Mississippi Battalion, one of the few units in Lovell's division to have seen any serious action: "Corinth could have been taken in one more hour's fighting had our division marched down the Memphis and

Charleston Railroad track. We lay quietly for orders to attack, but no such orders came."[26]

Lovell never gave a reason for his inaction. Perhaps to deflect criticism from himself, Lovell afterward said publicly that he had opposed the whole operation from the beginning. As despicable as was Lovell's duplicity, Van Dorn must bear a share of the blame for his inertia. Throughout the day Van Dorn had stuck close to the center of Price's corps, so near the action, in fact, that after the capture of Oliver's hill he was largely unaware of events on the right. He knew nothing of Moore's fight with McArthur near the entrenched camps or his subsequent engagement with Crocker. Not until he had abandoned the notion of attacking Rosecrans's inner works did Van Dorn communicate with Lovell, and then merely to tell him to feel his way cautiously "to develop the positions of the enemy."[27]

Van Dorn's ignorance of affairs beyond Price's narrow front nearly had tragic consequences. Neither the Mississippian nor his staff had an inkling that an entire Federal division lay in the deep forest east of the Mobile and Ohio Railroad, nearly a mile behind them. Van Dorn's left rear was completely exposed, with no one prepared to protect it.

*   *   *

Colonel Ducat ran the gauntlet of Rebel skirmishers and found General Rosecrans where he had left him, in the field of fallen timber near Battery Robinett. It was well past 4:00 P.M., and the pressure on Davies was extreme and growing. Rosecrans was beside himself. He listened with thinly veiled fury while Ducat related both Hamilton's recalcitrance and his own efforts to clarify for him the meaning of Rosecrans's order for a flank attack. Yes, Rosecrans assured Ducat, he had interpreted the order correctly for Hamilton. Flipping the paper over he scribbled on the back, below Hamilton's response: "Ducat has been sent to explain it." Ducat suggested that new, unequivocal written instructions, as well as a sketch of what the commanding general intended, might be appropriate. Rosecrans agreed. Ducat drew the diagram, and Rosecrans signed the new order himself. Having lost faith in Hamilton's judgment, Rosecrans was careful to address every possible exigency — or excuse for noncompliance. The order read, in part, as follows: "Rest your left on General Davies and swing round your right and attack the enemy on their left flank reenforced on your right and center. Be careful not to get under Davies' guns. Keep your troops well in hand. . . . Do not extend too much to your right. It looks as if it would be well to occupy the ridge where your skirmishers were when Colonel Ducat left by artillery well supported. . . . Use your discretion. Opposite your center might be better now for your artillery."[28]

Ducat started on his fifth ride of the day. He had both the route and the location of the enemy skirmishers memorized and had reduced the trip to fifteen minutes. Nevertheless, Rosecrans dispatched two orderlies with similar orders in case Ducat did not make it.

Ducat reached Hamilton at 5:00 P.M. The two orderlies were killed en route. Hamilton found the order both verbose and insulting but at least agreed to obey it. Ducat stuck with him to ensure the order was carried out correctly.[29]

It was not. Error followed error with a darkly comic inevitability. Hamilton decided he could not use his artillery as Rosecrans had envisioned because "the ground was too uneven and the forest too dense." Ducat concurred but suggested that Hamilton at least send forward one battery with Sullivan's brigade. Hamilton relented and ordered the Sixth Wisconsin Battery to support Sullivan as closely as the ground allowed.

Hamilton summoned Sullivan to explain to him the plan of attack, but the Indianan was in no condition to listen. What was wrong with him is uncertain. His senior regimental commander, Colonel Holmes of the Tenth Missouri, said he seemed "very much exhausted and barely able to keep his saddle." Sullivan himself said he had received a severe contusion from a large splinter that unfitted him for further duty that day. Hamilton accepted Sullivan's explanation and excused him.

Several minutes of rapidly receding daylight were lost in tracking down Colonel Holmes. The Missourian reported to Sullivan before the general quit the field, acquainted himself with the situation quickly, and moved to the attack shortly before 5:30 P.M. Hamilton and Ducat watched the brigade step off into the timber east of the Mobile and Ohio Railroad, satisfied that Holmes had matters well in hand.[30]

Brig. Gen. Napoleon Bonaparte Buford, on the other hand, was doing his best to prove himself unworthy of his distinguished name. He had intended to obey Hamilton's order that he wheel southwestward across the railroad, but a band of Armstrong's roving cavalry and a single cannon firing into his right flank while he was changing front distracted him. Rather than form alongside Holmes, he marched off after the Rebel troopers and the gun; that is to say, instead of wheeling to the southwest, he moved toward the northwest. Ducat noticed the movement first and asked Hamilton if Buford's brigade "was going to Bolivar or to attack the enemy." Hamilton sent a courier galloping after Buford to recall him, but the fifty-five-year-old Kentuckian was determined to chase chimera. To the messenger bearing Hamilton's summons, he replied, "Tell General Hamilton, the enemy is in my front and I am going to fight him." With a half-mile gap between Buford's left and Sullivan's right, Hamilton postponed the advance until Buford returned.[31]

Writing after the war, Hamilton expressed deep regret that Buford's detour had delayed the flank attack. That feeling apparently was the inspiration of memory. Just before Buford set off, Hamilton had let slip to Colonel Sanborn, with whom he was on intimate terms (and whom he probably would have preferred to have seen still in command of the brigade), his honest appraisal of the situation. It is a confession that goes far to explain Hamilton's halfhearted actions that afternoon and, as such, bears retelling in full. Recalled Sanborn,

> At about four o'clock in the afternoon, when the command was out about three miles on the Purdy road, and the enemy's line of skirmishers appeared in front, General Hamilton, in confidence, informed the writer that he saw no way of saving the position at Corinth; that the enemy's center was near the town and our depots; that his lines extended across the road by which we marched out to our position — which, in fact, was our rear — and that he supposed that the army would retreat during the night and would try and cross the Tennessee at Pittsburgh Landing and try and effect a junction with Buell's army in northern Tennessee or Kentucky, and that in that event my force must act as rear guard and fight and hold the enemy as long as possible at all available points. This was a thunderbolt. I had formed no idea of the seriousness of the situation. I went into action feeling that all was lost except the army, and that we must fight with desperation to save that.[32]

So Sanborn led the Fourth Minnesota in at 5:30 P.M., determined to fight to save the army from what Buford feared was the "much larger concealed force" before his brigade. His first contact with the enemy seemed to bear out Buford's fears. As the Minnesotans marched through an overgrown field toward a heavy belt of timber near the railroad, they were hit by an enfilading volley from the trees. Sanborn looked to his skirmishers, who were scattered in the field 100 yards in front of his right. Their commander, Capt. Robert Mooers, beckoned to Sanborn with his sword as if he had something important to tell him. Sanborn galloped toward the captain. He had ridden but a few yards when Mooers collapsed, shot through the head. From the course of the bullets, Sanborn guessed that Mooers had meant to tell him that the enemy was passing around their right flank. Sanborn quickly conveyed his impressions to Buford, who ordered him to change front to the rear and charge the enemy. Sanborn executed the movement flawlessly, drawing to within 150 paces of the enemy's concealed line, when the "fire increased to a perfect shower of balls." Sanborn screamed, "Forward 150 paces, double-quick!" The regiment rushed ahead with a shout. With fifty yards to go, the Minnesotans watched the now visible enemy flee "to the rear with the greatest precipitancy." The Minnesotans paused to deliver two or three parting volleys, then fell back to rejoin the brigade. When the toll was taken, it was found that just five men beside Captain Mooers had fallen.

Sanborn's was the only regiment from the brigade to encounter opposition. Satisfied that whatever threat had existed to his flank had been cleared away, Buford at last relented and took up his proper place beside Holmes. But his grudging cooperation came too late. Said Hamilton, "A precious hour had been lost, the sun had gone down, and the attack having to be made through a forest of dense undergrowth, it was too late to execute the flank movement with any chance of success. The enemy's fire on Davies's division had ceased." Hamilton called off the attack.[33]

But Holmes had gone ahead. While Ducat and Hamilton were trying to draw in Buford, the Missourian had marched to the Mobile and Ohio Railroad. There he captured eighty-two startled Rebel skirmishers, one of whom confessed that the Confederates' left flank lay in tangled woods 150 yards beyond the railroad. A sudden blast of solid shot and canister confirmed the prisoner's confession. The railroad track ran along a five-foot-deep cut, and Holmes and his regimental commanders waved their men into it. The cut "formed a good shelter," said Maj. Nathaniel McCalla of the Tenth Iowa, "their balls passing over our heads, many of them lodging in the opposite bank so closely had they raked the ground."[34]

Holmes had stumbled upon the batteries of Landis and Guibor. Deprived of infantry support by the advance of Gates and Green, the cannoneers had to fend for themselves. Recalled Sam Dunlap of Landis's battery, "A body of the enemy appeared on our left about two hundred yards distant and we could almost count the buttons on their coats. We had to think fast and act likewise, immediately wheeled the guns to the left at the same time pouring a heavy fusillade of grape and canister into their ranks." The Federals disappeared into the railroad cut, "and with greater confusion than they made their debut, we rent the air and made the woods ring with shouts of our little success."

Hunt Wilson of Guibor's battery told a similar story. As the Federals drew near the railroad, "one of the liveliest cannonades took place that I ever witnessed. We cut our shells at one and a half seconds and began blazing away. The effect of our shells could be plainly seen. Their line halted and then began to waver." Unlike Dunlap Wilson did not cheer. Both batteries were still unprotected, and Wilson was convinced that a determined rush by the Yankees would overwhelm them: "If that line had opened fire and charged—we were in easy rifle shot—they would have taken the last one of us." But no attempt was made. It was growing dark, and the Federals seemed unaware that they faced only artillery. Major McCalla contemplated a sortie against a battery—probably Wade's—that had unlimbered in front of his right flank to enfilade his line, but it was recalled the instant his men rolled over the railroad embankment. It was just as well, because Price and Maury had used the time bought by the artillerymen to bring up Colbert's and Cabell's brigades. They formed a hasty line of battle fronting the railroad that closed the way to the Confederate rear. Desultory skirmishing continued until nightfall.[35]

Hamilton's abortive flank attack was thus over before it had really begun. Far from destroying Van Dorn, as Rosecrans had hoped, Hamilton caused no more than "a great commotion." "Had the movement been executed promptly . . . we should have crushed the enemy's [left] and rear," Rosecrans wrote two decades later. Time had not mellowed his fury. "Hamilton's excuse that he could not understand the order shows that even in the rush of battle it may be necessary to put orders in writing, or to have subordinate commanders who instinctively know or are anxious to seek the key of the battle and hasten to its roar." Rosecrans declined to confess that he had not discovered "the key of the battle" until mid-afternoon or—to Davies's way of thinking—ever "hastened to its roar."[36]

19

The Men Would Do All They Could

In the blue dusk General Davies shuffled among his troops like a night-
mare-racked somnambulist. His division was all but destroyed. Nearly a
third of the men of Oglesby's and Hackleman's brigades were lost: dead,
wounded, missing, or incapacitated by the heat. All three brigade commanders were wounded—Baldwin lightly, Oglesby and Hackleman presumed mortally. Davies made no effort at restoring order that evening.
The men begged for water, and he had water hauled to them in barrels.
They demanded food, and he saw to it pork and crackers were issued. They
begged for whiskey, and he turned a blind eye while commissary ser-
geants confiscated liquor from sutlers and brought it to the front line.
Buckets of whiskey passed among the soldiers, but most—knowing the
morning would bring a renewal of the battle—drank sparingly.

Night came, clear and cold. The temperature plummeted quickly after
dark. Campfires were prohibited. Opposing skirmishers rolled into ditches
or crouched behind tree stumps and traded shots. The moon rose, full
and bright—"the brightest moon I ever saw," swore Samuel Byers of the
Fifth Iowa—and the stars gleamed with an incongruous luster.[1]

Davies wandered off to see Generals Oglesby and Hackleman. He
found them lying in the ladies' parlor on the ground floor of the Tishomingo Hotel. The lightly wounded Colonel Baldwin sat nearby. Above
them the rickety ceiling creaked and moaned from the weight of hundreds of wounded soldiers and scores of moving surgeons and orderlies.

Archibald Campbell, the army's medical director, had not intended to use the hotel as a hospital. At daylight he had appropriated a nearby commissary depot for his surgeons. It was large and well built and lay in a slight depression, which made it the safest place in the vicinity. Campbell put a quartermaster detail to work and by mid-morning had the depot filled with medicines, surgical instruments, cots, and buckets of water. As the fight grew in fury, Campbell knew he would need more space. He took over the Tishomingo Hotel and hastily made it ready. By the time General Davies arrived, the wounded were spilling over into yet a third building, the Corinth House.

Baldwin mumbled to Davies that his torn hand had left him weak and nauseated. Davies probably paid him scant attention. When Davies arrived, General Hackleman was taking his last, gasping breaths. Surgeons had stanched the bleeding from his neck but could do little else, and Hackleman died in Davies's presence. General Oglesby writhed about on his cot "in the most excruciating pain." Davies remained with him a moment, then left the room. He made his way down the narrow corridor choked with the dead and dying of his division, stepped outside, and, tired and angry, set off for army headquarters.[2]

The less illustrious, too, had their visitors. Colonel Hancock of the Fourteenth Wisconsin went to see Capt. Samuel Harrison, a brave company commander who had been one of the first Union officers to fall. "I remained one shot too long," Harrison told him. Hancock visited him every night until his death two weeks later.

Charley Colwell parked his wagons and entered the commissary depot, where the wounded of the Ninth Illinois had been brought: "Some were dying, some were groaning with pain of wounds, while others were talking and laughing. The doctors were all busy amputating their limbs and dressing their wounds." Colwell snaked among the cots and operating tables, looking for his friends. A familiar voice called out, "Say, there comes Charley." Another answered, "I wonder if he will come down here. . . . Be sure he won't leave 'til he sees us all."

Colwell spoke with each man. One begged, "Charley, I am so dry, won't you bring me a drink of water." Colwell passed water among them, as he had that afternoon near the White House. "I shook hands with all of them, covered them all and fixed their wounds as well as I could," recalled Colwell. "They all seemed like brothers to me." At midnight he left to lay down in an ambulance for a few hours of fitful sleep.

Sam Byers of the Fifth Iowa slipped away from his bivouac and hurried to the Tishomingo Hotel to see a lieutenant from his company who had been shot through the chest. "Never will I forget the horrible scenes of that night," he wrote. "The town seemed full of the groans of dying men. In one large room of the Tishomingo House surgeons worked all the night, cutting off arms and legs. . . . I saw the floors, tables, and chairs covered with amputated limbs, some white and some broken and bleeding. There were simply bushels of them, and the floor was running blood."

Byers's friend was beyond help. Death would come shortly. "Go back to the regiment," the lieutenant told Byers, smiling weakly, "all will be needed." Byers obeyed. "It was a relief to me to get back into the moonlight and out of the horror, yet out there lay thousands of others in line, only waiting the daylight to be also mangled and torn like these."[3]

*   *   *

Neither Van Dorn nor Rosecrans gave a thought to sparing the living their chance at agony, and Van Dorn, at least, considered renewing the fight by moonlight. Generals Bowen and Rust certainly hoped he would allow them to atone for their inactivity that afternoon by ordering a night attack. Rust acknowledged the hazards of doing battle at night but believed circumstances warranted the risk: "I think the enemy were whipped that night, and I would have attacked with more hope of success before the enemy had received re-enforcements than after they were there."

Van Dorn gave the matter serious thought. He, too, was anxious to deny the enemy reinforcements, knowing that already he was waging an offensive against a foe of numerical parity. But the Mississippian, in a rare display of caution, rejected the idea on the grounds that his men were too tired and that in the dark he could not precisely locate the Federal lines.[4]

That Rosecrans intended to resume the struggle after dark is doubtful. The sole evidence he did came from the pen, and perhaps the imagination, of Hamilton. Writing twenty years afterward, Hamilton claimed Rosecrans directed him to attack at midnight over the ground Holmes had crossed. Hamilton said Ducat delivered the order at 9:00 P.M., an order Hamilton quoted in his article but that never made it into the *Official Records*. Hamilton said the order astounded him and that with much difficulty he talked Rosecrans out of it, thereby saving his division from almost certain destruction in the dark forest.[5]

It was a fine story from Hamilton's perspective; unfortunately no one came forward to endorse it. Ducat denied having delivered any order for a night attack. Rosecrans makes no mention of it in his report, and neither do Hamilton's own brigade commanders. Surely such an assault would have been coordinated with Davis and Stanley, yet both were silent on the matter.

Rosecrans was hardly capable of planning anything so complex as a night attack. He was tired and bewildered, certain only he was badly out-numbered—at least three to one by his reckoning. Such odds alone would have rendered a counterattack unthinkable. Rosecrans believed his only option was to consolidate his army behind the inner fortifications—that is to say, the five fortified artillery redans that formed an arc around the west and north sides of Corinth—and await the enemy's assault. Rosecrans maintained he arranged his divisions deliberately. "McKean's division was to hold the left," he wrote later,

> the chief point being College Hill, keeping his troops well under cover. Stanley was to support the line on either side of Battery Robinett, the three-gun redan surrounded by a five-foot deep ditch. Davies was to extend from Stanley's right northeasterly across the flat to the Purdy road. Hamilton was to be on Davies's right with a brigade, and the rest in reserve on the common east of the low ridge and out of sight from the west. Colonel Mizner with his cavalry was to watch and guard our flanks and rear from the enemy.

But Rosecrans's orders effecting the concentration were slow in coming and suggest he improvised his lines during the night.[6]

General Davies got no hint of the commanding general's purpose when he visited headquarters at 8:00 P.M.. In any case Davies was more in-terested in speaking his mind than in executing orders. Said Davies, "I re-ported to General Rosecrans and stated to him that the services of my three brigadier generals were lost, many of my officers were killed and wounded and the men worn out with fatigue, and that he must not de-pend upon my command on the following day, although the men would do all they could." Rosecrans raised no objection, nor did he mention the frontline role he claimed to have decided upon for Davies's battered com-mand. Rather, he told him to take his division to the east side of town, where Rosecrans assured Davies he would be held in reserve on Octo-ber 4. A grateful Davies returned to the front, roused his men, and got them on their way at 10:00 P.M.

McArthur partly filled the gap with his own and Oliver's brigades, both of which were under the Scotsman's command. Rather than guarding the army's left with his entire division, as Rosecrans said he intended for McKean to do, the incompetent old Pennsylvanian was being quietly shunted to make way for McArthur. He was left with only Crocker's four Iowa regiments to defend the ground around Corona College.[7]

In the narrow triangle of felled timber between the railroads, General Stanley deployed Fuller's Ohio brigade to fill the remainder of the line vacated by Davies. The last of Fuller's regiments had shuffled into town at sunset, and at 9:00 P.M. the reunited brigade marched from Corona College to its assigned position. Rosecrans and Stanley were on hand to greet the men and to hear their plaints for water. Most had marched all day along dusty roads without a drop to drink, and they were nearly as exhausted as the troops they replaced. Water was ladled out, along with a liberal ration of whiskey mixed with quinine, the surgeons' prescription for heat exhaustion.

Rosecrans watched Fuller deploy with great interest. The ground between the railroads, watched over by Battery Robinett and its three twenty-pounder Parrotts, commanded the direct approach to the railroad junction and, as such, was sure to be the fulcrum of the fight on the morrow. Stanley understood this, too, and so saw to the dispositions personally. Battery Robinett sat atop a low ridge that sloped most abruptly to the east. Stanley placed the Forty-third Ohio, which was led by Col. Joseph Lee Kirby Smith, whom many thought the best regimental commander in the army, near the crest of the ridge on the left side of Battery Robinett. Smith's line fronted west and ran from the cut of the Memphis and Charleston Railroad to the edge of the redan. Fuller, meanwhile, put the Sixty-third Ohio on the right side of Battery Robinett, fronting north. Its left rested on the Memphis road, thirty yards from the redan. The Twenty-seventh and Thirty-ninth Ohio completed the line in the direction of the Mobile and Ohio Railroad.[8]

Mower's brigade was behind Battery Robinett, where it had reformed after its late afternoon tangle with Green's Missourians. Stanley sent it, less two regiments retained as a reserve for Fuller, to cover the ground between Batteries Williams and Phillips, which McKean, left with only Crocker's brigade, was unable to fill. Mower's best unit, the Eleventh Missouri, Stanley held directly behind Battery Robinett. Even in the moonlight the Missourians could see it was an unenviable position, as the battle line on the crest blocked their field of vision. Stanley held the Fifth Minnesota, which had reached Corinth two hours earlier, on the northwest edge of town, 400 yards behind Fuller's right flank.[9]

Fuller and Smith certainly appreciated the gravity of their surroundings—the Rebels could be plainly heard felling trees and rolling artillery

into position less than four hundred yards to their front, and their skirmishers greeted the Ohioans with a fusillade — but they did not seem unduly troubled by them. The two were old friends, and Fuller asked Smith to make the rounds with him. They chatted pleasantly. Smith was his usual, cheery self. He "joked in low tones with as much unconcern as though the rebels were miles away," remembered Fuller. "Colonel," Smith said, "where did you get forage for your horses tonight? I don't know whether mine smells the battle afar off, but he keeps singing out 'Ha(y)! Ha(y)!' and I think he made a remark about oats."

His brother officers knew they could always rely on Smith for a kind word, a good joke, or — in battle — solid leadership. Brig. Gen. Alpheus Williams, under whom the twenty-six-year-old Ohioan served early in the war, came to admire him deeply: "He was my beau-ideal of a young man. . . . There was a daily beauty in his life that won the hearts of all who knew him. . . . He was so capable, so brave, so self-reliant without vanity, so patient and so persevering in the line of duty, that I have looked confidently — though not without apprehensions for his personal safety — for splendid services and rapid and well-earned advancement." While Fuller and Smith joked, their men drank their water and whiskey and then, too tired to throw up breastworks, lay down to sleep beside their muskets.[10]

It was midnight before Hamilton drew in his division. Even then he failed to assume his precise assigned position. He anchored his left on a redoubt of freshly turned soil named Battery Powell (also known as Battery Richardson) that contrabands had completed that day. Rather than face his entire division toward the northeast to complete the army's concave arc around Corinth, as Rosecrans wished, Hamilton cajoled the Ohioan into allowing him to place Dillon's Sixth Wisconsin Light Artillery and the Eightieth Ohio and Tenth Iowa regiments from Sullivan's brigade on a ridge to the right of Battery Powell, fronting west so as to command the Purdy road. The rest of the division formed behind and at right angles to Dillon and his infantry supports, which were thus abandoned with their flank "in the air." The Twelfth Wisconsin Battery went into position 400 yards southeast of Dillon, fronting north. Next came Lt. Junius MacMurray's Battery M, First Missouri Light Artillery. The Eleventh Ohio Battery, which Lt. Henry Neil had refitted after Iuka with surplus guns and men drafted from infantry regiments, unlimbered on what was to be the right of Hamilton's line. The two remaining regiments of Sullivan's brigade — the Fifty-sixth Illinois and the Tenth Missouri — filled in the ground from the left of the Twelfth Wisconsin Battery to the right rear of Davies's division. Buford's brigade deployed in support of MacMurray's and Neil's batteries. It took Hamilton until nearly daybreak to complete his preparations, and no time was found for his weary men to throw up breastworks. Laying aside gross incompetence, the most charitable explanation for Hamilton's unorthodox dispositions is that he intended to fashion a defense in depth, with the Eightieth Ohio and the Tenth Iowa to pivot rearward on Battery Powell and pass through the second-line regiments as the enemy neared.[11]

Though his hand was unsteady, Rosecrans kept in the saddle deep into the night, to give at least the appearance of guiding his army's deployment. He dismounted at 3:00 A.M. to catch perhaps three hours of sleep, but not before issuing one final order that deprived Davies's division of any chance at rest. As Hamilton moved into line, Rosecrans reconsidered the wisdom of leaving Davies in reserve on the east side of Corinth. The Ohioan decided to move him back through town to reoccupy part of the ground Davies had abandoned four hours earlier. Davies had the men roused, a task that alone took nearly an hour, and got them going at 2:30 A.M.

They filed past Rosecrans headquarters and moved into line between Battery Powell and the Mobile and Ohio Railroad. What remained of Oglesby's brigade, now led by Col. August Mersy, formed in reserve. Behind Battery Powell, Capt. Henry Richardson unlimbered the three twenty-pounder Parrott guns of his Battery D, First Missouri Light Artillery, and the single twelve-pounder howitzer and six-pounder gun of Lt. John F. Brunner's section of Battery I. Davies posted the remainder of his cannon on either side of Battery Powell. Col. Thomas Sweeny lined up Hackleman's brigade along the summit of a gentle slope that fell off toward the Purdy road, facing west. Davies inserted what remained of the Union Brigade on Sweeny's left, fronting north. Col. John Du Bois, an insufferable but competent Missourian who had been serving on the army staff, was detailed to command the 757 men left in Baldwin's brigade. With them he completed the line but lacked the numbers needed to reach the railroad, leaving a gap of 250 yards between the brigade left and the track. The ground immediately in front of Davies was flat and open. Strewn here and there were bits of an abatis. From 325 to 600 yards away the flat yielded to forest, in the depths of which the Confederates could be heard assembling for battle. Soldiers of the Sixty-fourth Illinois Sharpshooters felt their way into the woods; an unattached unit that had missed the first day's fight, the Sixty-fourth had drawn the unenviable duty of skirmishing the division front.[12]

General McArthur withdrew his brigade and that of Oliver to make room for Davies and returned them to the nominal command of McKean at Corona College. McKean placed McArthur on Crocker's left, between Batteries Phillips and Tanrath, and kept Oliver in reserve. At 4:00 A.M. on October 4, nearly ten hours after the fighting had sputtered out, Rosecrans at last had a battle line assembled to resist the attack that most were sure would come with the dawn.[13]

*   *   *

To the Confederate lines the night wind brought the rumble of wagons and artillery carriages and the tread of marching soldiers. What these sounds meant, no one could tell. Van Dorn speculated the Federals were evacuating Corinth. Unable to hear the noise from his bivouac, Price accepted Van Dorn's guess. Martin Green thought otherwise. "What made me doubt they were evacuating was the chopping of timber," he said; retreating soldiers do not fell trees. Nor did many of the men think the Yankees were departing. On the contrary, the sounds of moving wagons and marching men suggested to them that the enemy was receiving substantial reinforcements. Some convinced themselves they heard locomotives — presumably laden with troops — chug into town.[14]

Van Dorn had no means of confirming his suspicions. Opposing lines were too close to permit a reconnaissance of the enemy's position. Not that the lack of hard intelligence on Federal defenses troubled Van Dorn much. Careful planning was contrary to his nature. Van Dorn was a hostage to his own success; the gains of the first day made it impossible for a man of his temperament to consider withdrawing, even with the prospect of Federal reinforcements close at hand. He knew that Rosecrans had at least as many troops as he and that the Ohioan's lines would grow stronger the closer he was pushed toward town: "The line of attack was a long one, and as it approached the interior defenses of the enemy that line must necessarily become contracted," Van Dorn observed. Yet he chose to attack at dawn with everything he had.[15]

An artillery barrage at 4:00 A.M. would signal the attack. The three batteries of Dabney Maury's division were selected for the duty, with orders to run their cannon forward to a partially cleared ridge that overlooked Corinth, 500 yards northwest of Batteries Robinett and Williams. Hébert would begin the attack with his division. The Louisianan had performed miserably on the first day, issuing no orders to Gates or Green until sunset, when he told them to withdraw from their advanced position and regroup along the Memphis and Charleston Railroad. "The Missourians, and indeed all his division, were exceedingly denunciatory of the conduct of this officer, the many truculent remarks made not needing repetition here," remembered Lieutenant Colonel Bevier. But as his objective was to turn the Federal right, and as it was Hébert's division that lay opposite it, Van Dorn had to rely on Hébert to redeem himself.

Van Dorn's instructions were explicit: "Hébert, on the left, was ordered to mask part of his division on his left; to put Cabell's brigade en echelon on the left also, Cabell's brigade being detached from Maury's division for this purpose; to move Armstrong's cavalry brigade across the Mobile and Ohio Railroad, and if possible to get some of his artillery in position across the road. In this order of battle he was directed to attack at daybreak with his whole force, swinging his left flank in toward Corinth and advancing down the Purdy Ridge."[16]

To Dabney Maury and Mansfield Lovell went supporting roles. As Maury explained his instructions, "The orders given me were to charge the town as soon as I should observe the fire of the Missourians, who were on my left, change from picket firing to rolling fire of musketry." In light of Lovell's torpor of the day before, Van Dorn's orders to him were too discretionary. He was to form his division with two brigades in line of battle and one in reserve, his left flank resting on the Memphis and Charleston Railroad, and then either wait in place or feel his way slowly forward, as he saw fit, until Hébert was engaged. Once Hébert had attacked, Lovell was "to move rapidly to the assault and force his right inward across the low ground southwest of town."[17]

The thoroughness of the Confederate preparations varied with the character of their commanders. Price summoned Hébert upon receiving the attack order from Van Dorn; the Louisianan, however, neglected to brief his brigade commanders. Van Dorn informed Maury directly of his role. The Virginian drew in J. C. Moore's brigade and aligned it with that of Phifer behind the embankment of the Memphis and Charleston Railroad.

Beyond showing them a crude sketch of their sector, Lovell had little to offer his brigade commanders; he could not say whether the Yankee lines rested on two redoubts or one or whether they had been reinforced. Lovell detailed Bowen's brigade as the storming party but told him to stop for further orders after he came in sight of the enemy redoubt, or redoubts, as the case might be.

Van Dorn counted on a swift and vigorous attack to drive Rosecrans from Corinth before he could definitely be reinforced. That urgency he failed to instill in his subordinates. Maury could be relied on to do his part, but caution would again be the watchword on Lovell's front, and Louis Hébert, who was to direct the main attack, kept his own counsel that night.[18]

*   *   *

Of course Rosecrans had not been reinforced, but troops were on the way to him, and in greater numbers than Van Dorn or his lieutenants probably imagined. Communications between Grant and Rosecrans were problematic. Rebel troopers had cut the railroad and telegraph lines into Corinth, so that all messages between the two headquarters had to be sent by courier from Bethel, Tennessee. With the Confederate army astride the direct road to Corinth, couriers had to detour through Farmington, meaning that eight hours were needed to get a note from one general to the other. Worse yet, from the evening of October 3 until well after dawn on the fourth, there was no telegraph operator on duty at Grant's headquarters in Jackson to receive the messages that couriers brought to Bethel for transmission.[19]

Despite the paucity of information, by the afternoon of October 3 Grant knew Rosecrans was engaged. He had received Rosecrans's morning message relating Lovell's demonstration against Oliver, and scouts at Bethel had reported heavy firing from the direction of Corinth. On the basis of these reports, Grant scraped together units to reinforce Rosecrans. He directed his chief engineer, Brig. Gen. James McPherson, to march four regiments from Bethel "with all speed" to Corinth. Later that evening he ordered General Hurlbut to start for Davis Bridge no later than 3:00 A.M. on October 4 with those units of his division not on outpost duty and two regiments from Leonard Ross's command. Should he find the enemy retreating, Hurlbut was to destroy the bridge and contest their crossing of the Hatchie River. If the enemy were still threatening Corinth, he was to press on and strike them in the rear. "Rush as rapidly as possible," Grant urged him. Before retiring for the night, Grant penned an empty exhortation to Rosecrans: "General Hurlbut will move today towards the enemy. We should attack if they do not. Do it soon. . . . Fight!"[20]

20

Death Came in a Hundred Shapes

The early morning hours of Saturday, October 4, were unseasonably frigid, a cruel contrast to the unrelenting heat of the day before. Sweat-soaked uniforms sharpened the chill of the night. Few could sleep. Those not marching into new positions cradled their rifle-muskets and drew themselves tight to rest, but the ground was simply too cold; no sooner did a man doze off than he awoke shivering. Then there were the sounds of impending battle to distract even the most exhausted. In front of Battery Robinett the sounds were especially ominous. Shortly after midnight the bumping and scraping of artillery limbers replaced the shuffling beat of marching men. For three hours the soldiers of Fuller's brigade listened to cannon being wheeled into line along the edge of the forest 400 yards away.

The noise especially troubled Fuller, and he walked to his outposts with Colonel Smith to get a better feel for what it portended. Earlier General Stanley had cautioned the Ohioan to watch Davies's skirmishers, who were to cover his front, for out of the forest beyond them the Confederate attack was certain to come. To Fuller's amazement his outposts told him that Davies's skirmishers had crept away after Fuller formed his brigade. Pickets from Dabney Maury's division took their place — Fuller could see them plainly under the bright moonlight. Behind the Rebel pickets came the cannon.

Fuller took steps to dislodge them. He told Maj. Zephaniah Spaulding of the Twenty-seventh Ohio to take skirmishers "cautiously through the fallen timber, if possible, to gain and hold the edge of the woods." At the same time he sent two companies from the Sixty-third Ohio under Capt. Charles E. Brown to sweep the Memphis road as far as the forest.[1]

Three hundred yards away, on a small rise beside the White House, Lt. Thomas Tobin placed the four guns of Hoxton's Tennessee Battery. Nearby, Col. William Burnett, Maury's chief of artillery, helped align Sengstak's and McNally's batteries. Together these batteries would provide the predawn barrage Van Dorn had ordered.

Lieutenant Tobin had more courage than common sense. First he ran one gun forward of his infantry support to within 200 yards of Battery Robinett. Then he decided to reconnoiter a bit farther himself. With his bugler he walked off the ridge and down the Memphis road, straight into the leveled rifle-muskets of Captain Brown's Ohioans. Tobin and his bugler were hustled off to division headquarters, and Brown waved his men into tall grass beside the road, there to await daylight. Likewise intent on bringing down enemy artillerymen at dawn, skirmishers from the Sixty-fourth Illinois crawled forward on Brown's right.[2]

It was 4:00 A.M. Tentative streaks of orange fingered the eastern horizon. Above, the stars shined brightly. Here and there a few campfires, started by Union soldiers more troubled by the cold than the prospect of attracting enemy fire, flickered furtively. Suddenly the night erupted. Wrote a startled Colonel Fuller, "What a magnificent display! Nothing we had ever seen looked like the flashes of those guns. No rockets ever scattered fire like the bursting of those shells!" General Rosecrans had returned to his headquarters thirty minutes earlier to get some badly needed rest. The first salvo awakened him, and he was back in the saddle before servants could bring him breakfast.

The shells from Colonel Burnett's barrage rained down. They "were flying through the air like mosquitos of a summer night and I thought they were the prettiest thing a bursting in the air I ever saw, most especially when they did not come close," said Charles Colwell, in town with Davies's trains. The first shot landed in the bivouac of the Twenty-seventh Ohio. "Well do I remember that first shot," recalled Pvt. W. W. Adams, "fired from a rebel gun from the edge of the timber right in front of Robinett—a fuse shell that struck a stack of guns and knocked them down right close to me, scaring me out of my sleep." Adams sat up with a start and watched the burning fuse and black shell streak toward a group of soldiers huddled in the rear making coffee. It toppled their camp-kettle but left the men unhurt. They doused the fire and scampered away.

Most of the Rebel rounds passed harmlessly overhead, and losses on the front line, where the men lay down behind logs or shielding folds in the ground to wait out the storm, were light. But in the rear all was a maddening confusion. The lighted windows of the Tishomingo Hotel and a roaring bonfire started to warm the wounded left outdoors were inviting targets. Soldiers of the Fifth Minnesota ran from their bivouac to extinguish the blaze, but not before Rebel gunners got the range of the hotel. A solid shot crashed through a window and eviscerated a man being carried down the stairway on a stretcher. Outside, Charles Colwell watched a shot ricochet off the ground and strike a fleeing woman in the back. There was no one to give Colwell orders, so he took it upon himself to lead the train out of range. When he returned, after the firing had ceased, surgeons and orderlies were dragging wounded from the hotel. Colwell set about gathering ambulances to help carry them off.[3]

For nearly an hour Burnett's gunners had the field to themselves, while Federal infantrymen puzzled over the failure of Batteries Williams and Robinett to respond. Their silence was intentional. Lieutenant Robinett was waiting for dawn to betray the source of the cannonade. Capt. George Williams, commander of the army's heavy artillery and directly responsible for Batteries Williams, Robinett, and Phillips, also was loath to fire in the dark. The evening before, he had noticed a Federal field battery unlimber between Batteries Williams and Robinett; he feared firing in that direction until he was sure it had left.

By 5:15 A.M. the darkness had dissipated enough to disclose both the Rebel cannon and the departure of the friendly battery. Lieutenant Robinett opened on Tobin's exposed battery from in front. Captain Williams joined in with an enfilading fire of thirty-pounder Parrotts from Batteries Williams and Phillips. In less than thirty minutes the Confederate guns ceased firing. At that instant Captain Brown's Ohioans rose from the grass and opened a short-range fusillade on the crews of Hoxton's battery, who were scrambling to limber their guns. Sharpshooters from the Sixty-fourth Illinois joined in. Horses fell by the dozen, and five cannoneers crumpled. Brown charged. He overran a James gun, which was brought in by men from Battery Robinett, and captured a caisson and five artillerymen. Skirmishers from the Second Texas Infantry dashed forward to prevent Brown from penetrating farther into the woods, and Colonel Burnett was able to get the rest of his pieces away.[4]

*   *   *

The field fell silent. The sun rose at 6:00 A.M. in a clear and cloudless sky, burning away the dew quickly—a sure sign October 4 would be as hot as the day before. General Rosecrans rode the length of his lines, hoping to get in one final inspection before the enemy attacked. The soldiers hardly noted his passing; all eyes were trained on the forest in front. An hour passed. The mounting tension demanded a resolution. "If they're goin' to take us, why don't they come and do it in the cool of the morning? It'll be hot after a while," snarled a soldier in the Eighty-first Ohio. Another hour slipped away. Hungry Federals dropped their guard long enough to gulp down a hastily boiled cup of coffee and a few slices of bacon.

No one needed food and rest more than General Rosecrans. For forty-eight hours he had eaten little and slept less. His nerves had worn thin, and only the seeming certainty of an attack kept him in the saddle. When none came, Rosecrans grew impatient, and at 8:00 A.M. he told General Stanley to find out why the Rebels had not stirred.[5]

General Van Dorn asked the same question. When dawn came and went without a sound from the left, the Mississippian sent a staff officer galloping to Hébert to learn what had delayed his attack. But Hébert could not be found. Van Dorn dispatched a second staff officer, then a third. Like the first, they returned empty-handed. Finally, at 7:00 A.M., Hébert came to headquarters. He reported himself sick and asked to be relieved of duty. Van Dorn excused him, and Price put Brigadier General Green in command of the left wing.

Whether Hébert truly was ill and, if so, it was illness that had made him invisible on the field of battle the day before will never be known. Sick or not, he did nothing to redeem himself before reporting to Van Dorn. He had neglected to inform his brigade commanders of the plan of battle or the hour of the attack. When the courier bearing the order placing him in command found Green, he was breakfasting behind his brigade, which he had withdrawn 100 yards from the Mobile and Ohio Railroad so that his men might also eat unmolested by Yankee skirmishers. Only moments earlier a messenger from Hébert had brought him word of the Louisianan's illness. Green was a "kind-hearted, unostentatious man," thought Lt. Col. Columbus Sykes of the Forty-third Mississippi, but not really up to division command. Lieutenant Colonel Bevier agreed: "General Green, upon whom the authority then fell, was hopelessly bewildered, as well as ignorant of what ought to be done." He placed Col. William H. Moore in command of the brigade and set about familiarizing himself with the situation; in the time spent, wrote Bevier, "we lost a score of our most worthy comrades" to Yankee skirmishers.[6]

As the morning wore on, skirmishers on both sides stepped up their deadly game of cat and mouse, in part to ease the strain of waiting. Cpl. Ephraim Anderson of the Second Missouri lost a messmate in the exchange. Will Ray long had been convinced he would come through the war unscathed, but that morning he awoke with a cold premonition of death. At dawn he wrote a farewell diary entry to his wife, asking those around him to make sure the diary, his personal effects, and a lock of his hair were sent home after he fell. Ray was shot through the heart and died beside Anderson, who stuffed the articles in his jacket.[7]

General Stanley saw to the execution of Rosecrans's reconnaissance order personally. Five men from each company in the division were selected for the duty, making a force equal to two regiments. They moved into the field on either side of the Memphis road. Stanley chose Colonel Mower to lead those to the left of the road; he would direct those on the right side. Stanley had erred in picking Mower; the Vermonter had kept warm with a bottle during the night and now was deeply in his cups.

The long blue skirmish line started forward at 9:00 A.M. Stanley's band managed to reunite with Captain Brown's Ohioans, but sharp and accurate fire from the Second Texas Infantry brought a swift end to their reconnaissance. Mower, meanwhile, had veered to the left. He crossed the Memphis and Charleston Railroad and pursued pickets from Lovell's division, who seemed as loath to fight as their commander. Too drunk to distinguish his own skirmishers from the Rebels, Mower gave chase for a mile before his small party stumbled upon the main line of Villepigue's brigade. The Rebels opened up at close range, but in the deep forest their aim was bad. A scattered volley from behind his left flank caught Colonel Mower's attention; assuming the fire to be friendly, he rode foggily toward the source. Coming into a clearing, Mower found himself in the midst of a small party of Rebels. That sobered him, and he spun his horse and dug in the spurs. A bullet toppled the animal, and Mower hurt his neck badly in the fall. The Southerners hustled him away as his now leaderless skirmishers retreated.

Stanley's reconnaissance neither clarified the situation for Rosecrans nor provoked an assault, but it did add to Van Dorn's anxiety. Any more such aggressive Federal probing might derail his plan of attack and cause the battle to open spontaneously. Already the skirmishing on Maury's front was reaching a dangerous intensity, and it was with difficulty the

Virginian restrained his frontline brigades. He, too, was impatient: "For hours we listened and awaited our signal. . . . I have never understood the reason for so much delay."[8]

* * *

Martin Green was having a hard time. Upon assuming command he inspected his lines, which rested behind the embankment of the Mobile and Ohio Railroad. His own brigade, now led by Col. William H. Moore of the Forty-third Mississippi, lay to the left of Phifer's brigade of Maury's division. Gates's brigade rested on Moore's left. Col. Robert McLain, who had taken charge of the Fourth Brigade after Martin was wounded, held the division left. Colonel Colbert's brigade was in reserve.

Green's inspection was appropriate under the circumstances, and the delay engendered was excusable. But what Green did next was not. Perhaps because it had been so roughed up in the White House fight, he pulled his brigade from the front line and replaced it with Colbert's. No sooner had he done this than Colonel McLain complained that the much longer Federal lines would overlap his left flank. So Green reversed himself, ordering Colbert to move to the left of McLain and reinserting Moore in his original place on the division right. These peregrinations cost an hour.

Green did not know what to do with Cabell's brigade, which Van Dorn had assigned to Hébert the night before to support his attack. Green dismissed him with ambiguous orders: although Cabell was to hold his brigade within supporting distance of William H. Moore, he would be subject to orders from his own division commander, General Maury. The incompatibility of these instructions would fast become evident.

Green at least appreciated what his division was up against. He understood that the concave nature of the Federal defenses meant Colbert and McLain would have far more ground to traverse before making contact with the enemy (in this case Hamilton's division) than would Gates and Moore, who were directly opposite Davies's division and Battery Powell. Consequently he directed Colbert and McLain to "move forward en echelon, throwing their left forward, so as to come to a charge at the same time as the right." Just before passing the word to attack, Green remembered Cabell: "I sent for re-enforcements, believing that we would need them, for I could see the enemy had two lines of fortifications, bristling with artillery and strongly supported by infantry."[9]

"Finally, about 10:00 A.M., somebody concluded we had better charge, and the order was given," said Lieutenant Colonel Bevier. "With a wild

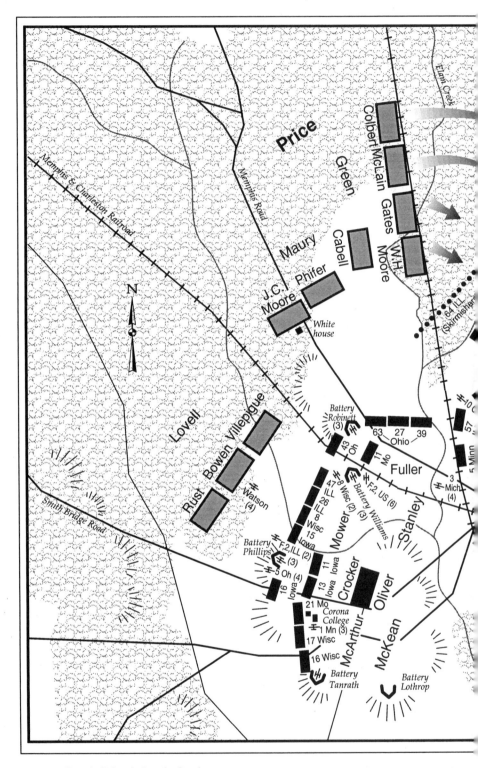

MAP 9. *Green's Belated Attack, October 4, 10:00 a.m.*

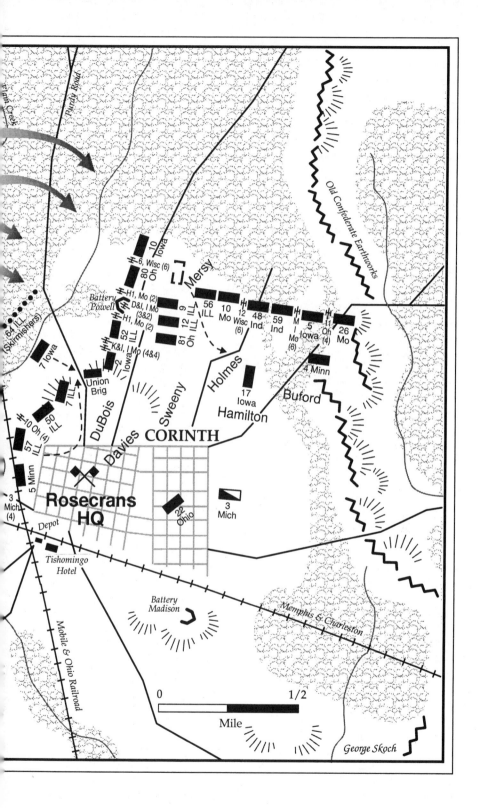

Tuka Creek

Purdy Road

64 ILL
(Skirmishers)

7 Iowa

7 ILL

50 ILL

10 Oh (4)

57 ILL

5 Minn

3 Mich (4)

Depot

Tishomingo Hotel

Mobile & Ohio Railroad

Battery Powell

H1, Mo (2)
D&I, I Mo (3&2)
H1, Mo (2)

52 ILL

K&I, I Mo (4&4)

2 Iowa

Union Brig

DuBois

Davies

Sweeny

Holmes

Hamilton

CORINTH

10 Iowa

6, Wisc (6)
80 Oh

Mersy

56 ILL

81 Oh ILL
12 ILL
9 ILL

10 Mo
12 Wisc (6)

48 Ind

59 Ind

M Mo (6)

5 Iowa

11 Oh (4)

26 Mo

4 Minn

17 Iowa

Buford

22 Ohio

3 Mich

Rosecrans HQ

Old Confederate Earthworks

Battery Madison

Memphis & Charleston

0   1/2

Mile

George Skoch

shout, our whole brigade jumped swiftly across the railroad, and charged toward the enemy's line," Lieutenant Colonel Hubbell of the Third Missouri recalled. Moore swept across the railroad on their right. But, as Bevier was shocked to see, they charged with their left in the air: McLain and Colbert had lagged behind and were lost to sight. And they attacked without a reserve. Before Cabell started, Maury diverted him to fill the gap on Phifer's left that Moore's advance occasioned.

"Stopping but a moment in the edge of the woods, to reform our companies, slightly disarranged by the fallen timber, our brave brigades pushed right ahead," continued Bevier. "The shot and shell from more than half a hundred guns crashed and whistled around us incessantly. No orders could be heard." [10]

None were needed. As the Missourians emerged from the forest, their objective was obvious: Battery Powell, alive with the flashes of a dozen cannon. Between them and the redoubt lay a half-mile of open ground and a dense curtain of blue-clad skirmishers. As he had Stanley, General Rosecrans had enjoined Davies to reinforce his skirmish line to feel out the enemy. A few minutes before 10:00 A.M. Davies dispatched the Seventh Iowa toward the left of the Illinois sharpshooters, who had crept to the fringe of the forest. Col. Elliott Rice was hardly beyond the breastworks before he spotted Green's Confederates crossing the railroad and making for the division right. Rice relayed the news to Davies, who recalled the Iowans at once.

The sharpshooters of the Sixty-fourth Illinois were too far out to hear the recall, but they could not have obeyed the order had they heard it. Lying in the dirt to escape shells from Battery Powell that whizzed dangerously low overhead, the Illinoisans watched thousands of Confederates pour forth from the forest in front of them and hundreds more double-quick around their flanks. Capt. John Morrill tried to shepherd his command back to the cover of a low ravine, but no one paid him any attention. The sharpshooters sprinted past, and the retreat he had hoped to direct disintegrated into a rush for safety. In fifteen minutes the Sixty-fourth lost sixty-nine men. [11]

As the Seventh Iowa and survivors of the Sixty-fourth Illinois cleared their front, the cannoneers in and around Battery Powell intensified their fire. The effect was devastating. Attested Sergeant Payne of the Sixth Missouri,

A sheet of flame leaped out from fronting rifle pits and showers of iron and leaden hail smote the onrushing men from Missouri with terrible and deadly effect. Great gaps were torn in their ranks, to be filled as soon as made. . . . Not for a moment did they halt. Every instant death smote. It came in a hundred shapes, every shape a separate horror. Here a shell, short-fused, exploding in the thinning ranks, would rend its victims and spatter their comrades with brains, flesh, and blood. Men's heads were blown to atoms. Fragments of human flesh still quivering with life would slap other men in the face, or fall to earth to be trampled under foot.[12]

The slaughter was the same in Gates's brigade. At first William Kavanaugh of the Second Missouri found the charge to be "a sublime sight . . . that magnificent body of men moving majestically forward in regular battle array." But then they came into the open and the shelling started. Men began to fall rapidly. Lines unraveled, and "unavoidably we became scattered." Colonel Francis Cockrell rode among his men, encouraging them forward, but the smoke had become so dense no one could see him.

Two of his messmates were dead, but Private Kavanaugh kept on. Beside him was Wallace Martin, his closest friend. Both were exhausted. They had lain awake most of the night, nervously relating their battle experiences to each other and speculating on what the morning might bring. Martin closed the conversation. Tucking up his blanket, he rolled away from Kavanaugh, saying, "Will, let's go to sleep, tomorrow some of us have to be killed, but we do not know who it will be." One hundred fifty yards short of the Yankee breastworks, Martin fell, shot through the brain. Kavanaugh was unhurt, but thirty-five of the forty-five men of his company already were down.

Many more Missourians would have died on the open field had Rosecrans not blanketed his front with skirmishers. The Seventh Iowa and the Sixty-fourth Illinois Sharpshooters blocked the fields of fire of most of the batteries around Battery Powell until the Confederates reached Elam Creek, between 300 and 400 yards away. At that range the gunners simply could not get off enough rounds to halt the Rebels before they closed on the Federal lines. They drew a hundred yards nearer, and as Davies had predicted, his men wavered. The New Yorker watched one man fire his rifle madly in the air, duck his head, then jump up from behind the breastworks and dash for the rear. "A very few of those who had fired followed his example, and I only regret that I was not near enough to the cowards to have them shot down, as I had shot at two the day before on leaving the line under similar circumstances."[13]

Davies egregiously understated the defections from his ranks. Two hundred yards short of the Federal works, Moore and Gates paused to dress ranks. The Confederates let go a volley, then came on at the double-quick, yelling, "On to the battery! Capture the Battery!" That broke the will of the Federals. Whole units collapsed. The entreaties of officers were useless. The order of their going is uncertain, but the inescapable truth is that nearly the entire division fled while the enemy was yet a hundred yards off.

The pitiful remnant of the Union Brigade present that day was among the first to fold. The Seventh Illinois, on the Iowans' left, gave way with them. General Rosecrans happened to be near the Union Brigade. He had a hearty disdain for the Iowans that was returned in equal measure. When they joined his command for outpost duty two months earlier, Rosecrans berated them in a letter to Grant: "The Mackerel—I mean Union—brigade, reported to General Granger 520, 300 for duty; advanced as far as Danville where . . . they attacked the pigs of Danville deploying skirmishers for that purpose, who opened a sharp fire and brought eight of the hairy rascals to the ground before Colonel Tinkham, commanding the station, arrived and informed the commander of the brigade that these natives were non-combatants as loyal as possible considering their limited information."

Rosecrans rode among the fleeing Iowans and Illinoisans. Forgetting Davies's prophecy, he screamed and swore at them: They were a set of cowards and old women; they would have no military standing in his army until they won it back in battle; it was no wonder the Rebels had thrown most of their force against Davies's division. The men kept running.[14]

Sweeny's brigade gave a better account of itself. On the brigade right the Fifty-second Illinois held its ground even after the batteries on either side of it "limbered up and galloped off in wild confusion," as Colonel Sweeny remembered. The limbers crashed down on Sweeny's reserves, the Twelfth Illinois and Eighty-first Ohio regiments of Mersy's brigade, which were stacked up in column, crushing several men and scattering the rest. General Davies was beside himself. "This communicated a stampede in the ammunition wagons in the hollow in the rear of the line, and they too started on the run to the rear." The general and his staff galloped after them. Davies lost himself to the madness of the moment. When one officer refused to turn around and return to the front with his men, Davies drew his revolver and shot him dead. A stunned private bent over to examine the body. "Let him alone," snarled Davies before riding off.[15]

The cannoneers in Battery Powell followed the examples of the batteries to their left. There were five cannon with their crews, horses, and limbers all pressed into the small redoubt. Smoke, the thud of guns, the shouting of gun commanders, the odor of powder, and the stench of dead and dying horses converted Battery Powell into an inferno of the senses. No one could stay there long. Lt. John Brunner fired double charges of canister from the two guns of his section of Battery I, First Missouri Light Artillery, until half his men ran off or were shot down and only five horses were left standing. Brunner waved the animals away and, leaving the guns behind, started rearward with the rest of his cannoneers. Captain Richardson's gunners kept firing until the enemy was twenty yards away. His drivers had all but lost control over their terrified animals. They danced about madly, making it impossible to limber up the cannon. The limbers and caissons made good their escape, as did most of Richardson's cannoneers, but his three twenty-pounder Parrotts remained in the redoubt. Abandoned with his and Brunner's cannon were two twenty-four-pounder howitzers from Welker's Battery H, First Missouri Light Artillery.[16]

Gates's one Arkansas and four Missouri regiments hit the rapidly emptying Union lines a few minutes ahead of Moore's brigade. Lieutenant Colonel Bevier's Fifth Missouri struck Battery Powell from in front. His men swarmed over its low earthen walls to find a score of mangled horses and five stilled cannon. "Unfortunately, we had nothing to spike [the guns] with," Bevier lamented. The inside of the lunette was a charnel house, "one of the bloodiest places I ever saw," said Missourian William Ruyle. "On all sides of me I could see both the enemy and our men lying in the bleaching sun, in the dust which was about six inches deep. Every second it seemed I could see some comrade fall dead or wounded."

Colonel Sweeny called on his own Fifty-second Illinois to change front and "meet the enemy boldly." Lieutenant Colonel Wilcox tried, but before he could arrange his lines, the Second Iowa began to give way. With the enemy just twenty yards from his right flank, Wilcox hurriedly barked the order to withdraw. The Third Missouri poured over the breastworks and delivered a volley that wrecked the Fifty-second; before Wilcox could rally them, the Illinoisans scattered. But they had hurt the Third Missouri badly before breaking. Its lieutenant colonel lay in the field wounded. Col. J. A. Pritchard dismounted at the breastworks. As the Fifty-second retreated, he waved his sword to urge his men forward in pursuit. A minie ball struck him in the left shoulder, "literally crushing the bones," said a captain who witnessed the scene. "My God, I am shot!"

said Pritchard. "Boys, take me off the field—don't let me fall into the hands of the Yankees." Grasping his left wrist with his right hand, his features contorted in pain, Pritchard stammered as he was carried away, "Boys, do your duty." Major Hubbell assumed command; Pritchard died sixteen days later.

"Waving his sword and pointing it to the grim mouths of the artillery," Col. Francis Cockrell led his Second Missouri against the only cannon remaining on that part of the field, the six guns of Capt. Henry Dillon's Sixth Wisconsin Battery. "Forward, my boys; we must capture that battery," he shouted repeatedly above the din.[17]

Dillon had only thirty-nine men to serve the battery, but they fought admirably, staying with their pieces until the Missourians were among the cannon. Several were bayoneted, and one cannoneer knocked a Rebel down with his ramrod in order to send a final charge home. Dillon escaped with his limbers and caissons but lost all six guns and twenty-four men. He halted the teams in front of the Fifty-ninth Indiana and sought out its commander, who was a friend. First Lt. William Bartholomew watched Dillon approach. "I shall never forget his expression. The tears rolled down his cheeks as he said: 'Colonel, my battery is gone. I did the very best I could. They shot down my men and my horses, and took my battery.'"

Dillon's infantry support had failed him miserably. Indeed, the Sixth Wisconsin Battery lost nearly as many men as either the Eightieth Ohio or Tenth Iowa Infantry regiments. On his left the Eightieth Ohio came apart at about the same time as the Fifty-second Illinois. When Bevier's Missourians poured into Battery Powell on their left flank and those of Cockrell closed in on their right, the Ohioans gave back. Maj. Richard Lanning was killed trying to rally them, and command devolved on the senior captain, who could do nothing to stem the panic.

The Tenth Iowa retired from its indefensible position after losing thirty-four men. With their right flank in the air, the Iowans had no recourse but retreat when the brigades of McLain and Colbert, having caught up with Gates and Green, slipped around them. The Iowans fell back 300 yards to the protection of the Twelfth Wisconsin Battery.[18]

By 10:45 A.M. the only regiment left to confront Gates's Missourians was August Mersy's hard-luck Ninth Illinois. The regiment was in closed column beside a large, white house on a hill behind Battery Powell when Sullivan's regiments collapsed and the Sixth Wisconsin Battery was taken. The limbers and caissons of the battery bounced onto the rear companies of the Ninth, throwing them into disorder. Out of the dust came the

Confederates. They poured a volley into the flank of the startled Illinoisans, who milled about the yard until Mersy rallied them. He somehow got the regiment into line to meet the Missourians, and the Ninth held on long enough to allow the gun crews from Battery Powell and the last fugitives from the Fifty-second Illinois to slip away safely.[19]

The moment was critical in the extreme. Gates had punched a quarter-mile-wide hole in Rosecrans's inner defenses and taken thirteen cannon. On his right William H. Moore had rolled over the Second and Seventh Iowa regiments of Sweeny's brigade and the Fifty-seventh Illinois of Du Bois's brigade. Nothing stood between Moore's men and the town but several hundred frightened Federals, most of whom had given up all thought of resisting.[20]

As stunning as their success had been, Gates and Moore could not carry the day alone. The Missourians had breached the Federal front line, but the effort had exhausted them. Key officers had fallen, and the surviving Confederates were nearly as disorganized as the Yankees. They needed immediate help to exploit their gains, but none was forthcoming. To their left the brigades of McLain and Colbert were faltering before Buford's brigade and Hamilton's massed artillery. To their right Dabney Maury had yet to begin his attack. Nor was support forthcoming from behind; Cabell's brigade remained beside Phifer, glued to the Memphis and Charleston Railroad. Martin Green had not called on Cabell, and Price apparently was not near enough to the action to recognize Green's need for reinforcements.

Inside Battery Powell, Lieutenant Colonel Bevier contemplated the probable fate of his men. The Federals were rallying and pouring a murderous fire into the flanks of his Fifth Missouri, and the cannon of Battery Robinett and the rifles of Fuller's Ohioans, who had been spectators to the contest, joined in to rake the lines of Moore. Said Bevier, "The battle was furiously raging, a fatal cross-fire was enfilading us, and more than half our men were killed or wounded. Still, we are inside the breastworks of the foe, and hold a portion of their guns; and if Cabell only would come! Why don't they help us on the right? A division hurled in that direction would attract that terrible cross-fire which is turned upon us. But we expected in vain."[21]

Even before their brigades crossed the breastworks, many Missourians had seen the inevitable outcome of the struggle and had acted accordingly. In open defiance of orders, they lay down when the command was passed to charge Battery Powell. Especially egregious was the case of the First Missouri Dismounted Cavalry, which had lost its commander early

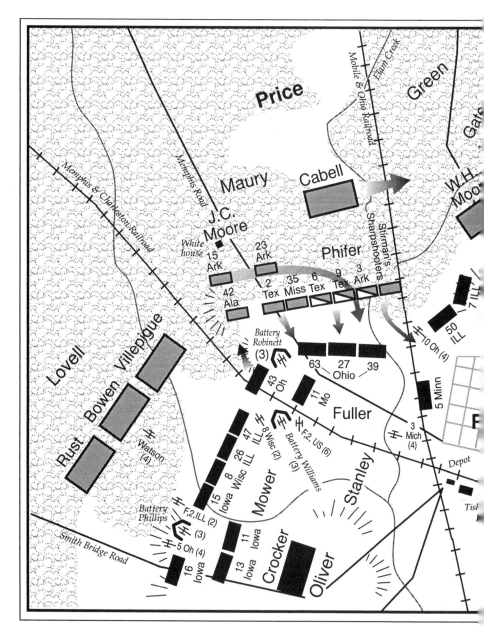

MAP 10. *Batteries Powell and Robinett, 11:00 a.m.*

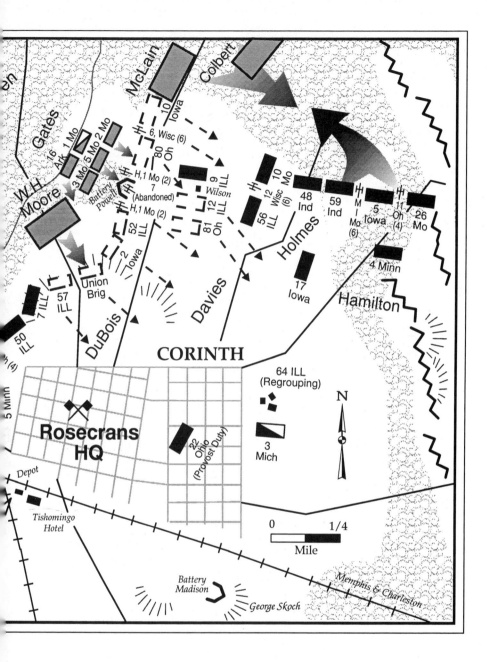

McLain

Colbert

Gates

Ark. 1 Mo

16

W.H. Moore

3 Mo 5 Mo 2 Mo

Battery Powell

H, 1 Mo (2)

7 (Abandoned)

H, 1 Mo (2)

10 Iowa

6, Wisc (6)

80 Oh

9 Wilson ILL

12 Oh ILL

81 Oh

56 ILL

12 Wisc (6)

10 Mo

48 Ind

59 Ind

M I Mo (6)

5 Iowa

11 Oh (4)

26 Mo

4 Minn

Holmes

17 Iowa

Hamilton

52 ILL

2 Iowa

Union Brig

57 ILL

7 ILL

50 ILL

5 Minn

h (4)

DuBois

Davies

CORINTH

64 ILL (Regrouping)

3 Mich

N

Depot

Tishomingo Hotel

Rosecrans HQ

22 Ohio (Provost Duty)

0   1/4

Mile

Battery Madison

George Skoch

Memphis & Charleston

in the action. Confessed Pvt. Will Snyder in a letter home, "Within three or four hundred yards of a very large fort, we laid down. . . . I think it was the intention for us to charge it by some means — we did not do it." James Fauntleroy was among those of the regiment who shirked their duty while others "rousted the Yankees out of their breastworks." But the field offered the laggards little protection. "We lay there for three-quarters of an hour under a blinding shower of shot and shell, wounding a good many," recalled Snyder. Fauntleroy lost two friends and took a bullet through his hat and one that burned his neck.[22]

Most of Gates's Confederates clung to the line of captured breastworks. A few dozen pressed on, through the yard of the Wilson house and up the slope of a low hill. On the opposite side, hidden from view, was the Twelfth Wisconsin Battery, posted on a slightly lower rise. To the left of the battery was the Fifty-sixth Illinois; to its right, the Tenth Missouri. Both regiments were intact, thanks to the clear thinking of their commanders, and had changed front to meet the impending onslaught. Reported Col. Green Raum, commander of the Fifty-sixth Illinois, "The front line began to waiver and fall back. Their retreat soon became a rout, and they came down pell-mell upon us, running over my men in every direction. The caissons and a number of loose horses came thundering down and passed through the interval between the Tenth Missouri and my regiment." Maj. Leonidas Horney of the Tenth Missouri took a more aggressive approach. He told his men to fix bayonets and run through anyone who tried to trample them.[23]

As soon as their front was clear, Raum and Horney ordered their men to stand up and fire at the unfortunate Rebels exposed on the hillcrest 100 yards away. The Twelfth Wisconsin Battery joined in with a rapid fire of shell and canister. Sgt. S. E. Jones, in command of the left section, had his one serviceable cannon spew double-shotted canister at the remarkable rate of six shots a minute. Jones's performance caught the attention of others in the battery. "Lord, boys, look at Jones," shouted a sergeant commanding a nearby gun. "Three cheers for Jones." First his crew, then the entire battery, and finally its infantry support took up the cheer. "In an instant the whole aspect of the situation was changed," said the battery commander, Lieutenant Immell. "Without knowing the cause of the cheering our retreating troops took courage and rallied on our left, while the enemy appeared paralyzed at they knew not what."[24]

Their indecision was fleeting. Through the smoke a Wisconsin gunner watched the Rebels, "who halted for a moment, then turned and fled, leaving their dead and wounded behind. When they started on the retreat they ran like frightened sheep, and at last accounts were still running." The Fifty-sixth Illinois and the Tenth Missouri charged down the slope after them with a shout. "On we went, yelling at the top of our voices, every man trying his best to reach the top of the next hill first down which the late advancing foe was now retreating on the 'thrible quick,'" quipped a Missouri Yankee.[25]

Back in Battery Powell, Lieutenant Colonel Bevier watched the nearing tide of blue: "While looking at them in dismay, and fruitlessly trying to pierce the misty atmosphere in the rear for some sight of more 'bold boys in gray,' my horse was shot through, and the ball flattened against my ankle. As he fell heavily and rolled in the dust, I was sure three-fourths of my leg was gone; but on arising, found it all safe and sound, though somewhat bruised. I was engaged in feeling and shaking it . . . when the order came to fall back — and it was time."

Few of Major Hubbell's Missourians waited for the order to retire. "A panic seemed to seize all the men on the right and left, until I stood alone, with only fifty of my own brave boys, who all offered to die with me," said Hubbell. "But I thought it would be sacrificing their lives to no purpose, and finally gave the painful order to fall back, which was obeyed."

The Missourians' retreat proved nearly as costly as their advance. The guns they had been unable to spike were turned against them with murderous effect. Major Horney's Tenth Missouri retook the Sixth Wisconsin Battery and blasted the fleeing enemy of their home state. In Battery Powell a scramble ensued for the chance to shoot Rebels in the back. Several companies of the Fifty-second Illinois had rallied and insinuated themselves into the counterattack of the Fifty-sixth Illinois, which retook the redoubt. As soon as it was secure, General Davies's adjutant general, Capt. Julius Lovell, jumped from his horse and, with the help of a soldier from the Fifty-second and a bugler from Richardson's battery, turned a twenty-pounder Parrott on the enemy. The trio got off ten rounds before Richardson's crews returned to their guns. Company C of the Fifty-sixth Illinois, which had trained as artillerists, manned the remaining pieces in the interim. It was 11:30 A.M. before Gates's brigade ran the gauntlet of shot and shell and returned, horribly bloodied and exhausted, to the shelter of the embankment of the Memphis and Charleston Railroad. But not all had obeyed the order to withdraw; 100 chose to surrender at the Federal breastworks rather than risk death recrossing the field. Some 300 more lay dead or wounded around Battery Powell.[26]

* * *

McLain and Colbert fared no better than Gates. During the hour Gates clung to the Yankee breastworks — from 10:30 to 11:30 A.M. — they fought Napoleon Buford's brigade inconclusively in open, rolling timber 400 yards east of Battery Powell. Behind a cloud of skirmishers they descended on Buford from the north. At a range of 600 yards the Federal artillery opened with shell. Mounted beside the cannon of his reconstituted Eleventh Ohio Battery, Lt. Henry Neil strained to make out the approaching enemy through his field glass. The breeze blew open their colors. Neil shivered with the shock of recognition; his gun barrels were trained on the Third Texas Dismounted Cavalry and the Third Louisiana Infantry. "Boys, there are the same troops that fought us at Iuka; are you going to let them touch our guns today?" he screamed. His artillerymen responded in kind. Recalled Neil, "The yell of rage that went up was more ominous than a rebel yell ever tried to be. The men worked like tigers in their desperate resolve that their beloved guns would never again feel the insult of a rebel touch."

Nor would they. At a range of seventy-five yards Buford's 1,800 infantrymen rose and delivered a near-simultaneous volley that brought the Confederates to an abrupt halt. The wait had been excruciating. "We were in a field of high weeds," said Samuel Byers of the Fifth Iowa. "The orders were to lie down, as the enemy was about to assault us directly in front. We lay there in the weeds for an hour without speaking. What a chance for strange thoughts! And the men, thinking of their comrades dead in the ditches of Iuka, did meditate." The sun was high and the heat blistering. Continued Byers, "The suspense, lying there in the weeds, every moment expecting a crash of musketry in our faces, was something intense." The strain was too much for Private Billy Bodley, still lost in grief over the death of his only brother at Iuka. Byers felt a light touch on his shoulder. "I am not afraid, but I am too sick to fight — you are the captain's friend; ask him to let me go back," implored Bodley. Byers crawled to the captain, who obliged Bodley. The heartsick Iowan crept away, only to be killed on another field of battle.

Byers had turned to watch Bodley go when someone yelled, "Rise and fire." Byers jumped up. He had been on detail far in the rear at Iuka; this was his first taste of combat. The moment was seared in his memory: "I was burning up with excitement, too excited to be scared. I was in the rear rank. I raised my musket and blazed away at nobody in particular. A comrade in front of me afterward said I 'nearly shot his ear off.' He glanced back once, he said, and I was only laughing. That was my first shot in an open, stand-up battle."[27]

Lieutenant Neil teetered between revenge and death. The Ohioan rode forward of his battery to taunt the Rebels. He waved his hat to attract their attention and yelled, "Come on! Come on! If you think you can play Iuka over again." Three times the Louisianans and Texans closed on their colors as if to charge; three times Lieutenant Neil signaled to the soldiers of the Fourth Minnesota, posted behind his cannon, to rise and fire. Each time the Minnesotans stood up, he motioned them back to the ground with the words, "No, no, they have broken again."[28]

For forty-five minutes the Confederates endured the pounding. The slaughter was terrific. Colonel McLain had his leg sliced off by a solid shot. The badly depleted Third Louisiana lost thirty-two men — one-third of those engaged. Company K of the Thirty-seventh Mississippi lost every officer.[29]

The fall of McLain took the fight out of his brigade. His Mississippians wavered, inspiring the Federal artillerymen to intensify their fire. The Forty-eighth Indiana delivered a particularly heavy volley at the same instant, and the leaderless Confederates broke and fled. The Forty-eighth swept forward on the run after them, firing as they went. The huge Fifty-ninth Indiana Infantry, 636 strong, joined the pursuit.

McLain's brigade got off without a man being taken prisoner. Colbert's brigade, which tarried until the Indianans were behind it, had 132 seized. Buford's Federals kept up the chase for nearly a half-mile before recall was sounded at 11:30 A.M. The attack on the Union right, in which Van Dorn had placed his greatest hope for victory, had ended in a rout.[30]

*   *   *

William H. Moore was still advancing when Gates, on his left, withdrew. There were no reserves behind the Seventh or Second Iowa of Sweeny's brigade or Du Bois's shattered Fifty-seventh Illinois, so Moore's men were able to chase the Yankees into Corinth. As they passed among the houses, fragments of Federal regiments rallied and began a door-to-door struggle with their pursuers.

Rosecrans was in the thick of the battle, but his presence was hardly inspiring. The Ohioan had lost all control of his infamous temper, and he cursed as cowards everyone who pushed past him until he, too, lost hope. Near the Tishomingo Hotel were the trains of Du Bois's brigade. With them were many of the brigade wounded. Both were under the charge of the chaplain of the Fiftieth Illinois. The crash of combat drew steadily nearer, when from a side street appeared a retinue of mounted officers. One rode forward, pointed at the wagons of the Fiftieth Illinois, and asked who was responsible for them. The chaplain spoke up. The officer declared that the army was whipped and told the chaplain to burn his baggage. "We are not whipped, sir," retorted the chaplain. The officer and his retinue galloped away, and the chaplain guided his train to safety a half-mile beyond town. Only later did the chaplain learn that the forlorn officer had been Rosecrans.

Rosecrans's histrionics nearly cost him his life. "On the second day I was everywhere on the line of battle," he wrote with disingenuous pride. "Temple Clark of my staff was shot through the breast. My sabre-tache strap was cut by a bullet, and my gloves were stained with the blood of a staff-officer wounded at my side. An alarm spread that I was killed, but it was soon stopped by my appearance on the field."[31]

Van Dorn was as removed from the action as Rosecrans was swept up in it. He was unable to correct the piecemeal deployment of Green's division, and Maury's failure to take up the attack in a timely manner escaped his notice. Rosecrans was too close to the fight to fashion a coherent response to the Rebel breakthrough; Van Dorn was too distant to exploit it.

21

My God! My Boys Are Running!

The delay in Green's assault may have perplexed Dabney Maury, but when it came his turn to attack, he proved equally delinquent. A desire to keep his command out of harm's way could hardly have been the reason; just 400 yards from the guns of Batteries Williams and Robinett, Maury's men had been taking a pounding since dawn.[1]

Shortly after daybreak Lt. Charles Labruzan of the Forty-second Alabama volunteered his company to reinforce the brigade skirmishers. Nearing the skirmish line Labruzan realized the danger to which he had exposed his Alabamians. "We got behind trees and logs, and the way the bullets did fly was unpleasant to hear," rued Labruzan. "I think twenty must have passed within a few feet of me, humming prettily. Shells tore off large limbs and splinters struck my tree several times. We could only move from tree to tree, bending low to the ground while moving. Oh how anxiously I watched for the bursting of the shells when the roar of the cannon proclaimed their coming!"[2]

Labruzan's company took cover beside the Second Texas Infantry. The Second was a sterling unit—"one of the finest regiments I have ever seen," proclaimed General Maury. Its colonel, William P. Rogers, was perhaps the most gifted regimental commander in Van Dorn's army. A burly, square-faced man with black hair and an iron gaze, Rogers was forty-four years old at the time of Corinth. He had led with distinction a company in Jefferson Davis's Mississippi Rifles during the Mexican War, and Rogers was said to be the second man to have scaled the fortress walls of Monterrey. Following the war he served as American consul in Vera Cruz, a post he resigned when his wife refused to follow him across the border. But she consented to go as far as Texas, where Rogers opened a successful legal practice. Rogers was an intimate friend of Sam Houston. Like the eminent Texan, Rogers deplored secession but felt duty bound to fight with the Confederacy, and he accepted a commission as lieutenant colonel of the Second Texas upon the outbreak of hostilities.

Rogers's Texans revered him. "A better or a braver man never led troops anywhere," averred one. Rogers returned their affection, and loyalty to his men overcame Rogers's desire to leave the army. His brother officers felt no less strongly toward him. Before Iuka, twenty field-grade officers from Arkansas and Texas regiments joined in recommending him for a major general's commission.

That Rogers had not risen in rank was a result of a deep personal enmity between him and President Davis. During the Mexican War the two had a quarrel resulting in a challenge to a duel, which only the intervention of Gen. Zachary Taylor prevented. The cause of their dispute is obscure—Rogers reviled the lieutenant colonel of Davis's regiment, with whom Davis apparently sided against Rogers, despite protestations of goodwill—and the antipathy between Davis and Rogers lingered. On the night of October 3, as he was visiting the wounded of his regiment, Colonel Rogers said to his senior company commander, "Captain McGinnis, tomorrow Jeff Davis must allow my promotion, or tomorrow I die."[3]

Rogers obviously was not a man to shirk battle. While Maury waited, the Texan pushed forward his regiment, reinforced by two companies of the Thirty-fifth Mississippi, to skirmish aggressively with the Yankees. From the edge of the forest the Texans exchanged fire with skirmishers from Fuller's Union Brigade, who were deployed among the fallen timber and partially completed abatis between the forest and Battery Robinett. So sharp was Rogers's fire that General Van Dorn feared he would bring on a general engagement before Green was ready to attack.[4]

The morning passed to the thunder of artillery and the crackle of skirmish fire. At 9:30 A.M. Lieutenant Labruzan came off the skirmish line with his Alabamians and took up behind a log with the major of the Forty-second. Thirty minutes later Labruzan was stunned by the "heavy volleys of musketry" coming from far to the left that told of Green's encounter with the enemy. That was to have been the signal for Maury to take up the attack, but no call to attention came. Not until 10:30 A.M. was the

Col. William P. Rogers
(courtesy of T. Michael Parrish)

command given to form for an assault. After donning an armored vest and pinning to his shirt a short note with his name, rank, and the address of friends, Colonel Rogers delivered the attack order to the Forty-second. Riding up to its supine officers, Rogers said simply, "Alabama forces," and the regiment came to life. Throughout the division the response was the same. "When the order to advance was given that fine body of soldiers obeyed as unhesitatingly as if the impulse to move had been that of a single man," said a member of the Second Texas. Generals Phifer and Moore formed their brigades in densely massed regimental columns. Each column was five lines deep with a front two companies wide. One hundred yards separated the assaulting columns. Moore was behind and to the right of Phifer, so that Phifer's right regimental column masked Moore's left.[5]

From the high ground on either side of Battery Robinett Colonel Fuller's Federals watched the Confederates mass for the assault. Capt. Oscar Jackson of the Sixty-third Ohio, a twenty-two-year-old former schoolteacher from Hocking County, looked on in awe as "the rebels began pouring out of the timber and forming strong columns. All the firing ceased and everything was silent as the grave. They formed one column of perhaps two thousand men in plain view, then another, and crowding out of the woods another. . . . I thought they would never stop coming."

Captain Jackson's fear caused him to exaggerate Rebel numbers, but even the most hardened veterans were dumbstruck. Colonel Fuller stood with his friend, Col. Joseph Kirby Smith, behind the Forty-third Ohio, to the left of Battery Robinett. Smith's lighthearted humor had left him, and he nervously ordered his Forty-third to change front forward on its right company to better meet the impending attack (that is, conduct a right half-wheel so as to face north). Smith's regiment held a key part of the line, the 200 yards of open ground between Batteries Robinett and Williams, but Colonel Fuller's thoughts were elsewhere. He feared for the Sixty-third Ohio, which lay on the far side of Battery Robinett, astride the Memphis road. Smith's Forty-third Ohio at least enjoyed the protection of breastworks; the Sixty-third was in the open. Assuming the Confederate columns guided on the road, Fuller deduced they would converge on Battery Robinett precisely in front of the Sixty-third. A regiment so exposed, he remarked to no one in particular, could not long withstand such an onslaught. Presupposing that the Sixty-third would collapse, Fuller placed the Eleventh Missouri twenty-five yards behind them. General Stanley shared Fuller's concern, and he placed himself and his staff behind the Sixty-third.

It was 11:00 A.M. before Dabney Maury set his brigades in motion. Captain Jackson studied their approach. "As soon as they were ready they started at us with a firm, slow, steady, step," he remembered. "In my campaigning I had never seen anything so hard to stand as that slow, steady tramp. Not a sound was heard but they looked as if they intended to walk over us. I afterwards stood a bayonet charge when the enemy came at us on the double-quick with a yell that was not so trying on the nerves as that steady, solemn advance."[6]

Before the Federals fired a shot, they saw the attacking columns reduced by a third. Perhaps repenting his tardiness, Maury released Cabell's Arkansas brigade to Green; that the Arkansans were needed was beyond question. As Cabell started his brigade toward Battery Powell, a courier from Gates's galloped up. Drawing rein in front of the lead regiment, he screamed, "Colonel Gates has captured forty guns, but cannot hold them unless you reinforce him at once — follow me!" The courier wheeled his horse and started back.

Cabell's Arkansans followed the courier through the forest at the double-quick until Cabell judged his brigade was directly behind Gates. Facing to the front, the Arkansans moved out onto the flat before Battery Powell, expecting to find the Missourians. What they saw astonished them. Said Cabell, "The Missouri Brigade had fallen back, taking a road on my extreme right. . . . Instead of meeting the Missouri Brigade, as I had been informed I would, I found the enemy in line of battle just outside of the timber and about three hundred yards in front of their breastworks." Cabell's left regiments became engaged before he could take in the situation. Bowing to the inevitable, Cabell ordered a charge.[7]

The Arkansans had run into the Fifty-sixth Illinois and the Tenth Missouri, which had chased Gates's Missourians beyond the breastworks. Flushed with victory, the Federals were ready for Cabell. The Twentieth Arkansas came up against the Fifty-sixth Illinois. "Every musket of the Fifty-sixth was turned upon the head of the column," said its commander. The Illinoisans fired with terrible effect. Col. H. P. Johnson toppled from his horse dead, and the lieutenant colonel and the major of the Twentieth also fell. Suddenly deprived of their leaders, the soldiers of the Twentieth Arkansas fled without firing a shot.[8]

Cabell's remaining regiments pressed on, forcing the Federals to retire into the breastworks. There, however, the Arkansans found a reformed Fifty-second Illinois and the remanned guns of Battery Powell. Coming up opposite the redan, the Twenty-first Arkansas was decimated in a matter of minutes. To the left of Battery Powell, the Eighteenth and Nineteenth Arkansas regiments held on a bit longer against the Seventh Iowa, and a few Arkansans made it over the breastworks; but ultimately the result was the same. The colonel of the Eighteenth was killed, and General Cabell fell from his dead horse, badly bruised. In the Nineteenth Arkansas, 15 officers and 105 men were wounded in twenty minutes.[9]

Cabell's belated assault had cost him more than 600 men and gained nothing. By 11:45 A.M. the struggle for Battery Powell was over. Van Dorn's grand flanking movement had played itself out in a series of fierce but poorly coordinated charges against a foe numerically equal and supported by seven batteries of artillery. At least 2,000 Confederates remained in Corinth itself, thronging the streets in the face of feeble resistance. Unless they were quickly and heavily reinforced by a decisive breakthrough somewhere along the Federal front line, their gains would prove as illusory as those of Gates had been.

*   *   *

Out on the low ridge beside Battery Robinett, Fuller's Ohioans braced themselves for battle. Inside the redoubt Lieutenant Robinett's gun crews loaded their three twenty-pounder Parrotts with canister and waited. It was a few minutes before noon. The "steady, solemn advance" Captain Jackson described was becoming unbearable. General Stanley rode back and forth behind the Sixty-third Ohio. Too nervous to be imaginative, he stole a line from history and yelled out to the men to hold their fire until they saw the whites of the Rebels' eyes. Few paid him much notice. All attention was given to the dense enemy columns, which were then marching with seemingly superhuman will into the abatis before Battery Robinett. Their pace was still measured, their lines yet compact. Captain Jackson walked behind his company. "I could see the men were affected. . . . They were in line and I knew that they would stand fire, but this was a strong test," said Jackson. "I noticed one man examining his gun to see if it was clean; another to see if his was primed right; a third would stand a while on one foot then on the other; while the others were pulling at their blouses, feeling if their cartridge boxes or cap-pouches were all right, and so on, but all the time steadily watching the advancing foe." Jackson tried to steady his men. He thought of all the customary battle cries that commanders uttered, such as "Remember some battle (naming it)," "Fire low," or "Stick to your company," but the best he could utter was "Boys, I guess we are going to have a fight." A bit embarrassed at having stated the obvious, Jackson quickly added, "I have two things I want you to remember today. One is, we own all the ground behind us. The enemy may go over us but all the rebels yonder can't drive Company H back. The other is, if the butternuts come close enough, remember you have good bayonets on your rifles and use them." [10]

When the enemy closed to within 250 yards, Fuller gave the order to lie down. Jackson's company was behind a small roll in the earth, which shielded it from the enemy's view. The guns of Batteries Robinett and Williams boomed their greeting, and the Rebel columns shivered. The first shell from Battery Williams exploded amid the Forty-second Alabama, knocking down or scattering some forty men. Wrote a soldier of the Thirty-ninth Ohio, "The shells from our battery made complete roads through them. You could see the poor fellows throw up their arms and leap into the air, and fall down. Still they pressed on." Thomas Duncan of the Second Texas described the scene from the Confederate perspective: "When they encountered the abatis—an obstruction of felled trees, with sharpened and interwoven branches—the formation was necessarily somewhat broken, just as the enemy's artillery began to blast and wither the moving mass of men; but each man, though but an atom of the fiery storm, moved with a separate though strangely cooperative intelligence, advancing with remarkable rapidity toward the common objective, Fort Robinett." [11]

The Confederates cleared the fallen timber and paused briefly to reform their columns. Fuller's skirmishers stumbled before them toward the protection of the Federal main line. The order to charge was given, and the Rebels swept up a low ridge—a "bluffish bank," Captain Jackson called it—fifty yards in front of Battery Robinett.

Fuller's four regiments opened fire almost simultaneously. Lieutenant Labruzan crested the ridge as the first Yankee volley shattered the air. "The whole of Corinth with its enormous fortifications, burst upon our view," said the Alabamian. "The United States flag was floating over the forts and in town. We were met by a perfect storm of grape, canister, cannon balls, and minie balls. Oh God! I have never seen the like! The men fell like grass. I saw men, running at full speed, stop suddenly and fall upon their faces, with their brains scattered all around; others, with legs and arms cut off, shrieking with agony." Labruzan commended his soul to God and kept on running.

The slaughter stunned the Federals. Captain Jackson's regimental commander, Col. John Sprague, said the enemy's "first line was literally shattered to fragments. The survivors rushed pell mell on the second line, throwing it into confusion." Matters certainly looked that way to Jackson: "It seems to me that the fire of my company had cut down the head of the column that struck us as deep back as my company was long. As the smoke cleared away, there was apparently ten yards square a mass of struggling bodies and butternut clothes." [12]

Five Confederate regiments—three from Moore's brigade and two from Phifer's—had come up against Fuller's Ohioans and the cannon of Battery Robinett. West of the Memphis road the Forty-second Alabama tangled with Smith's Forty-third Ohio. Colonel Rogers's Second Texas charged along the road itself. The unreliable Thirty-fifth Mississippi felt its way through the abatis east of the road. To their left the Sixth and Ninth Texas Dismounted Cavalry regiments of Moore's brigade came up against the Twenty-seventh and Thirty-ninth Ohio regiments. [13]

Fuller's first volleys drove them all back. The Rebels retired twenty or thirty yards. The survivors of the decimated frontline companies tumbled behind stumps and fallen trees and opened fire, while their regimental commanders regrouped the remainder for a second try at the Yankees.[14]

Now it was the Ohioans turn to suffer. The fight had become a contest of infantry. The cannoneers in and beside Battery Williams had ceased fire for fear of hitting their own men, as had the two guns situated behind the flanking walls of Battery Robinett; only one piece remained in action. The Confederates shot with greater accuracy than had the Ohioans, and Yankee losses were staggering. Although behind three-foot breastworks, nearly a quarter of the soldiers of the Thirty-ninth Ohio were gunned down. Colonel Smith sat behind them on horseback. "Those fellows are firing at you, Colonel," yelled one of his men solicitously. "Well, give it to them," answered Smith. An instant later a bullet smashed into his nose. It plowed through his head and emerged from his left ear, and Smith fell to the ground, dead. His "constant companion" and adjutant, the "accomplished young" 1st Lt. Charles Heyl, galloped up and was shot too. He sat upright for a moment, grabbed for his horse's mane to steady himself, then rolled off and fell beside Smith.

The death of Smith, whom Colonel Fuller called the "most accomplished officer in the brigade," incapacitated the Forty-third Ohio. Forgetting the danger, those nearest Smith gathered around his body. Word of Smith's fall spread along the line like an electric current. "The best testimony I can give his memory is the spectacle I witnessed myself, in the very moment of battle, of stern, brave men weeping as children as the word passed, 'Kirby Smith is dead,'" said General Stanley, who galloped up behind them. Colonel Fuller was too far away to help; he could only look on with alarm at a regiment that to him "seemed dazed and liable to confusion." The presence of Stanley settled the men, and Lt. Col. Wager Swayne stepped forward to take command. Together they steadied the ranks before the next Rebel charge.

The Forty-third Ohio may have been crippled, but on the opposite side of Battery Robinett the Sixty-third Ohio was nearly obliterated. Almost half the men in its left companies were hit, and casualties in the rest of the regiment were dangerously high. Only Captain Jackson's company seems to have escaped damage. The enemy in front of Jackson "fired too low, striking the ground, knocking the dirt and chips all over us, wounding not one man in my company."[15]

Fifteen minutes after the first assault ended, the Rebels came on a second time, pouring over the low ridge, through a shallow ravine, and up the gentle slope before Battery Robinett. Again they were thrown back.[16]

The Confederates re-formed a second time among the fallen timber. Colonel Rogers became a galvanizing force for the disordered survivors of Moore's brigade. He rode from regiment to regiment, challenging the men to greater efforts. Rogers's presence was especially needed in the Forty-second Alabama. A relatively green command, the regiment had lost its commander in the first charge. Rogers teased the Alabamians back into action. Turning his back to them, he unsheathed his sword and cried to his own regiment, "Forward, Texans." George Foster, the senior captain of the Forty-second, took the bait. "They shan't beat us to those breastworks," he remonstrated, then, raising his sword, yelled to his men, "Forward Alabamians!"

Moore's sharpshooters had cleared the way considerably during the lull. They had grown more daring after the second repulse and would crawl through the undergrowth to within a few dozen yards of Fuller's position to pick off officers. Jackson lost ten of his thirty-four men to sharpshooters' bullets and narrowly escaped death himself. Three bullets passed through his clothes, one of which burned a red streak across his ribs, but he was otherwise unhurt.[17]

Colonel Fuller was riding behind the Sixty-third Ohio when the enemy came on for the third time. The destruction the sharpshooters had wrought was apparent. "The Sixty-third, which had suffered greatly from a cloud of sharpshooters, seemed the principal target for the enemy, and its ranks were so riddled and broken that I could see the enemy's column as well as if their line had never intervened," he reported. With nine of thirteen line officers down, Fuller knew the regiment could not resist a third attack, and he yelled to the commander of the Eleventh Missouri to be ready to charge the moment the Sixty-third faltered. The ubiquitous General Stanley was on hand also, ready to rally the fainthearted.[18]

The first time they attacked, the Confederates had advanced deliberately, at the quick step. The second time they accelerated to the double-quick. This time they charged on the run. Still their losses were horrendous. Lieutenant Labruzan was at the head of his company, negotiating the carpet of bodies in front of the redan and bending low against the hail of bullets. "One ball went through my pants, and they cut twigs right by me. It seemed by holding out my hand I could have caught a dozen," said the Alabamian. "They buzzed and hissed by me in all directions, but I still pressed forward. I seemed to be moving right in the mouth of the cannon, for the air was filled with grape and canister. Ahead was one continuous blaze." Private McKinstry of Company D, the colors company, was also in the front rank of the Forty-second Alabama. He closed on the colors and plunged ahead toward Battery Robinett, just forty yards away. The flag had fallen once; now it fell again. A man named Crawford threw down his rifle and snatched it up. "On to the fort, boys," screamed Crawford. Nine buckshots riddled his torso, and he and the flag fell. Huddled in the ditch around Battery Robinett, neither Private McKinstry nor the handful of men left in the company had the courage to pick up the colors. The defenders began tossing hand grenades over the wall, but the Alabamians were too frightened to respond; they merely crouched lower and waited for the grenades to explode. Then someone yelled, "Pick them up, boys, and pitch them back into the fort." McKinstry grabbed one, lobbed it over the earthen wall, and had the satisfaction of hearing it explode among the defenders. "Over the walls, and drive them out," somebody screamed next, and up the steep embankment clambered McKinstry and five or six other Alabamians. At the top they were met by a volley of musketry. The man on McKinstry's right rolled into the redan; the soldier on his left was struck in the forehead. As he fell backward, the dying soldier clenched McKinstry savagely around the neck and carried him to the bottom of the outside ditch. There McKinstry lay, badly stunned.

Lieutenant Labruzan was hunched a few yards away, clutching the dirt. He and another man had tried to scale the wall of Battery Robinett, with ghastly consequences. The man beside him had "put his head up to shoot into the fort, but he suddenly dropped his musket and his brains were dashed in a stream over my fine coat, which I had in my arms, and on my shirt sleeves."[19]

Several squads from the Second Texas Infantry and two companies of the Thirty-fifth Mississippi fell in with the Forty-second Alabama in front of Battery Robinett. The remainder of the Second Texas swept around the redan's northern flank to grapple with the Sixty-third Ohio. At the head of the small band that came up against Battery Robinett was Colonel Rogers, still mounted on his black mare. He drew rein near the ditch and fired his revolver frantically through the embrasure. Next to Rogers, on foot, was Captain Foster of the Forty-second Alabama.[20]

Inside the redan all was bedlam. Shot in the head, Lieutenant Robinett fell beneath one of his cannon. Most of his cannoneers decamped out the rear end of the earthwork. Before the Confederates in the ditch realized their brief advantage, the right-wing company of the Forty-third Ohio spilled into Battery Robinett to man the cannon.[21]

On the northern side of Battery Robinett the full weight of the assaulting column of the Second Texas, less the few dozen men who came up against Battery Robinett with Colonel Rogers, struck the thin ranks of the

Sixty-third Ohio. The Second Texas fired into the front rank of Ohioans. From the ditch beside Battery Robinett Rogers's band contributed a volley that decimated the left-flank company of the Sixty-third. Captain Jackson's was the next company in line. Colonel Sprague yelled at him to take the place of the annihilated unit. "I will do it," replied Jackson with a salute. His twenty-two men faced to the left and started for the gap. "It was like moving into dead men's shoes, for I had seen one company carried away from there on litters, but without a moment's hesitation we moved up," wrote Jackson. Instantly the Ohioans were attacked from the front and flank by five times their number. So near were the Rebels that Jackson distinctly heard their commanding officer's exhortation: "Boys, when you charge, give a good yell." "It almost made the hair stand up on my head," confessed the Ohioan. "The next instant the Texans began yelling like savages and rushed at us without firing." Jackson took in the situation in an instant. "Don't load boys; they are too close on you; let them have the bayonet." Then, for the first and only time in his service, Jackson found himself in the midst of a hand-to-hand melee.

Fear narrowed Jackson's focus, and he fought in an adrenaline-fed rage. Jackson squeezed off rounds from his revolver into the faces of the enemy until one Texan knocked the pistol from his hand with his musket, then swiped him across the face, cutting Jackson's cheek to the bone. Never feeling the wound, Jackson picked up his revolver and kept on shooting until the Texans fell back into the ravine. As the blood poured down his blouse, the Ohioan surveyed his company. Of the thirty-four men he had taken into the fight, only eleven remained standing. Then Jackson glanced to the front, just as one of the Rebels turned to fire a parting shot. "I saw the fire was aimed at me and tried to avoid it, but fate willed otherwise and I fell right backwards. I was struck in the face," said Jackson. Corporal Savely saw him fall. Jackson's limbs quivered convulsively and the blood spurted from his face in a stream several inches high. Savely thought him dead, and someone cried, "The captain is killed!" So, too, thought Jackson: "I felt as if I had been hit with a piece of timber, so terrible was the concussion and a stunning pain went through my head. It was my impression that I would never rise, but I was not alarmed or distressed by the thought I was dying; it seemed a matter of indifference to me."

Jackson stood up and staggered a few paces to the rear. A fallen tree blocked his progress. He was too weak to cross it. Pvt. Frank Ingmire of his company stood nearby, staring widely at nothing. "Ingmire, help me over." "Yes," he whispered, "let me help you across." Ingmire offered

Jackson his left hand; only then did the captain notice Ingmire's right arm was dangling at his side, his hand dripping blood. A bullet had shattered Ingmire's wrist. Ingmire helped Jackson over the log. The captain clung to his left arm with both hands and pleaded with Ingmire not to leave him. Jackson took a few more staggering steps before he felt his hands slip from Ingram's arms and his knees buckle. Jackson sank to the ground, to welter unconscious in his blood until litter bearers carried him off. Two days later he awoke in a field hospital.[22]

Neither Jackson nor any of his men would have lived to tell their story had not Colonel Fuller and General Stanley been nearby. When the Second Texas struck the Ohioans, Fuller called the soldiers of the Eleventh Missouri to their feet. Their bayonets already were fixed. In a few moments Colonel Sprague ordered his left and center companies, less Captain Jackson's command, to fall back. The Missourians opened their files to let the Ohioans by, then fired. They reloaded and fired again. The enemy kept coming. When the Rebels were thirty yards away, the Missourians raised a yell and charged.[23]

Colonel Fuller was astonished to see General Stanley with them. He had dismounted and was "rushing in between the file closers and the line of battle of the Eleventh Missouri, his arms outstretched, to touch as many men as he could reach, pushing them forward to strike the head of the rebel column. I wondered how he got there; for, only a minute or two before, he was with the Forty-third Ohio, making it hot for the rebels to the left of the battery."

Considerations other than military led Stanley to risk his life to rally the Sixty-third Ohio. A good part of the regiment had been recruited near his old home in Wayne County, and several of the men were familiar to him. While Stanley passed among the dead and dying of the Sixty-third, a young lieutenant, who had received his promotion just a few days before, hailed him. "General, come here; I want to say good-bye; I am mortally wounded." Stanley was incredulous: "He spoke so naturally I could not believe it, and tried to encourage him, but he died in half an hour. He was born within two miles of my home."

Stanley gave the closest thing to an order the Eleventh Missouri received. As they swarmed past his horse, he clapped his hands and yelled to the Missourians, "Go in boys, go in, they are running, go in, go in."[24]

The charge of the Eleventh Missouri broke the momentum of the Texans' attack. They gave way stubbornly, killing or wounding sixty-three Missourians before quitting the field. As the Texans stumbled back into the ravine, the Missourians wheeled to the left to challenge Colonel

Rogers's band before Battery Robinett. Private McKinstry saw them coming: "Before giving the situation a thought, I immediately raised my gun and fired full into the breast of a Federal sergeant, who was in front of the column, and only a short distance from us."[25]

Captain Foster saw the futility of further resistance. "Cease firing, men! Cease firing!" he screamed while waving a white handkerchief. Colonel Rogers also judged surrender the only alternative to certain death. He bent down in the saddle and grabbed a ramrod from Pvt. T. B. Arnold of the Thirty-fifth Mississippi. To it he tied his handkerchief. In the smoke and confusion the Federals either did not see it or misunderstood his gesture. The entire Eleventh Missouri let go a volley into Rogers's fast-dwindling band. "Oh, we were butchered like dogs," mourned Lieutenant Labruzan. He escaped the volley by bending low in the ditch. Less fortunate was Private McKinstry. A minie ball crashed through his left hip, spinning him halfway around. Another tore through his right shoulder, the impact spinning him back to the front. A third bullet crushed his left shoulder, and he dropped his rifle. "I looked, and lo! every one of the fifteen men who were standing with me had fallen in a heap." The survivors broke for the safety of the timber 300 yards to the rear. Although bleeding profusely from three wounds, McKinstry stumbled off before the Federals could capture him. Miraculously, after laying on his back unable to move for three months, he recovered in time to fight at Lookout Mountain.[26]

Among the dead was Colonel Rogers. The volley that killed him may have come from the Sixty-third Ohio. Colonel Sprague had re-formed a few dozen of his men and with them joined the charge of the Eleventh Missouri. As they came around the side of Battery Robinett, Sprague said he noticed "a man on horseback, leading the charge in a most gallant manner." What he apparently failed to notice was Rogers's white token of surrender. Sprague jumped forward. He shook the shoulders of several of his men and shouted at them to shoot the man on horseback. They did, and horse and rider collapsed beside a large stump, a few yards in front of one of the embrasures of Battery Robinett. Rogers had been hit at least seven times at such close range that his body armor was useless.[27]

Captain Foster survived the volley. "Boys, you had better get away from here," he yelled, then started for the rear with Lieutenant Labruzan. A second Federal volley ripped the air, and a cannon in the redan belched a blast of canister. Lieutenant Labruzan dove behind a large stump. Foster was less fortunate. Recalled Labruzan, "Just then I saw poor Foster throw up his hands, and saying 'Oh, my God!' jumped about two feet from the ground, falling on his face. The top of his head seemed to cave in, and the blood spouted straight up several feet. I could see men falling as they attempted to run, some of their heads torn to pieces and some with blood streaming from their backs. It was horrible."

Labruzan tried to feign death, but the Yankees found him out. "Our boys . . . all around me were surrendering. . . . I was compelled to do so, as a rascal threatened to shoot me. I had to give up my sword to him. He demanded my watch also and took it; but I appealed to an officer and got it back. I had no means of defending myself. For the first time in many years, I cried to see our brave men slaughtered so," said the Alabamian. "I have never felt so bad in all my life."[28]

That any of the attackers survived was miraculous. The better part of five regiments had poured a converging fire on them. While the Eleventh Missouri raked their left, the Forty-seventh Illinois delivered a plunging fire into their right from near Battery Williams. Part of Colonel Sprague's Sixty-third Ohio contributed to the slaughter, as did the right companies of the Forty-third Ohio and part of the Twenty-seventh Ohio. Even the Regular infantrymen who manned the guns in Battery Williams, when they found the fields of fire of their cannon blocked, had picked up their rifle-muskets to join in the killing.[29]

North of Battery Robinett the fighting had been just as desperate. There the Sixth and Ninth Texas Dismounted Cavalry regiments fought point-blank with the Twenty-seventh Ohio. "There were numbers of men knocked down with fists and butts of muskets and trampled to death," averred an Ohioan. Col. Lawrence "Sul" Ross of the Sixth Texas was thrown from his white mare when the frightened horse bolted, giving rise to a rumor he had been killed. Ross was unhurt, but his regiment was decimated. After Ross went down, Maj. Robert White yelled at the men to fall back any way they could. A cannonball sliced off the hind legs of his horse, and White fell hard. Those of his Texans who got away did so on their hands and knees.[30]

Pvt. H. S. Halbert was trying to escape when his friend Jerome Kerr stopped him. Kerr's right shoulder was ripped open, and he begged Halbert to take off his belt and cartridge box to ease his burden. Halbert began to remove them, but the tongue of the belt buckle would not slip out of the buckle hole. While he tried to work it loose, a squad of Yankees opened fire from twenty yards. "I could do nothing further for Jerome, as the crisis was so pressing that everyone had to provide for his own safety," lamented Halbert. "There was a log lying near us, I suppose four feet off. In a few hurried words I told Jerome I could do nothing further for him, but to throw himself down behind this log, where he could be protected from bullets, and there would be some chance of saving his life. I there left him and made my own retreat the best I could," Halbert told Kerr's father. "We were at that time exposed to such a terrible fire, that I am satisfied that Jerome was shot down before I got six feet from him, and before he had time to get to the log."[31]

Between the Sixth Texas Dismounted Cavalry and the Twenty-seventh Ohio, the fighting had been nearly hand to hand. The Texans came within eight yards of the left center companies of the Twenty-seventh before Pvt. Orrin Gould of Company G shot down the Confederate color-bearer. He ran forward to grab the flag. A Rebel officer commanded his men to "Save the colors," at the same time firing a bullet into Gould's chest. The Ohioan kept coming. He covered his wound with one hand and snatched up the flagstaff with the other. Gould darted back into the ranks; as the Texans retreated, he waved the colors defiantly after them. The next day Colonel Fuller paid Gould a visit in the field hospital. The young man was stretched out on a cot, clearly in great pain. But "upon seeing me, his pale face was instantly radiant with smiles and pointing to his wound, he said, 'Colonel, I don't care for this, since I got their flag.'"[32]

Not all the attacking Confederates were stopped so abruptly as the Alabamians and Texans. Following the wooded bank and shallow bed of Elam Creek, General Phifer's Third Arkansas Dismounted Cavalry and Stirman's Arkansas Sharpshooters had slipped around the right of the Thirty-ninth Ohio, which had turned its attention to the Texans on its left flank. The Fifteenth and Twenty-third Arkansas regiments of Moore's brigade, along with most of the Thirty-fifth Mississippi, followed them. With General Phifer leading them on foot, hat in hand, the Arkansans and Mississippians charged across the Mobile and Ohio Railroad toward the Seventh and Fiftieth Illinois regiments of Du Bois's brigade. The Illinoisans were as demoralized as their comrades in Sweeny's brigade had been, and the two regiments disintegrated into small squads, fleeing past Rosecrans's headquarters with the enemy in close pursuit. Rosecrans hurled obscene epithets and threats their way, but no one paid him the slightest attention.[33]

Phifer's and John C. Moore's Confederates surged around the Federal headquarters and continued toward the Tishomingo Hotel. Rosecrans let himself be carried back by the wave of fleeing Yankees, across the track of the Memphis and Charleston Railroad to the doorstep of the hotel. To the commanding general the day was lost. Some fifteen minutes earlier he had ordered the chaplain of the Fiftieth Illinois to burn his train; now he told

Confederate High Tide at the Tishomingo Hotel *by Keith Rocco (courtesy of the Siege and Battle of Corinth Commission, Corinth, Mississippi)*

the survivors of the regiment to apply torches to the vast commissary and quartermaster stores stockpiled near the Tishomingo Hotel. The Illinoisans continued to ignore him, and the stores stood untouched when the Confederates came charging across the tracks a few minutes later. There, in front of the Tishomingo Hotel, the Arkansans of Phifer and John C. Moore mingled with the Missourians and Mississippians of W. H. Moore for a brief, triumphant moment before the Federals rallied to repel them.

Colonel Du Bois launched the first counterattack, against part of W. H. Moore's brigade. Moore was shot down near the railroad depot, and Du Bois drove his men 150 yards back through town before his own ranks were convulsed by a bombardment of shot and shell from behind. Posted on the east edge of town, Battery B, Second Illinois Light Artillery, had opened a blind fire on ground they thought to be held by the Confederates. Instead they struck dozens from the brigades of Du Bois and Baldwin and almost succeeded in eliminating their division commander. "Many of our men were killed by the shell and shot. . . . Seven or eight of these passed directly over my head, and one very close . . . taking off the legs of two of my brave soldiers directly in front," said Davies. "It was all up with us," confessed Du Bois, whose men scattered to fight or run as their consciences dictated.[34]

Du Bois's abortive counterattack had reversed the momentum of the battle in the town. A few minutes before 1:00 P.M. the Seventeenth Iowa completed the work. After the threat to Battery Powell was eliminated, General Sullivan turned his attention to the fighting in Corinth itself. He personally led the Seventeenth Iowa, the only one of his regiments not to have been committed to the fight for Battery Powell, up onto a low rise overlooking the streets on the northern edge of town. From there the Iowans picked off Confederates with impunity. When the left flank of W. H. Moore's brigade—or, rather, the left side of the jumbled mass of men his brigade had become—passed out of range, Sullivan ordered the Iowans to charge into town. They struck the rear of the Rebel mass before they could recover from Du Bois's counterpunch and the death of their commander. What remained of unit integrity disappeared. Dozens of Confederates dispersed to grab Yankee horses, which stood hitched to a fence beside the Tishomingo Hotel, and ride them to safety. Scores of Missourians and Mississippians, including the color guard of the Forty-third Mississippi, surrendered rather than run the gauntlet back to their own lines.[35]

John C. Moore's and Phifer's columns were thrown out of Corinth at about the same moment. Again it was the ever-present General Stanley who helped reverse the tide. He had posted the large and relatively fresh Fifth Minnesota Infantry in the rear of Fuller's brigade before dawn. Now, at 1:00 P.M., he called on the Minnesotans to repel the column that had eluded the Thirty-ninth Ohio. Col. Lucius Hubbard executed the order impeccably. He faced his regiment about, from the west to the east, and moved to the edge of the town square. There he waited until the enemy had spilled into the square before giving the command to fire. "The effect was tremendous, instantaneous," said the regimental chaplain. "The Confederates fell, staggered, turned back. The Fifth, the brandishing sword of Colonel Hubbard leading the way, hastened in pursuit. Beyond, other Union regiments, rallying from the confusion into which the Confederate charge had at first thrown them, fell in with the Fifth. The rout of the enemy was complete. The victory was ours."[36]

General Stanley was thrilled. "Should God spare me to see many battles I never expect to see a more grand sight than the battlefield presented at this moment. The enemy had commenced falling back from town and batteries before our advancing infantry. The roll of musketry and the flash of artillery was incessant as the enemy tried in vain to form line under fire."

From the edge of the timber, on the east side of the Mobile and Ohio

Railroad, a Confederate artilleryman watched the return of William H. Moore's Missourians: "It was very discouraging indeed to see the men falling back dispirited and gloomy, but they had done their whole duty . . . in a manner for which they deserve great credit and Missourians should never cease to venerate them."

Also on hand to greet the survivors was General Price. It was the first time the Missourians had been bested in a charge, and Price could not credit his senses. "My God! My boys are running!" he exclaimed. Then, in a low tone, as if talking to himself, he added, "How could they do otherwise — they had no support — they are nearly all killed."[37]

22

We Must Push Them

General Price may have whispered vague imprecations upon an unnamed culprit, but every soldier in his corps knew Mansfield Lovell was to blame for their repulse. While their comrades died by the score, Lovell's front was cloaked in a culpable silence.

Lovell had obeyed to the letter Van Dorn's discretionary orders that he move cautiously until Hébert was engaged. Wrote Maj. John Tyler facetiously, "General Lovell, with scientific and skilled precision, occupied his brigades for nearly two hours in searching the woods and beating about, as duty bound by his orders." Finally, about 9:30 A.M., the division debouched onto a ridge 600 yards northwest of Battery Phillips.

What they saw gave Lovell's men pause. On the horizon, said General Rust, was a strong redoubt and "long lines of infantry behind formidable-looking breastworks with abatis again in front." The ground between them and the brigades of McArthur, Crocker, Oliver, and Mower hardly inspired hope that an assault would succeed. Recalled a Mississippi lieutenant, "The ridge on which we were had been cleared away, and there was nothing larger or better to hide behind than an occasional blade of grass, or a dead leaf." After scanning the ground Lovell concluded to ignore Van Dorn's order that he "move rapidly to the assault" after Green became engaged. As Major Tyler put it, "Finally defining the works near the Female College, [Lovell] calculated their relative strength and rested upon the conviction of his own inferiority of power. . . . An assault upon

[Battery Phillips] was not in accordance with West Point tactics, and General Lovell abstained from making the . . . effort."[1]

That suited Generals Rust and Villepigue, but not Bowen. He fumed over the division's inactivity. In plain sight a half-mile to the east, Dabney Maury's command was being massacred before Battery Robinett. Incredibly, Lovell had disappeared from the front. Three times Bowen sent couriers importuning Lovell to return and lead the attack General Van Dorn had ordered. Lovell never responded. In frustration Bowen decided to test the Yankee lines himself. He sent sharpshooters toward the enemy, ordered the remainder of his infantry to lie down on the crest of the ridge, and then ran the four guns of the Watson (Louisiana) Artillery forward to shell Battery Phillips.

Bowen's purpose was laudable, but he had gravely miscalculated the Federal strength. The guns of the Watson Artillery fired their first rounds. Before the crews rammed home a second charge, Battery Phillips and College Hill erupted with the flash of a dozen cannon. Said a Mississippian who lay nearby, "In less than five minutes there was scarcely a man, horse, gun carriage, or caisson left of the outfit." Fifty infantrymen were also hit before Bowen pulled his command off the ridge. "There was no complaint when orders to fall back were passed along the line," quipped Mississippian I. E. Hirsh.

Not all Bowen's men shared Private Hirsh's lightness of conscience. Gazing upon Battery Phillips, Lieutenant Holmes of the Twenty-second Mississippi was ashamed of his unit's inaction: "We unexpectedly stopped under a hill to be protected from the shells of this fort. . . . Here we stayed, not daring to advance one foot, till General Price failed in his attack on Fort Robinett, and we, of course, had failed too, for outside of the skirmishers not a volley had been fired by our whole division."[2]

*　　*　　*

Under the blistering rays of a midday sun the survivors of the attacks on Batteries Powell and Robinett retreated into the woods northwest of Corinth. An Iowa Yankee watched them melt away: "With no guns and coats and hats gone a scattering few reach the timber and escape from the jaws of death. . . . I could not help but pity these poor fellows who thus went into certain and sure destruction here. They had been cut to pieces in the most intense meaning of that term. Such bravery has never been excelled on any field as in the useless assaults on Robinett."[3]

Van Dorn conceded the contest at noon: "Exhausted from loss of sleep, wearied from hard marching and fighting, companies and regiments without officers, our troops — let no one censure them — gave way. The day was lost."[4]

Van Dorn could not have rallied the men had he tried. Lovell's division excepted, the army was wrecked. What remained of Dabney Maury's division had been transformed into a unit of silent but studied insubordination. "Our men fell back in disorder, but sullenly," Maury wrote. "I saw no man running, but all attempts to reform them under the heavy fire of the enemy, now in possession again of their artillery, were in vain. They marched towards the timber in a walk, each man taking his own route and refusing to make any effort to renew the attack."

Capt. James Greene of the Eighth Wisconsin watched the histrionics of Rebel regimental and brigade officers unwilling to accept defeat. Some managed to get their men into line, but that was the extent of their success. Said Greene, "Again and again they formed their lines and advanced to the edge of the woods, but their men would go no further. Officers swore, and appealed to them to go in just once more; but they had enough."

Van Dorn and Price watched the proceedings from the Memphis road. Van Dorn, said Maury, "looked upon the thousands of men streaming past him with a mingled expression of sorrow and pity." Sterling Price was devastated. He "looked on the disorder of his darling troops with unmitigated anguish," Maury remembered. "The big tears coursed down the old man's bronzed face, and I have never witnessed such a picture of mute despair and grief as his countenance wore when he looked upon the utter defeat of those magnificent troops."[5]

It took an unexpected stirring in the Federal ranks to rouse the generals to action. Yankee skirmishers had swarmed into the field between the lines, amid resounding cheers from their comrades behind the breastworks. Fearful of a counterattack against his broken division, Maury appealed to Van Dorn for help. The Mississippian, in turn, called on Lovell to cover a general withdrawal. At 2:00 P.M. Lovell pulled Rust and Bowen out of range and inserted Villepigue's brigade between Maury's mob and the Federals. Villepigue formed astride the Memphis road, 1,000 yards from Battery Robinett. "These troops were in fine order, they had done no fighting," sneered Maury. Behind them the army streamed over the Chewalla road in retreat.[6]

*   *   *

Van Dorn and Maury overestimated their enemy. Pursuit was the last thing on the minds of most Yankee soldiers or their generals. Like their enemy, the Federals had endured two days of no food, little sleep, and de-vitalizing tension. "When the excitement of battle was over," said General Stanley, "they lay down exhausted on the ground."

General Rosecrans sank down with them. Confederate cavalry probes south of town momentarily convinced him the attacks on Batteries Powell and Robinett had been but a prelude to greater efforts against his flanks, and he scrambled to meet the threat. Rosecrans told Hamilton to regroup and watch for an attack from the east, and he had McArthur reconnoiter the Kossuth road to look for Confederates southwest of Corinth. At 2:00 P.M., faint from fatigue, Rosecrans sought the shade of a tree.[7]

Three distant explosions calmed his fears. To his staff the Ohioan remarked that Van Dorn must be blowing up his ammunition wagons, a sure sign of a retreat. "We must push them," Rosecrans added vaguely. But apart from sending Sullivan to skirmish with the fleeing Confederates, Rosecrans let Van Dorn alone. Not until 6:00 P.M., when Sullivan reported the enemy well on his way toward Chewalla, did Rosecrans rouse himself. In the company of his escort and staff he rode into the forest beyond Battery Robinett. Satisfied Van Dorn indeed had gone, Rosecrans returned to address the troops.

The battle had been won, he announced. After two days of battle and two sleepless nights of marching, the men should "replenish their cartridge boxes, haversacks, and stomachs, take an early sleep, and start in pursuit by daylight."[8]

Most welcomed Rosecrans's words. But some soldiers, able to look beyond their exhaustion, questioned the wisdom of allowing Van Dorn to steal nearly a day's march. Lt. William McCord of the Sixty-fourth Illinois wondered why McKean's division, which had done no real fighting the second day, was not sent after the enemy at once. So, too, did Colonel Chetlain. Sgt. Charles Hubert of the Fiftieth Illinois had the same thought. Had McKean or Hamilton moved out immediately, "a rich reward would have resulted. But," he lamented, "a contrary course was adopted, and thus a breathing spell was given to the broken and disheartened Rebels."

Some started out spontaneously after the enemy, and they were richly rewarded for their enterprise. Said Lieutenant McCord, "Squads of soldiers from our lines without orders did rush forward and made prisoners of large number of the enemy, in some instances one man capturing from five to twenty and in another instance twenty men capturing three hundred." Among those the Federals ran into as they started after the enemy

The Aftermath of Battle, Battery Robinett *by Keith Rocco*

was Joseph Mower. He had sustained a flesh wound in the back of the neck before surrendering. In their haste to depart, his captors neglected to post a guard on him, and he slipped away.[9]

Sleep came hard that evening. The dead lay in windrows among the living, and under the hot sun their bodies decomposed quickly. The stench was horrible. Commissary whiskey was ladled out liberally to keep the men from vomiting. Before Batteries Powell and Robinett the scene was particularly gruesome. Recalled a Federal, "The mangled bodies of living and dead before those forts should be seen, if one would have any adequate idea of them; heads carried off so that no trace of them could be found—so with limbs—others having all the flesh torn off the bones, leaving them white and bare." Col. Augustus Chetlain rode to Battery Robinett with a party of officers from Davies's division. Climbing the parapet, they counted thirty-six Rebels in one pile near the ditch.[10]

An instant of joy relieved the sorrow around Battery Robinett. Word reached Colonel Fuller that Smith had regained consciousness and that the surgeons had pronounced the wound not mortal. "I jumped upon a fallen tree in rear of the Forty-third Ohio and sang out to them that Colonel Smith was not killed, but would recover," Fuller said. "This was repeated by Swayne and others, and the cheer which followed, taken up by the men of other regiments also, would have gladdened Kirby's heart."

Fuller and Stanley rushed to Smith's bedside. They found him awake but too badly shot up in the mouth to speak. Smith offered a smile and a salute from his cot. Stanley sat down beside him. Smith motioned his desire to write something. Fuller handed him his memorandum book and a pencil. Smith scratched out a question: "How did my regiment behave?" Stanley took the notebook from him and began to write a reply before a quizzical look from the wounded colonel reminded Stanley and Fuller he could hear well enough, and Stanley answered, "Most gallantly." "This seemed to please Smith greatly, and he at once acknowledged it with one of his graceful salutes," said Fuller.

When Stanley rose and passed along to speak to others, Fuller took his place at Smith's side. Rather awkwardly Fuller told Smith he would do anything the junior colonel wished. Would he like to have him write to his mother? Smith nodded "Yes." Was there anyone else he wished Fuller to write? Smith "made no sign in response, but seemed hesitating about something he felt loth to drop, and kept looking at me with a steady gaze." Fuller understood. Should he write to Smith's fiancée? Smith smiled and nodded. With a promise to do so, Fuller returned to his command.[11]

About the time Stanley and Fuller called on Smith, Rosecrans rode into Corinth to find General McPherson on hand with the five regiments Grant had ordered from Jackson, Tennessee. McPherson handed him a dispatch from Grant. Expecting victory, Grant had ordered Rosecrans to pursue Van Dorn vigorously. He reminded him that Hurlbut would be within two miles of Davis Bridge by nightfall. If Rosecrans followed Van Dorn closely, he and Hurlbut might crush the Mississippian between them. But success depended on their cooperating: "If the enemy falls back push them with all force possible and save Hurlbut who is now on the way to your relief. The Corinth and Bolivar forces must act in concert," admonished Grant. "Hurlbut is not strong enough alone to handle the rebels without very good luck. Don't neglect this warning. I can reinforce you no more from this on — hence you will see the vital importance of your and Hurlbut's forces acting in conjunction."[12]

But neither Grant's peremptory order nor the presence of five fresh regiments changed Rosecrans's mind. He would start at dawn, with McPherson in the lead.

Rosecrans decided to pursue along two routes. McPherson was to march initially along the Memphis road, then pick up the Columbus road and continue toward Pocahontas. The divisions of Stanley and Davies would support him. McArthur would take his brigade, Oliver's brigade, and a squadron of cavalry out the Memphis and then the Chewalla road, also in the direction of Pocahontas. McKean was to follow him with Crocker's brigade and the rest of the cavalry. Hamilton would trail McKean. Rosecrans told his division commanders to issue three days' rations and 100 rounds of ammunition, then bedded down for the night.[13]

* * *

Earl Van Dorn was near collapse. He had counted on victory at Corinth to erase the stigma of Pea Ridge. While his lieutenants tried to rally their men, Van Dorn searched his troubled mind for a way to save himself.[14]

The troops thought mostly of what had been lost and who was to blame. Cavalryman Edwin Fay had very definite ideas on both questions. He wrote his wife angrily, "The Yankees killed a great many of our men and . . . our retreat was conducted with the greatest confusion. . . . Van Dorn was drunk all the time and Villepigue too and I expect Price too. Everybody was commander and Price did the fighting. We lost half of Price's army killed and *straggling*. Such demoralization was never seen in an army before. I think the cause of the Confederacy is lost in [the] West." Lt. Col. Hubbell of the Third Missouri agreed: "I fear the disaster will be a national one. But it certainly was at a terrible hazard we made the attack." Missourian Ephraim Anderson pondered losses more personal. Three of his closest friends were dead. One had had his brains blown out by a grapeshot in front of Battery Powell, and the other two had fallen inside the works. Anderson's company commander had been badly wounded, and hardly a messmate had escaped without some injury. "That night, when weary and worn, I stretched myself on a blanket, my feelings were very melancholy and depressed: the scenes and events of the day came up before me in all their dark and gloomy reality," recalled Anderson. Dabney Maury expressed the agony of Anderson and thousands like him, when he wrote, "The utmost depression prevailed throughout the army."[15]

Blind to the dejection so evident to Hubbell and Maury, that afternoon Van Dorn concocted a plan to renew the battle.

Assuming they would bivouac on the west bank of the Tuscumbia River — as far from the Federals as possible, Van Dorn's generals were startled to receive, at sunset, orders to make camp at Chewalla. "Why [the army] was stopped here was to everyone an enigma," said Major Tyler. "If retreating, and there seemed nothing else to do, we surely should have crossed the Tuscumbia River only four miles off, where our trains still reposed, if indeed, we did not continue over the Hatchie, requiring a march of only four miles more, while the bridges were yet in our possession."

When they learned why the army had halted, the generals were mortified. Van Dorn had no intention of retiring; rather, he would march rapidly south along the Mobile and Ohio Railroad to Rienzi, then double back and attack Corinth from the south.[16]

Price learned of the plan at midnight. It came, he thought, from "a mind rendered desperate by misfortune." Assembling his staff officers, all of whom agreed Van Dorn's scheme was preposterous, Price sought out Dabney Maury. Despite his continued admiration for its author, he, too, thought the plan absurd. "My division had marched from Chewalla to attack Corinth with four thousand eight hundred muskets the day but one before," said Maury. "We left in the approaches and the very central defenses of Corinth two thousand officers and men killed or wounded, among them were many of my ablest field and company officers." With what was he to carry out Van Dorn's orders? Maury wondered angrily.[17]

Price and Maury convinced Van Dorn to call a council of war. Price did most of the talking. How, with two of his three divisions wrecked, did Van Dorn think he was going attack Corinth? "Van Dorn, you are the only man I ever saw who loves danger for its own sake," Maury interjected. "When any daring enterprise is before you, you cannot adequately estimate the obstacles in your way." Van Dorn pondered a moment, then replied, "While I do not admit the correctness of your criticism, I feel how wrong I shall be to imperil this army through my personal peculiarities, after what such a friend as you have told me they are, and I will countermand the orders and move at once on the road to Ripley." The army would cross the Hatchie River at Davis Bridge, Van Dorn added, then turn south for the twenty-eight-mile march to Ripley.[18]

Welcome as was Van Dorn's return to reason, it appeared to have come too late. Before dawn on October 5, couriers from Wirt Adams reported the Kentuckian had clashed with Federal cavalry six miles west of Davis Bridge the day before. Following the Yankee horsemen was infantry of indeterminate strength.

A new crisis presented itself. No one in Van Dorn's military circle had expected a fight from behind; all assumed the only threat to be from Rosecrans. Consequently, Van Dorn had assigned his strongest division, Lovell's, to rear-guard duty and had placed his most depleted command, Maury's, in the lead. There was no way to reorder the march column, so Van Dorn improvised. He ordered Armstrong and Jackson, whom he had sent to Rienzi, to rejoin the army at once. At dawn he directed Lt. Col. Edwin R. Hawkins, whose First Texas Legion guarded the supply train two miles east of Davis Bridge, to join Wirt Adams on the Hatchie River.

Together they were to delay any Federal crossing until the main body came up.[19]

The army marched at sunrise. Van Dorn and his staff rode with the vanguard. At 10:00 A.M. a courier from Adams told them "the enemy in heavy force is moving from Bolivar to oppose the crossing of the Hatchie." Even as they spoke, the courier added, Adams was heavily engaged a mile west of the river. Van Dorn felt the noose tighten. As he confessed in his report, "Anticipating that the Bolivar force would move out and dispute my passage across the Hatchie Bridge I pushed rapidly out to that point in hopes of reaching and securing the bridge before their arrival, but I soon learned by couriers from Colonel Wirt Adams that I would be too late. I nevertheless pushed on with the intention of engaging the enemy until I could get my train and reserve artillery on the Boneyard road to the crossing at Crum's Mill."[20]

Van Dorn looked to Maury to hold off the Yankees until the army was across Crum's Bridge, which spanned the Hatchie River six miles south of Davis Bridge. "Maury, you are in for it again today," said Van Dorn. "Push forward as rapidly as you can and occupy the heights beyond the river before the enemy can get them."[21]

With his line of retreat in jeopardy and his options limited to one, for perhaps the first time in his military career Earl Van Dorn understood the odds against him.

23

I Never Saw Such Slaughter

Stephen Augustus Hurlbut was a volunteer soldier who brought to the army a talent for politics, a taste for the bottle, and a lack of ethics. A South Carolinian by birth, Hurlbut moved to Belvedere, Illinois, at age thirty and entered politics as a Democrat. Ten years later, while a state representative, he switched allegiance to the new Republican Party. Most people changed parties to combat slavery; Hurlbut did so for spoils.

Hurlbut's primary virtue was his ability to pick a winner. Early in Lincoln's political career he made himself a useful, if obsequious, ally. As president, Lincoln rewarded his loyalty with a brigadier general's commission. Hurlbut was assigned to duty in Missouri, where he embarrassed the president and himself by being frequently drunk on duty. The press pilloried him, and his political patrons in Illinois abandoned him. But Lincoln gave him a second chance, and Hurlbut fought well at Shiloh.[1]

At dawn on October 4 Hurlbut led the Fourth Division of the Army of the Tennessee, 5,000 strong, out of Bolivar. He covered twenty-three miles before dark and bivouacked along Big Muddy Creek, three miles from Davis Bridge. Hurlbut's cavalry pushed Wirt Adams's pickets to Metamora, a hamlet along a high ridge overlooking the Davis farm and the wooded banks of the Hatchie River, at the intersection of the State Line, Ripley, and Pocahontas roads. A late afternoon counterattack drove the Yankee troopers back toward Big Muddy Creek. From Adams's spirited defense Hurlbut guessed he would run into the Rebel army the next

Maj. Gen. Stephen A. Hurlbut
(Library of Congress)

morning along the Hatchie River, and he cautioned his brigade commanders, Brig. Gens. Jacob Lauman and James Veatch and Col. Robert Scott,
to move carefully.[2]

Veatch set off at 8:00 A.M. on Sunday, October 5, a warm and sunny
morning. The way forward was horrible. Deep hollows and ravines bisected the road, and dense thickets and woods impeded maneuver off it.
Veatch lacked a cavalry screen, as Hurlbut had diverted his cavalry to the
left of the State Line road in order to take Adams's troopers, whom Hurlbut presumed to be on the Metamora heights, in the flank. Expecting
trouble, Veatch deployed his lead regiment, the Twenty-fifth Indiana,
across the road and pushed a two-gun section of Capt. William Bolton's
Battery L, Second Illinois Light Artillery, behind it.

Hurlbut lost the chance to direct the coming battle to General Ord,
who arrived from Bolivar just as the Twenty-fifth Indiana set out.
Rather insouciantly, Ord approved of Hurlbut's arrangements and allowed Veatch to continue.[3]

For 800 yards Veatch's Indianans traipsed through a tangle of underbrush and vines without encountering a single Rebel. Not until they
reached the Robinson farm, two miles west of the Hatchie River, did they
run into opposition. There several dozen pickets from Wirt Adams's

cavalry had taken shelter, and they clung to the farm so tenaciously that Veatch thought he had run into enemy infantry. Captain Bolton came forward to shell the place. Six rounds sufficed to set the farm ablaze and scatter the Rebels. The Twenty-fifth Indiana surged over the knoll until Veatch stopped the Hoosiers in order to add the Forty-sixth Illinois and the Fourteenth Illinois to the front line for the last leg of the march.[4]

At Davis Bridge, Wirt Adams prepared for the Federal onslaught. Excitement got the better of him, and Adams parceled out his forces recklessly. Hoping their presence on the west bank would slow the Yankee skirmishers then spilling over the Metamora ridge, Adams threw Lt. Col. E. R. Hawkins's First Texas Legion, 360 strong, across Davis Bridge. His own skirmishers spread out on their right, in the huge Davis field, and peppered the Federals.[5]

The Confederates would have done better to stay on the east bank, where terrain made common cause with defenders to create an almost impregnable barrier to a lodgement. The Hatchie twisted and turned in a tortuous and baffling course near Davis Bridge, and maneuvering room on the east bank was at a premium. Three hundred yards north of the bridge the river bent sharply from the southwest to the south. Just south of the bridge its green and sullen waters again changed course, twisting abruptly to the east. The State Line road paralleled the Hatchie along this stretch, with only a half-acre between the road and the river bank. Badly overgrown with weeds and blackberry bushes, the road here was more a trail than a true thoroughfare.

Five hundred yards east of Davis Bridge the ground rose from spongy bottomland to a steep, timbered bluff—a position of great natural strength dominating Davis Bridge. Taken together, the bluff and the river banks formed the boundaries of a twenty-two-acre killing zone of scrub pine, sassafras, wildflowers, and clinging vines into which any force crossing the Hatchie would be channeled.

Dabney Maury had no time to ponder the terrain. General Van Dorn had charged him with occupying the Metamora ridge before the Yankees could carry it, and he personally led J. C. Moore's brigade and Captain Dawson's St. Louis Artillery across the rickety, rotten timbers of Davis Bridge, close on the heels of the First Texas Legion. Fatigue, a lack of water, and the morning heat had caused scores to fall out of the ranks, so that Moore had only 300 men with him when he crossed the bridge. Maury and Moore rode ahead, unaware the Yankees were on the reverse slope of the Metamora ridge.[6]

Maury's display impressed the Yankees. Remembered an Indianan,

"When we got upon the hill, we could see the Rebels crossing the Hatchie River on the bridge. They were about one-half mile away. The ground between us was clear, it being a farm. They were crossing their troops and forming a line of battle and running some artillery over and getting them into position." To some it looked as if the Rebels would never stop coming. "An awfully majestic sight was now afforded us," said Illinoisan James Dugan. "We could see thousands of 'Butternuts' file right and left, and take up their position in the margin of the woods on the opposite side of a field from us, preparatory to a series of brilliant movements, that they considered sure to result in our discomfiture." Dugan wildly overcounted the enemy. No more than 1,100 dismounted troopers and tired infantrymen faced Veatch, who had arrayed his entire brigade along the heights in a three-quarter-mile battle line.[7]

Veatch's artillery served him well. Dawson's Confederate battery had unlimbered west of the Davis house on a hogback, square in the middle of the field. The Rebels opened fire first, but their aim was bad. Most of their missiles burst harmlessly behind the Metamora ridge. Hurlbut's chief of artillery, Maj. Charles Campbell, saw to it the Federal return fire was more exact. At a range of 750 yards he opened on the head of Moore's column with two batteries. The first shot spattered sand on General Maury, who knew a losing proposition when he saw it. He commanded Moore's brigade to file off the road and take cover in a skirt of trees lining a wisp of a creek called Burr's Branch. Moore brought his left into contact with the right flank of Hawkins's Texans.[8]

After Moore disappeared into the timber, the Yankee batteries trained their guns on Dawson's exposed cannon. For forty-five minutes they pounded the Missouri battery, eviscerating horses and smashing limbers and caissons. Dawson kept up a valiant but pointless fire that emptied his ammunition chests but hurt few Federals.[9]

Generals Ord and Hurlbut watched the artillery duel intently. When, a few minutes after 9:00 A.M., Dawson's fire weakened, Ord commanded Veatch and Scott to fix bayonets and attack. The Federal cannon ceased firing, and the Northerners swept off the ridge, running and yelling. They leapt ditches, climbed fences, and shoved through brambles and hedges to close with the enemy. The sheer weight of Federal numbers overwhelmed Moore and Hawkins, and their efforts at resisting proved worse than futile. The Fifty-third Indiana struck hardest, crashing into the angle formed by the junction of Hawkins's right flank and Moore's left. Hawkins tried to fall back, but in the confusion his First Texas Legion broke apart. Hawkins guided half the legion to the bridge in fair

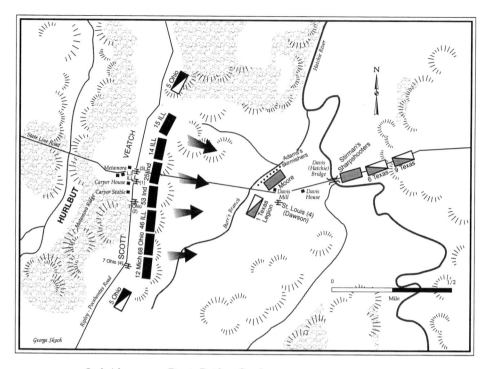

MAP 11. *Ord Advances on Davis Bridge, October 5, 9:00 a.m.*

order. His senior captain got away with most of the rest, but seventy-five Texans fell captive.

The defection of the First Texas Legion doomed Dawson's battery. With no ammunition left and too few horses to pull their pieces, the Missourians abandoned all but one gun to the Fifty-third Indiana.[10]

North of the State Line road, the Fourteenth and Fifteenth Illinois regiments completed the destruction of Moore's brigade. When Moore changed front to contend with the Twenty-fifth and Fifty-third Indiana regiments, which had edged around his left flank after Hawkins withdrew, the Fourteenth and Fifteenth slipped into a field of tall corn beyond his right. There they paused, aimed, and fired. Moore was stunned: "We had not fired more than two or three rounds before a perfect shower of balls was poured into our right flank from the direction of the corn field." The Illinoisans raised a cheer and charged past Moore's flank for the river. "We now saw that we must either fall back or be surrounded," said Moore, who gave the order to retreat.

It came too late. The Yankees had a head start, and they blanketed the bridge with deliberate volleys. Moore's men scattered. Some threw aside their rifle-muskets and plunged into the river, where dozens drowned. A few ran the gauntlet of Yankee fire and recrossed Davis Bridge, but most chose surrender. "As for my part I was so hot and tired that I didn't care about trying to swim with my clothes on and risk getting shot in the back," said Arkansan Albert McCollom, "so of course I surrendered." McCollom had plenty of company. Bunched into a river bend north of the bridge, 200 Rebels were, as one Confederate officer put it, "gobbled up." Scores more were tracked down in the tall grass and trees near the river.[11]

The Federals had a field day. William Garner and Charles Rafesnyder of the Fourteenth Illinois were told by their company commander to escort six prisoners back to Metamora. After the captain walked away, Garner remarked to his partner, "Charley, haven't we the power to take care of more prisoners than the captain has assigned us?" "I believe we could," agreed Rafesnyder. "Well, then, let's follow down the river and I think we'll find some of them hidden under the bank," said Garner. Combing the bank, they found a Rebel crouched under some bushes at the water's edge. Garner raised his rifle-musket and barked at the man to give up. The brush stirred, and the Southerner came out. The brush stirred again, and again. More Southerners stood up and showed themselves, until Garner and Rafesnyder found themselves surrounded by two captains, three lieutenants, and thirty-two armed soldiers. Garner got in the first word. "Boys, you are in a tight place, and I am sorry, yes, very sorry for you. The woods are filled with our scouting parties. There is no escape for you. Now, I will give you your choice, which is to surrender to me, and I will take you back to Metamora, where you shall be well treated and have plenty to eat and drink, or you can shoot us down and take the consequences when our men capture you, for capture you they will." Garner's bluster—and the promise of food—carried the day. The Rebel officers conferred, then told their men to stack their arms and get in line. Garner gave the command "Forward, march," and he and Rafesnyder led their company of captives down the State Line road.[12]

Moore's brigade had ceased to exist. "Scarce an organized company came out of the conflict," lamented Dabney Maury. That night the major of the Forty-second Alabama, who himself had been wounded in the chest, mustered only ten men present for duty. The rest, he told a staff officer from division headquarters, had been shot or captured or had straggled off. The only regiment to recross the Hatchie River intact, the Thirty-fifth Mississippi, dispersed into the woods. Their commander had been sent back to Corinth under a flag of truce to bury the dead. With no one to rally them, the leaderless Mississippians started for home. More than 300 deserted. In the brigade as a whole, only 600 officers and men were left of the 1,895 who had marched on Corinth two days earlier.[13]

Reinforcements arrived, but bad leadership almost condemned them to Moore's fate. Stirman's Arkansas Sharpshooters of Phifer's brigade came upon the field first and were blithely waved across Davis Bridge by Sterling Price, who yelled, "Boys, if there are many we will take them, and if there are but few we will take them the quicker." The Sixth and Ninth Texas Dismounted Cavalry regiments followed closely.

The acting brigade commander, Col. Sul Ross, was across with them before he could take in the situation. Only twenty-four years old, Ross was one of the hardest fighters in the army, having been seasoned battling Comanches on the Texas frontier during college vacations.

By the time the Texans were over, the fighting on the west bank resembled frontier bushwhacking. Ross was in his element but could contribute little; fugitives from Moore's brigade broke up his ranks before he was able to form a battle line. Moore was among them, yelling at Ross's men to run before the Federals trapped them. Colonel Ross hesitated, and General Maury gave the order to retreat for him. Most got back across the bridge safely, but Veatch's Federals gathered in 100 of Ross's men.

Ross took his men back to the bluff overlooking Davis Bridge. On the crest they lay down alongside the survivors of Moore's brigade and Hawkins's legion. Also on the bluff was a long line of cannon, placed there by the division chief of artillery, Maj. W. E. Burnet. While Price was foolishly enjoining the infantry to cross the river, Burnet had scouted the east bank for a good firing position. The bluff suited him, and Burnet, whom Maury praised as "one of the bravest and ablest artillery officers of our army," massed five batteries along its crest. Burnet had the cannon charged with double canister and trained on the bridge. Then, with his artillerymen, he awaited the Federals' move.[14]

*   *   *

Something about the fighting around Davis Bridge inspired foolhardiness. In sweeping the enemy from the west bank of the Hatchie River, General Ord achieved a singular success: he had closed off the enemy's line of retreat at the loss of fewer than 100 men. Ord need simply hold what he had won, and Van Dorn would be compelled to search for another crossing site farther downriver, which would consume the rest of the day. Rosecrans could use the time to close up behind Van Dorn.

Of course Ord was unaware he had struck the vanguard of a retreating army. For all he knew, the Rebels might have defeated Rosecrans and were turning to destroy the Bolivar column. In either case, however, logic dictated that Ord maintain a strong defensive position until the enemy revealed his intentions. But Ord was beyond the reach of rational thought. Like Adams, Maury, and Price before him, the Marylander lost himself to the excitement of the moment, and he told Veatch to cross the bridge and form his and Colonel Scott's regiments in line. Ord also sent word to Jacob Lauman to press forward, intending to commit his brigade on the east bank as well. He envisioned a front twelve regiments long, with six regiments on either side of the State Line road.

The orders to Veatch and Lauman horrified Hurlbut. He remonstrated passionately with Ord; there was only a half-acre between the south side of the road and the river, hardly enough for one regiment, much less six. Hurlbut spoke with authority. Earlier in the year he had camped on the very bluff Ord hoped to take. But Ord was adamant. Not only did he ignore Hurlbut's warning, but he also directed Bolton's battery to shell the Rebels on the bluff from the east bank. How he expected the gunners to aim uphill through the forest, Ord failed to say.[15]

It was the misfortune of the Fifty-third Indiana to be nearest the bridge when Ord ordered the crossing. The Indianans started across at 11:30 A.M. in column of fours. Canister and rifle volleys ripped the regiment apart. Lt. Col. William Jones herded the men off the road by the right flank into the soggy strip beside the river. Pandemonium ensued. Gunpowder smoke dispersed slowly in the dank, humid air, cutting visibility to a few feet. Canister continued to rake the Yankee ranks. To escape it, the men dove into the river or clung to the bank, refusing to rally.

The Fourteenth Illinois crossed next and filed to the left into an open field. The extra room to maneuver meant nothing. A shower of shell and canister sent the Illinoisans reeling to the river bank as well. Behind the Illinoisans came the Twenty-fifth Indiana. In keeping with Ord's vague orders, the Indianans filed to the right, straight into the milling mass the Fifty-third Indiana had become. In a matter of minutes the two regiments were hopelessly intermingled.

The Fifteenth Illinois negotiated Davis Bridge next. When Col. Cyrus Hall saw the regiment coming, he led his Fourteenth Illinois from the river bank to a thick forest. The Fifteenth filed to the left behind Hall's command, changed front forward, and came up on his left. There, sheltered by trees and fallen timber, both regiments lay down and awaited orders. Back on the west bank, General Veatch watched the slaughter and, with his own Forty-sixth Illinois and Scott's Provisional Brigade, also waited for Ord's next command.[16]

It proved a foolish one. A few minutes before 1:00 P.M. Jacob Lauman rode up to Davis Bridge at the head of his brigade. Lauman's men had run two miles to the bridge and were played out. But Ord gambled Lauman's command on what was, to those on the east bank, obviously a lost wager. He told Lauman to leave two regiments on the west bank and cross the other two to reinforce Veatch's troops in the pocket south of the State Line road.

The Fifty-third Illinois led the way at the double-quick time, plunging off the road into the confused swirl of blue on the far bank. Said their commander with decided understatement, "After crossing, and before I had formed in line of battle, I was met by men falling back, which staggered my men a little, but they recovered, formed in line, and commenced firing upon the enemy, who . . . soon opened a murderous fire with canister, shot, and shell, together with small arms."

A truer picture of affairs on the Illinoisans' front came from the pen of Sgt. Mark Bassett of Company E. The company, said Bassett, came apart on the bridge. Capt. Charles Vaughan fell like a rag doll in the road, his knees crushed by canister. First Lt. Armand Pallissard shepherded the men into the tall grass. When they tried to resist, he waved his sword menacingly and yelled, "Men, stand firm; we must not lose our ground . . . ." A canister shot ripped open his chest in mid-sentence. Pallissard fell dead, and the company broke up. Capt. John McClanahan, the acting regimental commander, noticed the confusion. "Who is in command of this company?" he demanded. "All the officers are killed or wounded," someone replied. McClanahan turned to Bassett. "Throw down that musket, take the sword off that officer and take command of the company," he said, pointing to Pallissard's body. Bassett and another man bent over Pallissard. Bassett watched as the soldier "turned him on his back so as to get at the belt fastening, unbuckled the belt and removed it, turned the breathless body back again on its face just as it fell a few minutes before, then removed the sword and fastenings and assisted in putting it on me. I assumed command."

The colonel of the Twenty-eighth Illinois, Amory K. Johnson, was painfully honest about his regiment's plight. His men stumbled into the rear rank of the Fifty-third, confessed Johnson. He tried to move them to unmask their front but fast ran out of ground. Those with a clear field of fire began to shoot. That they hit anything through the trees and brush, said Colonel Johnson, was unlikely.[17]

The Confederates, on the other hand, could hardly miss. Col. Ras Stirman had never seen such a wealth of targets. His Arkansas Sharpshooters fired with impunity. Said Stirman, "We would allow them to approach until we could see the whites of their eyes, then without exposing ourselves in the least, we would pour volley after volley into them, cutting them down like grass. I never saw such slaughter in my life."

The Confederates had been reinforced further. Simultaneous with Lauman's appearance on the east bank, Cabell's brigade came onto the bluff beside Ross. Cabell had just 550 men, but the natural strength of the position more than compensated for his meager numbers. As the Arkansans opened a rapid fire from the crest, Cabell's sole worry was that he would run out of cartridges before the Yankees broke off the fight.[18]

With four regiments huddled helplessly on a half-acre of ground, facing slow but certain annihilation, General Ord at last awoke to his error. But rather than recall them, he started across himself, determined to rally at least the Fifty-third Illinois. A blast of canister denied him the chance, and he toppled onto the bridge with an iron ball in his leg.[19]

Command again was Hurlbut's. The South Carolinian acted swiftly and wisely. Judging that the regiments on the east bank would suffer as greatly trying to recross Davis Bridge as they had charging over it, he elected not to recall them. But Hurlbut had no intention of reinforcing failure. Leaving the crowd south of the State Line road temporarily to its fate, Hurlbut ordered every available regiment across the river to the north side. By extending the line the Fourteenth and Fifteenth Illinois had begun, Hurlbut hoped to outflank the Confederates.[20]

The results were mixed. General Veatch moved the Forty-sixth Illinois and the two regiments of Scott's Provisional Brigade to the east bank at 3:00 P.M., extending the line northward by several hundred yards. But Lauman misunderstood Hurlbut's orders, and he personally led the Thirty-second Illinois into the deadly pocket south of the road, where four regiments were vying for the chance to be obliterated first. In his eagerness to lead the Thirty-second, Lauman neglected his second regiment, the Third Iowa. Commanded by a captain, the Iowans stumbled across the bridge with no idea where to deploy. Fifty-seven men and half the officers were cut down crossing. In their haste to clear the bridge, the Iowans ran into the rear rank of the Twenty-fifth Indiana. Capt. Matthew Trumbull reacted well: "I saw no way to extricate the regiment but by planting the colors in the middle of the road and ordering the men to rally to them and form a new line of battle. This was promptly done, nearly every man springing to his place."

Trumbull fairly appraised his men's response. The Confederate fire had fallen off to almost nothing. With no one shooting at them, the Yankees south of the road came out from beneath the river bank and rallied easily to their colors. For the first time since crossing three hours earlier, regimental officers were able to assemble something approximating a line of battle. Availing themselves of the lull, Veatch and Lauman set their commands in motion toward the bluff shortly after 3:00 P.M. Hurlbut was thrilled: "It is among the proudest moments of my life when I remember how promptly the several regiments disengaged themselves from their temporary confusion and extended to the left, and with what a will they bent themselves to conquer the hill." Hurlbut neglected to mention the agony their muddle had inflicted. Behind the advancing lines, heaped on the bridge, strewn along the road, or clustered in the half-acre-wide thicket south of it, were more than 500 dead and wounded Yankees.[21]

24

Van Dorn Has Done It, Sure Enough

Earl Van Dorn passed the afternoon of October 5 in an agony of his own making. In a few short hours he had been driven from the notion of again attacking Corinth to improvising a way to save his army. His train of nearly 500 wagons and most of his infantry lay in the fork of the Tuscumbia and Hatchie Rivers. Neither was fordable. The nearest bridge over the Hatchie River — Davis Bridge — was already in Federal hands, and Young's Bridge over the Tuscumbia lay in Rosecrans's path. Van Dorn's only hope was to hold the enemy on the far banks of both rivers while Armstrong's cavalry secured an alternate crossing site along the Hatchie.

Ironically, that morning Van Dorn had ordered Armstrong to destroy the very bridge that looked now to be his only path of escape. Spanning the Hatchie River six miles south of Davis Bridge near a gristmill, Crum's Bridge was small and rickety but sturdy enough to support infantry and wagons. It was reached by way of the Boneyard road, an overgrown, back-country trail that branched off from the State Line road two miles east of Davis Bridge. From there the Boneyard road ran southwest, over the Hatchie River at Crum's Bridge and on to the Ripley and Pocahontas road, which it bisected eight miles south of Metamora.

Armstrong had executed Van Dorn's order too rapidly. By the time the sounds of battle rolled up the river from Davis Bridge, troopers from the Seventh Tennessee Cavalry had yanked up the floor planks and ignited the supports of Crum's Bridge. "With a soldier's instinct," said General Maury gratefully, Armstrong "understood at once the condition of affairs. He sent a courier to Van Dorn to say that he might turn the train and army into the Boneyard road, and he would have the bridge repaired by the time they would reach it."

Armstrong was as quick to rebuild as he had been to demolish. Beside the burning frame of Crum's Bridge he erected a temporary bridge of cast-off logs and puncheons atop an old dam and smoothed the steep river banks leading up to it. Armstrong had logs piled in great heaps, to be ignited after dark to guide the army to the crossing site.

Maury withdrew from Davis Bridge at 3:00 P.M., after Armstrong informed him the dam bridge was ready. His retreat was unmolested, as Hurlbut's Federals were too disorganized to give chase. Maury passed the 1,200 soldiers left to him through Martin Green's Missourians, who had formed behind him. By 6:00 P.M. all were well on their way toward Crum's Bridge.[1]

For six hours Maury had withstood assaults by an enemy four times his number. He had inflicted 570 casualties at a cost primarily of the 300 men from Moore's brigade captured on the west bank of the Hatchie. Federal fire had struck few. Only seven men were killed and twenty-two wounded in Ross's brigade. Cabell had lost just two killed and eight wounded, a "thing unprecedented, considering the obstinacy of the fight," averred the Virginian. Most importantly, Maury had saved an army that Van Dorn had been ready to gamble away the night before.[2]

*   *   *

General Grant awoke October 5 with high hopes for the destruction of the Rebel army. He found no fault with Rosecrans for delaying the pursuit — "two days' hard fighting without rest probably had so fatigued the troops as to make earlier pursuit impossible," Grant wrote — but he expected great things of the Ohioan once he got started. At 8:00 A.M. he wired Halleck that Rosecrans was well on his way, and after news of the Davis Bridge fight reached him, Grant added, "At this distance everything looks most favorable and I cannot see how the enemy are to escape without losing everything but their small arms. I have strained everything to take into the fight an adequate force and to get them to the right place."[3]

Far removed from the fighting, Grant might easily entertain false expectations for the army at Corinth. But Rosecrans, who awoke to the stench of rotting flesh and lingering gun smoke, had to face squarely difficulties Grant blithely dismissed. His army was still exhausted; one night's fitful sleep on the field of battle did not change that. Horses were weak for want of forage. Sunday, October 5, had dawned as hot and dry as the day before. A thick carpet of dust covered the roads leading out of town. Streams and creeks were dry, and wells were empty. The soldiers would have no drinking water but that which they could carry in their canteens. Stanley neatly summed up the condition of the army that morning: "The heat was excessive and the men were worn out; they had narrowly escaped a most terrible defeat, and no one was anxious to crowd their late antagonists."[4] Not even McPherson, it seemed. He took literally Rosecrans's orders of the night before that he was to *be prepared* to move, and so he did nothing until Rosecrans accosted him just before sunrise.

"Why are you not under way?" Rosecrans demanded excitedly. McPherson handed him his written orders. Rosecrans told him to march, and McPherson started out. Colonel Ducat was disgusted: "Most men breathing the air of pursuit . . . would have moved at the first streak of dawn on such orders as McPherson already had."

To forestall further delays, Rosecrans drew up an elaborate special order for his division commanders that expanded greatly on the movement orders of the night before. Covering everything from interdivisional communications to march intervals, the order enjoined the generals to march rapidly: "The attention of leaders of columns in pursuit is called to the well-known principle of war that it is safe to pursue a flying enemy with a greatly inferior force, and they will take care, while conducting their advance with caution to guard against ambuscade, to push the enemy with vigor and firmness." For the time being, Rosecrans would remain in Corinth to coordinate the chase and keep in contact with Grant.

Rosecrans's didactic instructions were useless. General Stanley started hours late and became entangled with Hamilton, who had taken the wrong route. Although on the right road, Stanley retraced his steps, intending to "follow the beaten trail of the enemy and move toward the cannonading plainly heard in the west." He set out on the Chewalla road — the route Hamilton was to have taken.

Rosecrans disapproved Stanley's common-sense solution: "You should have taken the road to the right, this side of Cane Creek, which keeps north of the railroad. If you are not too far advanced it would be better for you to face by the rear and do it now, as you will reach Chewalla sooner."[5]

Rosecrans's remonstrance came too late, as Stanley had become tangled up with McKean's wagons near Cane Creek. Stanley had no use for McKean, whom he considered a consummate failure, and his opinion of the old soldier seemed quite correct. McKean was on the right road, but he supposed himself lost; worse yet, he had taken his trains with him. Nearly a mile long, they clogged the roads and brought both divisions to an abrupt halt.

While McKean and Stanley argued pointlessly, General Hamilton cantered up at the head of his division. It was past noon. The New Yorker had started the day full of fight but short on direction. "The enemy is completely stampeded," he had told his men at dawn. "Their baggage, guns, equipment, provisions and ammunition are thrown away. The enemy are running like hell." Unfortunately, Hamilton had no idea how to find them: "The division started at 7 o'clock without other instructions to me than to follow McKean's division."[6]

Hamilton had nothing to offer now except indignation. General Davies showed up next, and the four generals adjourned to the roadside. Rather than try to break the impasse as ranking officer, Hamilton merely apprised Rosecrans of the bottleneck and waited for an answer: "Much confusion and delay occurred from want of a commander [but] I deemed myself restrained by my instructions from assuming the command so long as the march was without resistance," said Hamilton. "Had we encountered the enemy I should not have hesitated to exercise my right of seniority in the absence of the general commanding."[7]

Rosecrans blamed McKean for the jam and scolded him for having marched with his wagons. When McKean ignored Rosecrans's order to clear them from the road, Rosecrans reprimanded him: "Hamilton says you are waiting for orders. . . . You have your orders to push ahead, follow your advance guard closely, and report frequently."

It was mid-afternoon before McKean moved, too late, Stanley thought, for a pursuit to amount to much: "The train accompanying the division was so long and cumbersome that any idea of making a successful pursuit must at once be dropped. . . . I will not say that my division could have overtaken and engaged the enemy, but I will say that we could have aided General McPherson."[8]

McPherson actually was doing quite well on his own. Despite the heat, his column made good time. Six miles south of Chewalla McPherson caught the sounds of Hurlbut's fight, and he "pressed on in the most lively manner" until, at noon, his cavalry escort ran into Van Dorn's rear guard a mile south of Chewalla. McPherson tried to give battle, but the Rebels vanished. He resumed the chase.

The next Rebels McPherson met were under a flag of truce. Ostensibly for humanitarian reasons, but more probably to slow the Yankee pursuit, Van Dorn had sent a party of 300 men under Col. William Barry to bury the Confederate dead at Corinth. Barry tried to detain McPherson, but the Ohioan angrily demanded that he yield the road. There was fighting going on, McPherson told Barry, and only a direct order from General Rosecrans could suspend it. Barry's party stepped aside, and the Federals pressed on.[9]

Four hours later McPherson encountered the enemy rear guard arrayed on Big Hill, above Young's Bridge on the east bank of the Tuscumbia River. This time, said McPherson, the Rebels looked ready to fight.

The Confederates atop Big Hill were under the command of Brig. Gen. John Bowen. One of the most vocal opponents of the attack on Corinth, the Georgian now found himself responsible for the integrity of the army's retreat. His men were fresh and they discharged their duty well. When McPherson sent forward one brigade at sunset to storm the hill, Bowen's Confederates met the Federals with deliberate, well-aimed volleys that sent most spilling down the slope. Bowen himself led the Fifteenth Mississippi in a counterattack that cleared away the remainder. As twilight fell, Bowen crossed the Tuscumbia River at "my leisure, tore up and burned the bridge, obstructed the ford near by, and joined the division about three miles beyond." McPherson licked his wounds and bivouacked on the east bank. Absent the rest of the army, he felt he had done all he could.[10]

The only other Federal commander who rightfully could claim to have spent the day productively was John McArthur. He marched far beyond the Cane Creek snarl before noon and would have passed McPherson had he not allowed himself to be detained by Colonel Barry and his detail. McArthur waited three hours for instructions from Rosecrans, who saw through the artifice and told McArthur to dismiss the Southerners with the promise their dead and wounded would be cared for properly. Just south of Chewalla McArthur ran into the rear of McPherson's division, which had come over from the Columbus road. He ceded the way and followed McPherson in close support, bivouacking with him on the Tuscumbia River.[11]

Everyone else came up short. Stanley went into camp two miles west of the Tuscumbia River. Hamilton bivouacked four miles short of the river. McKean settled in near Hamilton, and Davies made camp behind Stanley. The troops were thirsty and tired, but morale was high. Dispirited Rebels had surrendered by the score, and abandoned wagons, artillery carriages, and camp equipment—sure signs of an army near collapse—littered the road.

Although he had hoped for more, General Rosecrans had done little to accelerate the chase. Hectoring notes to McKean from Corinth were no substitute for the commanding general's presence at the front. Rosecrans had been too far away to exercise control over the pursuit. Had he been with one of the march columns, the Ohioan might have both resolved the bottleneck at Cane Creek and dismissed Colonel Barry in a matter of minutes.[12]

Puling telegrams to Grant during the day hinted at Rosecrans's fumbling. At noon Rosecrans exhorted Grant to have Hurlbut attack at once. "Where is Hurlbut?" he demanded. "Now is his time to pitch in." Later in the day Rosecrans confessed the Rebels had gotten a three-hour jump on him, the army having started late "through errors." At dusk Grant read a summary of the day's sorry efforts: "Leading divisions arrived at Chewalla. No news from McPherson since noon. Progress very slow. McKean in the way. Order us forage at once or our animals will starve." Grant pocketed the message and made a mental note to recommend McPherson's promotion to major general "above others who may be promoted for the late battles," a thinly veiled swipe at Rosecrans's stumbling division commanders. Though he said nothing then, Grant's opinion of their leader was fast declining.[13]

*   *   *

Sterling Price was at Crum's Mill, watching the army wagon train roll uneasily over Armstrong's makeshift bridge. The afternoon receded, and a chill caused by more than the early autumn twilight gripped the general and his soldiers. The woods were weird; a spirit of inevitable doom, heralded by the drum roll of Yankee cannon, marched through the countryside. Hope was at a premium. Everyone knew the army was hemmed in, their only way out a few planks laid over a dam, their only guide through the dark woods a young farmer from Chewalla.[14]

"Shall that night ever be forgotten?" mused a Mississippi private. "Will its vivid impressions ever be erased? Dust like a heavy impenetrable fog obscuring our comrades an arm's length was constantly stirred from the ashy soil. Burning, overturned wagons wrought lurid ghosts in dusty gleams through the forest. This is the darkest gloom that has ever been mine in struggling for freedom."[15]

With the army's fate riding on a rickety bridge, general's braid was worth less than a teamster's lash, and Price played wagon master and chief engineer. After each wagon passed, Price examined the bridge for damage. The strain told on him. To distract himself, Price helped with repairs. "Every few minutes something was knocked to pieces about the hastily constructed bridge," said Ephraim Anderson, whose brigade reached Crum's Mill at 9:00 P.M. "As we passed over, a wagon knocked down some of the puncheons, and the general, standing on the opposite bank, immediately called out for some of the boys to halt and assist in righting them. Several of us volunteered immediately, and replaced the slabs, and the general, himself, assisted in throwing the heavy rocks upon them, to hold them in place. His whole soul seemed to be in the work, and when it was done, he straightened himself from his stooping posture, remarking, 'Well done, boys—now stand back and let the train pass.'" Price kept at it all night. For his toils, he grew in the estimation of the army.[16]

Hardly anyone spoke well of Van Dorn, whose contribution to the crossing was limited to an admonition to Price to make haste. Rumors were rife that Van Dorn had been drunk at Corinth and that he had come unhinged on the retreat. Few trusted him to lead the army to safety. Thad Welch, the regimental wag of the Second Missouri, gave voice to the sentiment of most. On the march to Iuka Welch had said his motto was "victory or crippled." On the way to Corinth it had sunk to "victory or death." As the regiment shuffled along toward Crum's Mill, with the boom of Federal artillery echoing behind them, someone asked Welch what his motto was now. "Motto now," he growled, "there's no motto for this place—I can only say, we all thought Van Dorn had played hell at Elkhorn, and now he has done it, sure enough."[17]

Price saw the last of the army across the Hatchie at 1:00 A.M. on October 6. Once over, there was no pause. "No time for order in marching, no time to wait for the trains to stretch out into the road and to follow it then in twos," said a Texan. "We fell into the road pell-mell, and moved in any style we wished to, in among the wagons, or any way just so we moved along and kept out of the way of those behind us."[18]

Toward dawn the moon disappeared behind a scud of clouds. A warm rain tattooed the marching soldiers, easing their thirst and settling the dust. The army turned off the Boneyard road at daybreak and onto the Ripley and Pocahontas road. Well into the afternoon the men trudged on, too frightened to stop and too hungry to keep in ranks. They spread across the countryside by the hundreds in search of food. Farms were ransacked, but the plunder was meager. The occasional head of beef went fast, "the hungry men cutting the flesh from the carcass before the hide was off." Texan Sam Barron counted himself lucky. He had found a small pile of cornmeal spilled from a commissary wagon. Barron scooped it into a castoff cup, along with a generous portion of dirt, stirred in some river water, and warmed the mixture. "Surely it was the most delicious piece of bread I have ever tasted, even to this day," he wrote two decades later.[19]

Bread was scarce—nonexistent in the stores of Maury's division. Potatoes saved the army. The men foraged for them, and their commanders bought them wholesale from farmers along the way. "I never had seen such fine potatoes before," said Texan Joe Scott. "I scratched up as many as my haversack would hold and made my way back to the main line, where the army had stopped to feed the artillery horses. Someone had killed a cow, and my messmates had secured a small piece and without any salt were boiling it. We made a hearty meal of what was before us."

Most made do with potatoes, and then not always cooked. Lt. Col. Frank Montgomery of the First Mississippi Cavalry eagerly took a raw sweet potato that was offered him and at once regretted his voracity: "I was very hungry and I began on it as I rode along, but the first thing I knew I was choking, and would have choked if I had not had some water in my canteen. It took me a long time to eat that potato, but I at last got through with it. If any of my readers have never tried to eat a cold sweet potato and see how hard it is to swallow it, I recommend [they] try it."[20]

Potatoes and pursuing Federals could propel the army just so far, and that evening, seven miles north of Ripley, Van Dorn had to grant a respite. Willie Tunnard studied his companions as they sank to the ground. They were "in a terrible condition. Worn out with fatigue, sick, ragged, filthy, and covered with vermin, it was not strange that even their brave spirits should give way under the accumulated disasters, sufferings, and hardships which had so rapidly befallen them."[21]

*   *   *

Rosecrans left Corinth shortly before midnight on October 5. He was, as he told McKean, "coming out to Chewalla with a carload of water. . . . Baggage, I understand, has interfered with your progress, which certainly has not been remarkable." The Ohioan made the rounds of his generals' camps. He roused McPherson at 3:00 A.M. with orders to "push the enemy as soon as it is light." McKean he sent packing to Corinth with the wagons, and McArthur at last assumed formal command of the division.[22]

The Federals were on the road before 8:00 A.M., marching toward Crum's Mill past the flotsam of a broken army. "I have never seen such a stampede in all my life," said an Iowan. "Everything imaginable was strewn along the road—tents, bake ovens, corn meal, fresh beef, and many other things; some of their supplies they burned to keep from falling into our hands." Too tired to keep on, Rebels came in freely. Said Iowan Cyrus Boyd, "All along the roadside under the bushes, in the hollows, and behind logs the panting fugitives were found, glad to surrender. Glad to do anything to save all they had left and that was their lives. They all agreed in saying that no such terrible calamity had ever overtaken them in the west as the battle of the fourth."[23]

The pursuit was slow. An hour was lost to repairing Young's Bridge, a second to clearing felled trees from the Boneyard road, and four more to replacing the bridge over the Crum's Mill dam, which the Rebels had dismantled. Again in the lead, McPherson bivouacked for the night at Jonesborough, five miles west of the Hatchie River.

Notwithstanding the delays, Rosecrans's mood lifted during the day. His energy returned, and the Ohioan once again managed affairs with characteristic dispatch. He ordered 30,000 rations to Chewalla and eighty wagonloads to meet his advance. To ease congestion on the Boneyard road, he rerouted Hamilton's division to Rienzi. His messages to Grant were crisp and certain. At noon he wired, "The enemy are totally routed, throwing everything away. We are following sharply." At 6:30 P.M. he wrote from Crum's Bridge that the enemy seemed to be "aiming for Holly Springs. Bridge built; part of the troops across; we shall pursue them." Rosecrans closed with a request for orders.[24]

Grant had none to give. "You will avail yourself of every advantage and capture and destroy the Rebel army to the utmost of your power. . . . All news received cheering and all parts of the army have behaved nobly." Rosecrans retired for the night, ready to push the pursuit hard the next morning.[25]

Grant had been less than candid with Rosecrans. His breezy recital of cheering news and noble actions belied the truth concerning Hurlbut, whom Rosecrans counted on to aid him in the chase. After the fight at Davis Bridge, Hurlbut had counted his division too badly cut up to do more. In his report Hurlbut checked off his reasons for bowing out from the pursuit: "The total want of transportation, the loss of battery horses, the shortness of provisions, and the paramount necessity of burying my dead, taking care of my wounded, and securing the prisoners and captured munitions of war prevented my pursuing."

All were good arguments, but the real reason Hurlbut stopped was personal: he was too drunk to do otherwise. Perhaps the horrendous losses on the Hatchie led him back to the bottle; in any event, by sunset on

October 5 Hurlbut was in a fog. His most aggressive act that evening was to get between an angry teamster and the horse he was beating. The teamster knocked Hurlbut down with a punch to the jaw. The general got up and staggered to his tent.[26]

*　*　*

Despite the disorder in his ranks, Van Dorn was able to stay a step ahead of the Federals. Ironically, Lovell's division had become the glue holding the army together. At each report of the enemy's approach, Lovell faced his command to the rear and made ready to do battle. The men stood to the task, but the false alarms eroded morale. After they were called out at 3:00 A.M. on October 7, seven miles north of Ripley, to battle another rumor, Lovell's soldiers began to slip from the ranks, having decided to conduct their own personal retreat.[27]

While Lovell marched and countermarched on the outskirts of Ripley, Van Dorn groped for a way to extricate his fast-shrinking army. It was evident he could not remain at Ripley; at their present rate of march the Yankees would reach the town in two days, and Van Dorn knew his troops were in no condition to oppose them. Van Dorn settled on Holly Springs, a railroad town thirty miles west of Ripley, as the army's sanctuary.

Holly Springs had much to recommend it. A small Confederate force, around which the army could regroup, garrisoned the place, and a supply of rations had been gathered there. Neighboring farms had been spared the ravages of marching armies. Crops were in, and gristmills stood ready to make bread rations. More importantly, Holly Springs was the northernmost stop on the strategic Mississippi Central Railroad, which passed within forty miles of Vicksburg. Control of Holly Springs offered the best forward defense of the Mississippi River citadel.

Appropriate marching orders went out at midnight on October 6. A strong guard was posted in Ripley to prevent stragglers from lingering, and the army passed through after daybreak. Hunger proved a stronger master than their generals, and troops by the thousands dispersed to beg food from frightened citizens. Said a townsperson, "We were awakened at one o'clock with the heavy tread of cavalry and baggage wagons on their retreat, and by morning the town was full of soldiers, some wounded, all famished and begging for something to eat, if but a piece of bread. . . . All that miserable morning we were cooking to feed the famished men." The last Confederate left Ripley at sunset. Over the road and through farms and forests they traipsed "in small squads, making their way toward Holly

Springs," said Willie Tunnard. "They were worn out with fatigue, sadly depressed, almost demoralized."[28]

*   *   *

The Federals were on the road early on October 7, and Rosecrans drove them hard. The day was hot and dry. South of Ripley the army entered the Mississippi pine barrens, a sparsely settled, poorly watered wilderness. Rebels kept coming in, "exhausted and half-starved," but the Yankees were suffering now, too. Rations gave out, with no sign of the commissary wagons. Water was scarce; General Hamilton pushed his division twenty-three miles without finding a single well. Hundreds of men sneaked off in search of food and, with no enemy to threaten them, easy plunder. Efforts to prevent looting were mixed. General Stanley mandated stern punishment, but enforcement in his division was lax. "The field officers of the regiments all seconded my efforts to keep order and prevent straggling, but I am pained to say I find too many company commanders who are totally unconcerned as to whether their men march in ranks or go along the road like a flock of geese." By midday the roads were thronged with hungry and thirsty blue-clad stragglers.[29]

That afternoon Rosecrans made a serious tactical blunder — not on the field but in his communications with Grant. He suggested supply shortages might slow him and added he was sending rations from his own limited stock to Hurlbut to enable him to join in the chase. Grant answered promptly: "General Hurlbut took with him three days rations and I forwarded him three days more by wagons. If they are out now they must have wasted them. . . . You need not send him supplies as he will be back in Bolivar tonight. . . . We can do nothing with our weak forces but fall back to our old places. Order the pursuit to cease."

Rosecrans was incredulous. Nothing had prepared him for Grant's volte-face. He was handed the message at midnight in the field near Jonesborough, and he answered it angrily. Under the uncertain shelter of a huge oak tree, with a warm rain falling, Rosecrans held a candle and dictated to Colonel Ducat his response:

I most deeply dissent from your views as to the manner of pursuing. We have defeated, routed, and demoralized the army which holds the lower Mississippi Valley. We have the two railroads leading down toward the Gulf through the most productive parts of the State, into which we can now pursue them with safety. The effect of our return to old position will be to pen them up in the only corn country they have west of Alabama, including the Tuscumbia Valley, and to permit them to recruit their forces, advance and occupy their old ground, reducing us to the occupation of a defensive position, barren and worthless, with a long front, over which they can harass us until bad weather prevents an effectual advance except on the railroads, when time, fortifications, and rolling stock will again render them superior to us. Our force, including what you have with Hurlbut, will garrison Corinth and Jackson and enable us to push them. Our advance will cover even Holly Springs, which would be ours when we want it. All that is needed is to continue pursuing and whip them. We have whipped, and should now push to the wall and capture all the rolling stock of their railroads. Bragg's army alone west of Alabama River and occupying Mobile could repair the damage we have it in our power to do them. If, after considering these matters, you still consider the order for my return to Corinth expedient, I will obey it and abandon the chief fruits of a victory, but I beseech you to bend everything to push them while they are broken and hungry, weary and ill-supplied. Draw everything possible from Memphis to help move on Holly Springs and let us concentrate. Appeal to the Governors of the States to rush down some twenty or thirty new regiments to hold our rear and we can make a triumph of our start.[30]

Rosecrans probably was pleased with his answer, a missive of great strategic sweep that would compel cooperation by the soundness of its logic. But Grant thought otherwise. To him Rosecrans's telegram was poorly reasoned, presumptuous, and more than a little insubordinate. Grant had his reasons for recalling Rosecrans: "When I ascertained that the enemy had succeed[ed] in crossing the Hatchie I ordered a discontinuance of the pursuit. . . . This I regarded, and yet regard, as absolutely necessary to the safety of our army. They could not have possibly caught the enemy before reaching his fortifications at Holly Springs, where a garrison of several thousand troops were left that were not engaged in the Battle of Corinth. Our troops would have suffered for food and from fatigue."

Although certain of his own judgment, Grant submitted Rosecrans's objections to the War Department. No doubt he presumed Henry Halleck, whose pursuit of Beauregard after Shiloh was the epitome of extreme caution, would support him. Grant told Rosecrans to remain at Ripley until Halleck replied.[31]

While he awaited Halleck's answer, Grant received intelligence that disproved the premise of his order halting the pursuit of Van Dorn: far from being heavily fortified, Holly Springs had been abandoned. Ironically, it was Hurlbut who communicated the news to Grant. "I have just heard from Holly Springs," he wrote on the evening of October 8. "There are no forces there; all left on Sunday. . . . I am of opinion that the rout of Van Dorn's army is complete, and that Pillow's force, late at Holly Springs, has caught the panic."

Here was reason to go on. But Grant refused to budge. Not only did he not change his mind, but he also reneged on his promise to wait for Halleck's guidance and instead ordered Rosecrans to return to Corinth. Rosecrans ignored the order; he would take his chances with Halleck.

Grant had misread the general in chief. While he left the decision to Grant, Halleck wondered, "Why order a return of your troops? Why not reinforce Rosecrans and pursue the enemy into Mississippi, supporting your army on the country?"

Grant remained adamant, merely inventing new reasons to call off the pursuit. "An army cannot subsist itself on the country except in forage. They did not start out to follow but a few days and are much worn out," Grant told Halleck. Moreover, scouts reported the enemy had "reserves that are on the way to join the retreating column. . . . Although partial success might result from further pursuit, disaster would follow in the end." Nevertheless, if the War Department wished, Grant would press forward. "If you say so, however, it is not too late yet to go on, and I will join the moving column and go to the farthest extent possible."

Venturing an opinion was all Halleck dared do. A positive order to a commander in the field to undertake a vigorous pursuit—something Halleck himself never contemplated after Shiloh—was beyond him. Grant correctly took Halleck's silence as assent, and he told Rosecrans a third and final time to desist.[32]

Rosecrans obeyed grudgingly. He then believed, and would ever afterward maintain, that one of the best opportunities of the war had been sacrificed to undue caution. His troops were tired, but rations had arrived—eighty wagonloads met the army at Ripley. An autumn rainstorm promised cooler weather. The land southwest of Ripley was prime corn country, and the corn was ripe. Everything favored an advance; all that wanted was the will at department headquarters and the reinforcements that Rosecrans had implored Grant to send him. "If Grant had not stopped us, we could have gone to Vicksburg. My judgment was to go on, and with the help suggested we could have done so. Under the pressure of a victorious force, the enemy were experiencing all the weakening effects of a retreating army, whose means of supplies and munitions are always difficult to keep in order," argued Rosecrans. "We had Sherman at Memphis with two divisions, and we had Hurlbut at Bolivar with one division and John A. Logan at Jackson with six regiments. With these there was nothing to save Mississippi from our grasp. We were about six days' march from Vicksburg, and Grant could have put his force through to it with my column as the center one of pursuit. Confederate officers told me afterward that they never were so scared in their lives as they were after the defeat before Corinth."[33]

25

The Best We Had in the Ranch

Before slumping beneath a tree on the afternoon of October 4, General Rosecrans rode along the lines proclaiming victory, then wandered over the ground in front of Battery Powell. Dead and dying Confederates lay everywhere, shrouded with a quilt of gun smoke. An Arkansas lieutenant with a mangled foot caught his eye, and Rosecrans bent down to offer him water. "Thank you, General," the Arkansan politely declined, "one of your men just gave me some." "Whose troops are you?" Rosecrans asked. "Cabell's." Rosecrans kept up the conversation: "It was pretty hot fighting here." "Yes, General, you licked us good, but we gave you the best we had in the ranch."[1]

Many of the best had been lost. Federal casualties totaled 355 killed, 1,841 wounded, and 324 missing of some 22,000 engaged, or one man in ten. Another 570 were lost at Davis Bridge. Southern casualties were far worse; indeed, for the numbers engaged, they were staggering. Price contributed 13,863 troops to the battles, 10,498 of whom were infantry. He lost 428 killed, 1,865 wounded, and 1,449 missing. All but 71 of the 3,742 casualties sustained were infantrymen. In other words, nearly 35 percent of those who charged the Federal works at Corinth or defended Davis Bridge became casualties. More than half of Price's line officers fell. Maury's division existed in name only. The Virginian took some 3,900 men into the battles and lost 2,500. Another 600 melted away during the retreat. John C. Moore's brigade was obliterated, losing 1,295 of 1,895 men engaged.

Mansfield Lovell's losses were scandalously light, a fact not lost on Price's survivors. Lovell had gone into action with 7,000 troops. He reported 77 killed, 285 wounded, and 208 missing—almost all in the fight for Oliver's hill.[2]

Everyone but Van Dorn recognized the campaign for the fiasco it was. To his wife, Lt. Wright Schaumburg of Maury's staff lamented, "The Confederacy wakes today to the realization of its darkest hour. . . . The never before conquered Army of the West has been most signally defeated and routed in this expedition against Corinth. All is dark and gloomy now." The division inspector general, Capt. Edward Cummins, was equally downcast. At the first opportunity he wrote General Beauregard a "brief account of our recent expedition and disaster." Battle losses and straggling had destroyed Price's corps, said Cummins. He had no idea of the enemy's strength, but "when we got into Corinth he swallowed up seven brigades of as good fighting men as I ever saw in about twenty minutes. . . . God bless you, my dear general, and send us better days."[3]

Mansfield Lovell was the army's near-unanimous scapegoat. "With but little dissent," said Colonel Bevier, "the opinion of the participants . . . is that the loss of the battle of Corinth . . . was mainly owing to the misconduct of General Lovell and the inaction of the right wing." A disgusted Col. Ras Stirman wrote, "This was another victory lost the Confederates by the inactivity and failure of one man." Dabney Maury agreed. Had Lovell attacked vigorously, "it is altogether probable that . . . Van Dorn would have captured Corinth."[4]

Incredibly, Lovell was oblivious to the army's rancor. "If I am to believe all that I hear, my conduct throughout the operations has rebounded much to my credit all through the army," he told his wife. "Hundreds of persons have come to me and said that although they began the campaign with prejudices against me, they would now rather serve under me than anyone else." Lovell insisted his conscience was clean: darkness, rather than his indolence, had prevented the Confederates from taking Corinth on the first day. Neither did Lovell apologize for his actions on the second day.[5]

Unlike the army the Southern press spared Lovell. The culprit, most editorials agreed, was Van Dorn. He had been "terribly deceived"—lured into a trap and "made to fight against heavy odds without the hope of any advantage for this waste of blood and life," said a Mobile editorial. "The future is dark in this part of the country. . . . Other reverses are in store for us, including the loss of much ground now held by us," warned the *Atlanta Southern Confederacy*.[6]

Press comment was tame next to that of Van Dorn's fellow Mississippians. Senator James Phelan excoriated the general. "I doubt not you have ascertained . . . the unhappy condition of affairs in this state," he told President Davis. "The army is in a most deplorable state. . . . It is yet called 'Van Dorn's Army;' and the universal opprobrium which covers that officer, and the 'lower than the lowest depth' to which he has fallen in the estimation of the community of all classes, you cannot be aware of." Phelan played on the general's reputation for womanizing to demand his removal: "He is regarded as the source of all our woes, and disaster, it is prophesied, will attend us so long as he is connected with this army. The atmosphere is dense with horrid narratives of his negligence, whoring, and drunkenness . . . so fastened in the public belief an acquittal by a court-martial of angels would not relieve him of the charge."[7]

Van Dorn was too broken in spirit to confront the truth. He had not lost a battle, he assured his wife, Emily, but, rather, "was not entirely successful. . . . All that valor attempted was won but the enemy was too strong for us. . . . I have lost nothing. I will fight again soon, and if fighting will win. Do not be mortified at what they say. We cannot expect impossibilities."[8]

Van Dorn was denied a chance to win back what he could not concede he had lost. Anticipating disaster, the president had sent Maj. Gen. John C. Pemberton to Jackson, Mississippi, to await the outcome of Van Dorn's campaign. When he learned of the defeat at Corinth, Davis nominated Pemberton for promotion to lieutenant general and told him to take charge of the Department of Mississippi and East Louisiana, an area embracing both Van Dorn's theater of operations and Vicksburg. The Confederate Senate confirmed the nomination three days later.

Mortified, Van Dorn appealed to Secretary of War Randolph for relief: "I shall act for the best, but I am now an isolated body in the field in Mississippi, relieved of command of my department. I hope this will be corrected." It was not, and Van Dorn became subordinate to a general whom he had recently outranked. His despair grew. "I am weary, weary," he wrote Emily. "I sigh for rest of mind and body. If my death would give pain to no one I should court it. . . . I have struggled for others and they abuse me."[9]

As the abuse became more personal, Van Dorn sank deeper into self-pity. Mississippi had turned its back on him; he would turn his back on Mississippi. Schoolgirls from his hometown sent him a hundred pairs of socks to distribute to the Texans in his army. Van Dorn forward them to a Texas colonel with a sardonic note: "You will observe that these little angels identify me with Texas. They are right. I am a Texan. A Mississippian no longer except in my love for the pure hearted children of her evil who have not yet learned to make the name and fame of one of her sons the butt of malignant archery." Van Dorn was relieved to learn that his wife, at least, discounted rumors of his philandering. "Your kind letter came to me like a ray of sunshine in my cloud," he told Emily on October 30. "I was happy to learn that you have not believed all the villainous slander of my character. I scorn to answer the public accusations. . . . I live still and until it pleases God to take me from the vile race I shall continue to do so."[10]

From an unexpected source came a new challenge. As discontent within the army grew, Van Dorn replaced Lovell as the object of ridicule and censure. Price and Maury stood by him, but Price's staff schemed for Van Dorn's removal. When their conspiratorial dealings floundered, General Bowen stepped forward to accuse Van Dorn of "neglect of duty" and of "cruel and improper treatment of officers and soldiers under his command."[11]

Bowen's seeming perfidy astounded Van Dorn. Only two weeks earlier he had promoted Bowen to command Hébert's division. Van Dorn demanded a court of inquiry, which Pemberton granted.

The court convened on November 7. Pemberton selected Price, Maury, and Lloyd Tilghman to sit in judgment. The charges and specifications were read. Lacking a proper map of Corinth and having inadequate commissary stores, and "without due consideration or forethought," Van Dorn had marched the troops "in a hastily and disorderly manner, [and hurled] them upon the enemy with an apparent attempt to take a command by surprise whose outposts had been engaged with his advance for thirty-six hours before the attack." He failed to follow up the gains of the first day and had permitted the enemy to be reinforced during the night. And, Bowen concluded, during the retreat Van Dorn had subjected the army to "long, tedious, and circuitous" marches and had neglected the wounded.[12]

A parade of witnesses testified for the prosecution, but not as Bowen hoped. General Rust repudiated most of the charges his fellow brigade commander had leveled. General Green said his men were adequately supplied and had marched in good order. While he thought the attack should have been pressed on the evening of October 3, Green was unable to see how Van Dorn could have prevented Federal reinforcements from entering Corinth. Col. Robert Lowrey said his men, while hungry, were "marched as troops generally would be." [13]

Van Dorn called Sterling Price and Dabney Maury to the stand. Price thought that Van Dorn had acted prudently and with "great energy." Maury said Van Dorn was as well informed of the country around Corinth as any officer in the army. He considered Van Dorn incapable of "cruelty or inhumanity or intentional injustice to anyone."

Van Dorn closed the defense with a long and impassioned address, refuting each charge and specification and reaffirming his dedication to the cause: "Gentlemen of the court, I am a Mississippian by birth. My blood has always been ready for her, yet in the midst of my struggles for her my name has been blighted by her people. My trust is that the investigation of this court will vindicate it from dishonor."

It did. Sterling Price read the verdict: Every allegation made against General Van Dorn was "fully disproved." No further proceedings were warranted. Pemberton dissolved the court of inquiry on November 28. To celebrate his exoneration, Van Dorn had 1,000 pamphlets containing the court proceedings printed and distributed at his expense. [14]

Van Dorn's legal victory was hollow. Reading Van Dorn's pamphlet, President Davis agreed that the court's findings had cleared him, but by then events had swept Van Dorn aside. While Van Dorn was battling in the courtroom, Grant had quietly built up his army. By the end of October he had nearly 50,000 troops on duty at Memphis, Corinth, and intermediate posts. Rid of Van Dorn's army as an offensive threat and no longer a manpower pool for Buell, who had driven Bragg from Kentucky, Grant was ready to take the offensive. To Halleck he proposed to concentrate his forces at Grand Junction, twenty miles north of Holly Springs, and begin a drive down the Mississippi Central Railroad to Vicksburg. Grant ordered five divisions to assemble at Grand Junction and repaired there himself to await Halleck's reply.

Halleck not only endorsed his plan, but he also promised Grant 20,000 fresh troops from Illinois. Grant elected to suspend his overland march on Vicksburg until they reported in. In the meantime he directed McPherson to reconnoiter aggressively toward Holly Springs with two divisions. [15]

Van Dorn was in no position to contest even such a limited move. The exchanged Fort Donelson prisoners had arrived, and stragglers by the hundreds drifted into camp daily, swelling the army to a respectable 24,000 troops. But the men were indifferently armed, low on provisions, and in poor health. Winter struck early, blanketing northern Mississippi with snow the last week of October. Hundreds took ill. Pemberton at Vicksburg had no troops to spare. He could only suggest Van Dorn abandon Holly Springs in favor of blocking positions along the Tallahatchie River.[16]

Van Dorn slipped out of Holly Springs on November 9, four days ahead of Union cavalry, and took up position south of the Tallahatchie. His army was still in no state to resist an advance. Rations continued to be short, and the weather had worsened. Morale plummeted with the cold. When Grant moved southward on December 1, Van Dorn withdrew to Grenada, a mere 120 miles north of Vicksburg. Although Grant paused to consolidate, Van Dorn's lieutenants knew it was but a question of time before he would force them from Grenada. Disaster seemed inevitable. The army was in shambles. An ice storm struck during the march to Grenada, touching off an epidemic of typhoid fever and pneumonia. Desertions climbed. Van Dorn's detractors grew bolder, and his supporters wavered. After Van Dorn dispersed Maury's division to guard fords across the Yallabusha River, the Virginian complained: "In my opinion, we should all be concentrated about Grenada with our 'staves in our hand and our loins girt about,' and ready for a quick exodus. Why should we scatter our division in this way when no one division is strong enough to fight, and when it is not intended?"[17]

Van Dorn never had a chance to answer Maury. Secretary of War Randolph had resigned in mid-November 15 in disgust over the president's policies. Randolph's departure weakened Davis's hand with Congress, and he at last yielded to congressional demands that he appoint his old nemesis, Gen. Joseph Johnston, to orchestrate Confederate strategy in the West. Johnston's appointment occasioned a reshuffle in Pemberton's department. Van Dorn was reduced to corps command, on an equal footing with Price. A week later he fell farther. A Texas colonel had suggested to Pemberton that a cavalry expedition into the enemy's rear, properly led, might capture Grant's huge supply depot at Holly Springs, forcing him to fall back. Pemberton consented to give the scheme a try. He detached three brigades of cavalry for the expedition and selected Van Dorn to command them. Turning over his corps to Maury, Van Dorn left to join the troopers for what must have struck the ever-ambitious Mississippian as a ride into military oblivion.[18]

*   *   *

His recall to Corinth on October 9 had disgusted Rosecrans. To Colonel Ducat he grumbled, "Those people have ordered us back. The orders are imperative, and a great part of the fruits of victory will be lost. We must cross the Hatchie tonight and try to reach Corinth tomorrow."

"But why?" asked Ducat. "Is anything taking place that we don't know of? It's not possible for any troops in the center to be marching toward Corinth or wedging in anywhere between Chattanooga and Corinth."

"The whole rebel army of the west, and certainly the whole flower of it, is in our front, whipped and demoralized," continued Rosecrans. "We could drive them like a flock of sheep. It seems to me the time has come to win the war in the west."[19]

Rosecrans returned to a reception hardly befitting his success. During his absence Grant had issued General Orders Number 88. Although presented as a congratulatory message, the document was insulting to Rosecrans. At Corinth Rosecrans had inflicted losses twice as great as his own and repelled some of the fiercest attacks of the war. Yet Grant scarcely acknowledged his achievements. Said the journalist Whitelaw Reid, "Passing by the brilliant battle of Corinth with a single clause, [Grant] devoted most of the order to extravagant praise of Hurlbut for the brief onslaught he made upon the enemy during their retreat." Grant also intimated that relations between the commands of Rosecrans and Ord were strained: "Between them there should be, and I trust is, the warmest bonds of brotherhood."[20]

The insinuation offended Rosecrans. "The part that expresses the hope that good feeling will exist between Ord's command and my own amazes me," he wrote Grant. "So far as I know there was nothing even to suggest the fact that it might be otherwise. Under such circumstances the report is to be regretted, because our troops, knowing that there was no foundation in it for them, will be led to think there is some elsewhere."[21]

Rosecrans's allies in the press corps took up his cause. In his account of the Battle of Corinth, *Cincinnati Daily Commercial* correspondent William D. Bickham minimized Grant's role in the campaign: "And now, to whom is due the honors of the battle of Corinth? The verdict of the whole army is in favor of General Rosecrans. . . . It would seem from General Grant's dispatches that he claims the honors. . . . There is no doubt that the public will give the credit to General Rosecrans, where it belongs."[22]

Grant was incensed. He held Rosecrans responsible for the report and demanded he silence Bickham and members of his staff who had instigated "a distinction of feeling and spirit" within the department. He also upbraided Rosecrans for paroling prisoners incorrectly, a triviality to which Grant devoted a hectoring dispatch.

Grant's hostility to Rosecrans was an open secret at the headquarters of both generals. It most certainly derived from indiscreet chatter on the part of Rosecrans's staff, who "knew all their general knew" about Grant's failure to support him at Iuka. Querulous members of Grant's military family answered that Rosecrans, who had erred in bringing on the battle too soon and in leaving open the Fulton road, sought credit for the victory at Grant's expense. As Col. Mortimer Leggett told John Rawlins, "It [is] a gross outrage for the minions of a newly fledged major general, not only to attempt an exclusive appropriation of all the honors but, by irresponsible assertions, and mysterious insinuations, to attempt to awaken and deepen, former prejudices against the general to whom naturally and rightfully the first honors belonged. Major General Rosecrans is undoubtedly an excellent officer . . . but the evidence is such that . . . he must be at least privy to the whole devilish scheme."[23]

Nettled by Rosecrans's high-handedness after Corinth, Grant allowed himself to believe the conspiratorial ranting of his military circle, where three weeks earlier he had praised Rosecrans for vigor and good judgment at Iuka. Grant also began to complain that Rosecrans had missed the chance to destroy Van Dorn's army when he waited a day to give chase. Rosecrans behaved as badly as Grant. His brittle ego and hair-trigger temper got the better of him, and he enlisted his field commanders for a verbal duel with Grant. Those who refused to cooperate, Rosecrans slighted in his report of the Battle of Corinth.[24]

Grant and Rosecrans clashed openly on October 21. Their exchange began with an innocuous query from Grant regarding cavalry arms: "General, are you in receipt of or have you any rifles for cavalry on the way for use of troops at Corinth? If so how many? I remember hearing you say something on the subject and want to know so as to know how to distribute when they arrive."

Rosecrans bristled. A shipment of rifles indeed was on the way. The rifles should go to his cavalry regiments, which had done most of the fighting, rather than to those operating nearer Grant. Rosecrans's impertinence angered Grant, and he replied sharply: "Your remarkable telegram is just received. If the troops commanded by you are not a part of my command, what troops are? The Eastern District is the same to me and I have no partiality for any portion of it, over any other portion. General, I am afraid from many of your dispatches that you regard your command giving privileges held by others commanding geographical divisions. This is a mistake."

Rosecrans found Grant's missive equally remarkable, and he demanded an apology for its tone. Grant not only did not apologize, but he also obliquely accused Rosecrans of insubordination. He repeated his anger over the "leaky [nature] of some on your staff or in confidential relation to you as evidenced by newspaper correspondents and their attempt to keep up an invidious distinction between the armies of the Mississippi and the Tennessee," and over Rosecrans's improper disposition of Rebel prisoners, which "looked like ignoring higher authority."

Rosecrans could not let matters rest. He took exception to Grant's continued criticism of his handling of prisoners and denied anyone at his headquarters was fomenting discord; the problem rested with Grant's staff. Rosecrans closed with a challenge: "There are no headquarters in these United States less responsible for what newspaper correspondents say of operations than mine. This I wish to be understood to be distinctly applicable to the affairs of Iuka and Corinth. After this declaration I am free to say that if you do not meet me frankly with a declaration that you are satisfied I shall consider my power to be useful in this Department ended."[25]

A full-dress showdown between Grant and Rosecrans looked inevitable. Rosecrans tried to preempt it by demanding Halleck transfer him. Making it clear that "mousing politicians on Grant's staff" were responsible for the "feeling of jealousy" Grant had toward him, Rosecrans begged Halleck to assign him to "any other suitable duty."[26]

Rosecrans was right about Grant's subordinates, but he did Grant a disservice in accusing him of petty jealousy. Rawlins, McPherson, and Hurlbut all wanted Grant to relieve Rosecrans, and they appealed to Julia Grant to use her influence with her husband to that end. Julia spoke to the general, but Grant dismissed the matter. He intended to overlook Rosecrans's indiscretions. "Rosecrans's action was all wrong," he conceded, but Rosecrans was "a brave and loyal soldier with the best of military training, and of this kind we have none to spare at present. Besides, 'Rosy' is a fine fellow. He is a bit excited now but he will soon come around alright. Do not trouble yourself about me, my dear little wife," Grant assured her with a smile, "I can take care of myself."[27]

Halleck spared Grant the trouble. On October 23 he told Grant to have Rosecrans report to Cincinnati, where he would receive an onward assignment. "General Rosecrans is ordered to Cincinnati to receive orders," Grant confided to Ord. "I suspect he is going to take Buell's place. Have had no intimation of the fact but Buell's failure to come up with Bragg, whether his fault or not, will raise such a storm that he will probably have to give way."[28]

Grant guessed correctly. Bowing to pressure from the press and from

Republican governors, President Lincoln replaced Buell with Rosecrans as commander of the Army of the Ohio.

Both Rosecrans and Grant were delighted: Rosecrans to have an independent command, and Grant to be rid of a troublesome subordinate. He felt no need to castigate Rosecrans further. The pressures of his new assignment would try Rosecrans's nerves far better than could Grant. "It is a great annoyance to gain rank and command enough to attract public attention," Grant told Ord. "I have found it so and would really prefer some little command where public attention would not be attracted towards me." [29]

26

The Reckoning

Apart from poisoning relations between Grant and Rosecrans and wrecking the career of Earl Van Dorn, what had the Battle of Corinth accomplished? Federal generals removed from the squabbling of Grant and Rosecrans considered Corinth a signal defeat for the South. "The effect of the battle of Corinth was very great," wrote William T. Sherman. "It was, indeed, a decisive blow to the Confederate cause in our quarter, and changed the whole aspect of affairs in West Tennessee. From the timid defensive we were at once enabled to assume the bold offensive. In Memphis I could see its effects upon the citizens, and they openly admitted that their cause had sustained a death blow." David Stanley agreed Corinth was a disaster from which the Confederacy never recovered. A convalescent Richard Oglesby had no doubt he had been wounded in "one of the most decisive battles, in its immediate results and in its effects upon subsequent campaigns, of the early engagements of the war."[1]

Those close to Grant and Rosecrans emphasized lost opportunities rather than gains. Grant supporters blamed Rosecrans for letting Van Dorn slip away. Grenville Dodge, whom Grant placed in command of Davies's division after Corinth, summed up the consensus at district headquarters: "Van Dorn and Price retreated completely demoralized, and should have been relentlessly followed, and their trains and artillery captured; and although Grant urged this in dispatch after dispatch, for some reason there were delays."

Arthur Ducat disagreed vehemently. It was Grant who had permitted Van Dorn to escape. "If Grant had supported you . . . the Vicksburg campaign would never have been necessary," Ducat assured Rosecrans. "I regard the calling back of you at Corinth as an unexplained military crime, and shall so regard it while I live unless your superiors will admit that they were insane or jackasses."[2]

Overwrought language aside, Ducat's argument has merit. Contrary to Dodge's claim, Grant never urged Rosecrans on, and Rosecrans pursued aggressively — eventually in defiance of Grant's orders. Could Rosecrans, as he himself suggested, have gone all the way to Vicksburg, or were his troops too exhausted and their supply lines too attenuated for an extended pursuit, as Grant maintained?

Testimony from the men in ranks suggests they could have pushed on far beyond Ripley. They were tired but in excellent spirits. Demoralized Southern stragglers and abandoned Rebel equipment gave the Federals a feeling of being in on the kill. Straggling among the Yankees was a problem until rations met the army at Ripley; a brief rest there, and with the weather turned cool, the army certainly could have pushed on. Unlike the Rebels, Rosecrans's men were never starved; they found foraging quite fruitful. "We ran out of rations and foraging parties were sent out," said an Iowan. "They brought in sweet potatoes and fresh pork . . . and the boys have been pitching freely into cattle and hogs in this locality." "We feasted on hog and sweet potatoes," a member of the Fiftieth Illinois recalled fondly. "The regiment feasts on chickens, geese, and sweet potatoes," said a soldier of the Seventh Illinois.[3]

Had Rosecrans continued beyond Ripley, Grant feared he "would have met a greater force than Van Dorn had at Corinth and behind intrenchments or on chosen ground, and the probabilities are he would have lost his army." On the contrary, a Federal drive southward would have turned Van Dorn's flank, rendering Holly Springs indefensible. Van Dorn's army was in no condition to resist, nor were his subordinates inclined to obey him.

For the Union, victory at Corinth secured control of northern Mississippi, making possible Grant's offensive against Vicksburg. With Van Dorn's army near collapse, a protracted Union pursuit might have eliminated the principal Southern force in the state and made the Vicksburg campaign shorter and less costly. That Rosecrans delayed a day in giving chase is immaterial; it took Van Dorn weeks to restore his army to fighting trim. As Dabney Maury recalled, "At Holly Springs five thousand exchanged prisoners taken at Fort Donelson joined us, and many absentees and stragglers came in. The enemy remained supine, and for more than a month we were encamped about Holly Springs, and actively engaged in reorganizing, refitting and reinforcing our army. A vigorous pursuit after our defeat at Corinth would have prevented all this and effectually destroyed our whole command."[4]

Confederate defeat at Corinth had ramifications far beyond Mississippi. It fatally compromised Braxton Bragg's position in Kentucky at the climax of his campaign. Bragg had counted on Edmund Kirby Smith to cover his right flank and Van Dorn to protect his left during his drive on Louisville. Both failed him utterly. Smith considered his part in the campaign finished after he captured Lexington in early September. Rather than unite with Bragg or threaten southern Ohio to ease pressure on him, Smith scattered his army across eastern Kentucky. Van Dorn and Price, of course, never made it beyond the Mississippi-Tennessee state line. With both flanks exposed, Bragg stopped short of Louisville, yielding the initiative to Buell.

On October 1, while Van Dorn marched on Corinth, Buell advanced from Louisville to do battle with Bragg. At Perryville on October 7 and 8 the two armies fought to a standstill. Believing himself outnumbered, Bragg withdrew toward the Cumberland Mountains. On October 12 he called his generals and Smith to a council of war to consider future plans. Despite Smith's lack of cooperation and the lukewarm reception Kentucky had accorded his army, Bragg had not given up on the state. But while he waited for his generals, word came of Van Dorn's debacle at Corinth. Van Dorn's defeat left the Army of the Mississippi as the only force between the Appalachians and the Mississippi capable of contesting a Yankee invasion of the Deep South. And with Federal control of the Memphis and Charleston Railroad now undisputed and northern Mississippi and northern Alabama emptied of Confederates, Chattanooga was vulnerable to a rapid strike from the west. Bragg decided he must retreat at once to save his army and protect Chattanooga.[5]

Bragg later alleged the failure of Van Dorn and Price to launch an offensive into Middle Tennessee to be a deciding factor in the collapse of the Kentucky campaign. Only Bragg knew whether he honestly believed that or was merely dissembling to mask his own errors. Bragg had done everything possible to guarantee disappointment. He left no one in charge when he quit Tupelo, never made his purpose clear to Van Dorn, and phrased most of his appeals for help to Price ambiguously.

Unquestionably, a thrust across the Tennessee River by Van Dorn and Price in early September would have changed fundamentally the complexion of the Kentucky campaign. Strung out on the march across Middle Tennessee, Buell's army would have been vulnerable to flank attack and defeat in detail. At a minimum the time lost in concentrating his command to oppose Van Dorn and Price most probably would have cost Buell the race to Louisville. With only raw troops and state militia to oppose him, Bragg might have realized his desideratum of taking Cincinnati. But neither Van Dorn nor Price alone was strong enough to pose a credible threat to Buell, and it took the near-annihilation of Price's army at Iuka to bring them together. By then the chance to strike Buell a fatal blow from western Tennessee had passed.

Price and Van Dorn wasted September working at cross purposes. Price wanted to help Bragg but rightly judged as too great the risk of Grant striking his moving column from behind and cutting his lines of supply. He simply lacked the men needed to repel a Union counterthrust in western Tennessee.

Van Dorn disagreed with Bragg over the wisdom of a move into Middle Tennessee, preferring instead to clear the Yankees from the Mississippi River Valley, beginning with Corinth. But Van Dorn and Price had too few troops to give an assault on Corinth a reasonable chance of success. It is axiomatic that an attacker need have at least a three-to-one advantage over his opponent. Van Dorn and Price mustered only a few hundred more men than Rosecrans. The great nineteenth-century Prussian military theorist Karl von Clausewitz cautioned that an attack against an able opponent in a good position was a "risky business" that should only be tried after careful reconnaissance ruled out all other options for dislodging the enemy. As John Bowen correctly charged at Van Dorn's court-martial, the Mississippian conducted no reconnaissance and threw his tired army against a foe amply warned of his approach.[6]

The abrupt collapse of Rosecrans's outer line on October 3 suggests aggressive action by Lovell would have compelled the Federals to abandon Corinth. Only Crocker's lone brigade stood between Lovell and the town. Had Lovell pressed on and gained Rosecrans's rear, a Confederate victory would have been a near-certainty. Hamilton's division on the Federal right contributed nothing to the battle, and Stanley's brigades would have been chewed up on the march.

Angry survivors of Maury's and Hébert's divisions were convinced the poor showing of Davies's Federals on October 4 gave Lovell a second chance to turn the tide of battle. They were mistaken. Lovell's division stood opposite three well-positioned Federal brigades and the guns of

Battery Phillips. Had Lovell attacked, his division would have been mauled as badly as Maury's.

Without a doubt Lovell and Hébert undermined Van Dorn's plan of battle—Lovell by his inertness the first day, and Hébert by absenting himself with a sudden illness on the second day. Charges of insubordination were warranted, but none were brought. The War Department decided the potential damage to Southern morale from a public airing of the army's dirty laundry too great.

In a larger sense Lovell's and Hébert's failings were irrelevant. Van Dorn could not have held Corinth had he won it. The 14,000 Rebels who survived the battle could never have repelled the 50,000 fresh troops Grant would have been able to muster rapidly against the town. True to character, Van Dorn had gambled everything on a long shot for glory.

*   *   *

Rosecrans's army returned to a Corinth awash in suffering, its landscape blasted and torn. The forests were littered with broken branches and treetops lopped off by cannonballs. Unexploded shells and shell fragments lay everywhere. Fields were pounded to dust. The stench of bloated corpses permeated the town. Wounded men by the thousands lay in homes, hotels, schools, and stores. Under a flag of truce Sterling Price's assistant medical director, Dr. J. C. Roberts, and thirty-six Southern surgeons retrieved the Rebel wounded, but the dead remained, rolled into shallow graves or left to rot where they fell. Roberts had found only one wounded Confederate under a tent or in a room. The rest were scattered on lawns or on the streets of town or were still on the battlefield. The industry of the Sanitary Commission meant wounded Federals fared better; but their agony was nonetheless intense, and scores succumbed to fever and infection.[7]

No amount of care could save Pleasant Hackleman, who died in the Tishomingo Hotel on the night of October 3, a few hours after General Davies left him. General Oglesby clung to life, but no one expected him to pull through. He was moved from the hotel to a private residence, where a team of surgeons, under the supervision of the medical director of the Army of the Tennessee, Dr. John Holsten, examined his wound and pronounced it mortal. After probing without success for the bullet, they bandaged him up and waited.

Oglesby was still alive at dawn. The bleeding had stopped, and he was breathing a little easier. As hope for his recovery grew, Dr. Holsten telegraphed Grant to ask that Silas Trowbridge, a brigade surgeon on duty at Bethel, Tennessee, be permitted to come to Corinth to attend to Oglesby. Grant agreed but warned Trowbridge that enemy cavalry marauded the railroad between Bethel and Corinth, and he had no cavalry to spare for an escort. Trowbridge left Bethel with an assistant surgeon at once.[8]

Grant acceded to Holsten's request because of presidential pressure. When Lincoln telegraphed Grant after Corinth, he expressed more concern for Oglesby's well-being than for the state of the army. "I congratulate you and all concerned on your recent battles and victories — How does it all sum up?" the President began. "I especially regret the death of General Hackleman; and am very anxious to know the condition of General Oglesby, who is an intimate personal friend."[9] Trowbridge himself had a deeper reason for going. Oglesby had been his patient and friend before the war, when Trowbridge was in private practice in Decatur, Illinois.

The way to Corinth proved clear, and Trowbridge reached town on the evening of October 6. His portly friend sat painfully in a rocking chair, "pale, haggard, and in much distress, incapable of lying down, with a pulse of 136 per minute . . . expectorating small quantities of arterial and much larger amounts of venous blood . . . skin bathed in a cold perspiration, excretions from the kidneys and bowels almost suspended."

As Trowbridge gently probed the wound, careful not to disturb the perforated lung, Oglesby asked him what he thought of his chances. Trowbridge recited a medical axiom: if the patient survived the initial shock and loss of blood, there was a good possibility of saving him. Oglesby thanked Trowbridge for the first hopeful words he had heard, then fell into a deep, morphine-induced sleep.

The next morning Trowbridge consulted with Dr. Holsten, whose preferred treatment horrified the Illinois physician. Assuming Oglesby would die, Holsten had decided to ply him with liberal doses of opium, beef soup, and Catawba wine to ease his suffering. Trowbridge remonstrated sharply: "I told him I had hope of the recovery of General Oglesby . . . and that it was my duty . . . to treat the case with an eye [to] that result. I further told him that I could not adopt his views, and begged his pardon for not accepting those of men so much my senior in years and experience. But that I could see no chance for a recovery with opium sickness and a stomach crammed with beef-steak and the heaviest wine known to surgeons."

Holsten stormed off to Oglesby's room and began talking with the general. Trowbridge followed and yanked Holsten aside. Was he recommending the patient prepare for death? Holsten said he was. Trowbridge nearly struck him: "I peremptorily forbid his doing so, and only succeeded in preventing him by assuring him that in case he did so I should prefer charges against him."

Holsten left the house, complaining to all who cared to listen that Trowbridge was an ignorant "upstart." Left at last with his patient, Trowbridge stopped the wine, opium, and morphine and started Oglesby on a guarded diet. The general lost weight and became jaundiced, but the wound healed. His wife and sister joined him. They and Trowbridge removed Oglesby from Corinth, with its "unwholesome and disgusting odors," and from the pernicious presence of Dr. Holsten. After a brief stop at Jackson, Tennessee, where Grant came to the train station to congratulate Oglesby on his improved condition, they continued on to Decatur. Oglesby recovered to serve two terms as governor of Illinois, then went on to the United States Senate. "And [he] now," wrote Trowbridge in 1873, "holds a much better mortgage on a long and especially useful life than Dr. Holsten."[10]

*　*　*

Capt. Isaac Jackson awoke with a start. For two days he had lain unconscious, given up for dead by the surgeons. "When I aroused from my stupor, I could scarcely recollect what had happened. Both eyes were swelled completely shut from the wound, and although it was day time, I supposed it was night." Jackson's first thought was of his men. Corporal Harrison was at his bedside. "How are my men," Jackson asked hurriedly. "The company is badly cut up, captain," confessed Harrison. "For God's sake tell me who were killed," Jackson begged. Harrison recited the roll of the dead. Jackson thought he had named the entire company. "As soon as I got the news of how terribly my men had suffered, actually a feeling of gladness came over me that I had been wounded and had something to suffer."

Jackson recovered. Surgeons had left the minie bullet in his head so as not to destroy the right eye, on the off chance Jackson survived. His sight was permanently impaired, and the optic nerve pained Jackson intermittently; but the wound healed without disfiguring him. Jackson returned to the Sixty-third Ohio, closing the war as its colonel.[11]

*　*　*

Five months after Corinth Sterling Price got his wish — in part. On February 27, 1863, General Pemberton issued orders relieving Price of duty in Mississippi and directing him to report to Edmund Kirby Smith in the trans-Mississippi department. Two weeks later Price crossed the Mississippi River with his staff and a small cavalry escort. He reported for duty at Little Rock, Arkansas.

His Missourians never came. Price had warned Richmond they would all desert or mutiny after his departure. Few deserted, and none mutinied. The Missourians wanted to go home, but not through Arkansas, the scene of so much suffering. Remaining with Pemberton the Missourians fought gallantly until the surrender of Vicksburg. They were captured, exchanged, and fought with the Army of Tennessee from Atlanta to Franklin. In April 1865 the last remnant of Price's Missouri division, 400 men, surrendered at Mobile.[12]

*   *   *

Earl Van Dorn was a broken man. Stripped of senior command, he became reckless in his whoring. Van Dorn regained a measure of dignity in December 1862 when he surprised everyone by leading his new cavalry command on a successful raid against Grant's supply depot at Holly Springs. At dawn on December 20 Van Dorn and his troopers thundered into Holly Springs at a full gallop, yelling and shooting, the way Van Dorn liked it. They rode over the sleeping Federal infantry, capturing all but 130 defenders. Van Dorn's simple report of the most stunning triumph of his career, stripped of the conceit and overblown verbiage common to his correspondence, reflected a tired, chastened man: "I surprised the enemy at this place at daylight this morning, burned up all the quartermaster's stores, cotton . . . an immense amount; burned up many trains; took a great many arms and about 1,500 prisoners. I presume the value of stores would amount to $1,500,000."[13]

Van Dorn's raid paralyzed Grant. To protect his remaining bases of supply, Grant broke off his land advance on Vicksburg and retired to Memphis. Van Dorn had accomplished with three brigades of cavalry what he had failed to do with an army: severely disrupt Union operations in Mississippi.

For a time Van Dorn's star shined. In January 1863 Joseph Johnston transferred him to Tennessee along with 6,000 cavalrymen to raid the lines of communication of General Rosecrans's Army of the Cumberland. Organizing his command into two mounted divisions, Van Dorn started from Tupelo in the dead of winter. It was late February before he reached Tennessee. Rosecrans had bested Bragg at Stones River and was in control of Middle Tennessee. All Bragg could do was harass him with cavalry.

Van Dorn accepted the duty, cutting a swath of victories across the Union rear that badly damaged Rosecrans's supply network.

With the coming of spring, Van Dorn's spirits lifted. "I *am* a soldier," he wrote his sister after his mounted forays, "and my soul swells up and tells me that I *am* worthy to lead the armies of my country."

Only his family agreed. The long-coveted promotion to lieutenant general never came, and Van Dorn resigned himself to the job before him. Heartsick and bitter, on April 1 he wrote Emily an epiphany of his woes. They must not hope for his promotion; the stigma of Corinth was too great, he began. But he had been right to attack Corinth, Van Dorn reassured her. Reverting to ornate and frenzied prose, he told her "the attack on Corinth was the best thing I ever did in my life—if it had been successful I would have been pronounced the most brilliant General of the war. . . . I was not wounded in spirit by the powerful 'vox populi' at all—only when falsehood assailed my private character, and made me the butt of malignant archery in my own native state—expelling me after weary watchfulness and long mental pain for them from his soil."[14]

Van Dorn's passions were too great to confine to rhetorical outbursts. The time between raids passed slowly; to alleviate the tedium, said a member of Van Dorn's command, "the officers from the general down, found time for sport and amusement amongst the generally wealthy and hospitable citizens." Van Dorn went too far, and accusations of philandering and corruption reappeared. The *Chattanooga Rebel* said his drunken and licentious carrying-on had cost him the confidence of the people of Tennessee. While Van Dorn was a cadet at West Point, his strong libido was cause for good-natured teasing by his class-mates. The young Mississippian "was very sentimental and always in love with someone," remembered John Pope. But untethered passions in an army commander could have tragic results, and Van Dorn's officers worried about his fondness for women.[15]

Of particular interest to Van Dorn was Jessie Peters, the vivacious, twenty-five-year-old wife of Dr. George Peters, a middle-aged physician of some standing. She made Van Dorn's acquaintance during his lengthy stay at Spring Hill, Tennessee. Local gossips whispered that the Mississippian had his eye on Jessie, and a hostile journalist wrote that Van Dorn and Mrs. Peters often went riding in her carriage.

The rumors reached Dr. Peters, and he returned to Spring Hill to investigate. The doctor found the little town buzzing with talk of the affair. For three weeks he listened with increasing fury to tales of Van Dorn's lechery. When he caught a servant passing a note from Van Dorn to his wife, Peters exploded with an ultimatum: "I distinctly told him," the doctor later wrote, "I would blow his brains out if he ever entered the premise again." The same went for Van Dorn, if he ever set foot in the Peters's yard.

Perhaps convinced his threat was enough, Dr. Peters left for Nashville on business. When he returned several days later, meddlesome townspeople told Peters that Van Dorn had visited his wife every night while he was away.

The doctor set a trap. Feigning a trip to Shelbyville, he hid in the house and waited. In the early hours of May 6, as Peters told it, he caught Van Dorn "where I expected to find him." Peters confronted him at gunpoint. Van Dorn bade him shoot. Concerned for his wife's honor, Peters offered to spare Van Dorn's life in exchange for a signed statement exonerating his wife of unseemly conduct.

The next morning the doctor called at Van Dorn's headquarters. Peters claimed he came to pick up the promised statement. A sentry at headquarters said the doctor had come to ask Van Dorn for a pass through the lines. Whatever the pretext, Dr. Peters was armed and angry. Words were exchanged. Peters said Van Dorn dismissed him scornfully. The doctor threatened to shoot the general if he did not start writing. "You damned cowardly dog," Van Dorn purportedly replied, "take that door, or I will kick you out of it." Peters snapped. He drew a derringer and shot Van Dorn in the left side of his head. The Mississippian was dead before he fell. Peters escaped through the lines.[16]

The press bludgeoned Van Dorn in death as it had in life. "Van Dorn has been recognized for years as a rake, a most wicked libertine — and especially of late. If he had led a virtuous life, he would not have died — unwept, unhonored and unsung," scolded the *Fayetteville Observer*. "The country has sustained no loss in the death of Van Dorn. He was unfit to live, let alone having charge of such important trusts as he had."

The *Nashville Dispatch* was equally unforgiving. Said its Richmond correspondent: "My informant tells me that [Van Dorn] had degraded the cause, and disgusted everyone by his inattention to his duties and his constant devotion to the ladies. . . . He was never at his post when he ought to be. He was either tied to a woman's apron strings or heated with wine."

There was no compassion forthcoming from the army, either. "General Van Dorn was killed yesterday for tampering with a fellow's wife," a sergeant wrote home. "If that be the case he was served right."[17]

Earl Van Dorn died on May 7, 1863. His last, best chance for glory perished seven months earlier on the battlefield of Corinth.

APPENDIX

The Opposing Forces at the Battle of Iuka

The following list was assembled from *The War of the Rebellion: A Compilation of the Official Records of the Union and Confederate Armies*. With respect to officer casualties, (k) signifies killed, (mw) mortally wounded, (w) wounded, and (c) captured.

Union Forces

ARMY OF THE MISSISSIPPI
Maj. Gen. William S. Rosecrans

Second Division — Brig. Gen. David S. Stanley
First Brigade — Col. John W. Fuller
    27th Ohio; 39th Ohio; 43d Ohio; 63d Ohio; Battery M, 1st Missouri Light
    Artillery; 8th Battery, Wisconsin Light Artillery; Battery F, 2d U.S. Artillery
Second Brigade — Col. Joseph A. Mower
    26th Illinois; 47th Illinois; 11th Missouri; 8th Wisconsin; 2d Battery, Iowa Light
    Artillery; 3d Battery, Michigan Light Artillery

Third Division — Brig. Gen. Charles S. Hamilton
First Brigade — Col. John B. Sanborn
    48th Indiana; 5th Iowa; 16th Iowa; 4th Minnesota; 26th Missouri; 11th Battery,
    Ohio Light Artillery
Second Brigade — Brig. Gen. Jeremiah C. Sullivan
    10th Iowa; 17th Iowa; 10th Missouri; 80th Ohio; Company F, 24th Missouri;
    12th Battery, Wisconsin Light Artillery

Cavalry Division — Col. John K. Mizner
2d Iowa; 3d Michigan; Companies B and E, 7th Kansas

Unattached
Jenks's Company, Illinois Cavalry

Confederate Forces

ARMY OF THE WEST
Maj. Gen. Sterling Price

First Division — Brig. Gen. Henry Little (k), Brig. Gen. Louis Hébert
First Brigade — Col. Elijah Gates
    16th Arkansas; 2d Missouri; 3d Missouri; 1st Missouri Dismounted Cavalry;
    Wade's Missouri Battery
Second Brigade — Brig. Gen. Louis Hébert
    14th Arkansas; 17th Arkansas; 3d Louisiana; 40th Mississippi; 1st Texas Legion;
    3d Texas Dismounted Cavalry; Dawson's St. Louis Battery; Clark Missouri Battery

Third Brigade — Brig. Gen. Martin E. Green
   7th Mississippi Battalion; 43d Mississippi; 4th Missouri; 6th Missouri; 3d Missouri
   Dismounted Cavalry; Guibor's Missouri Battery; Landis's Missouri Battery
Fourth Brigade — Col. John D. Martin
   37th Alabama; 36th Mississippi; 37th Mississippi; 38th Mississippi

Cavalry — Brig. Gen. Frank C. Armstrong
Adams's Mississippi Regiment; 2d Arkansas; 2d Missouri; 4th Mississippi Cavalry;[1]
1st Mississippi Partisan Rangers

The Opposing Forces at the Battle of Corinth

Union Forces

ARMY OF THE MISSISSIPPI
Maj. Gen. William S. Rosecrans

Second Division — Brig. Gen. David S. Stanley
First Brigade — Col. John W. Fuller
   27th Ohio; 39th Ohio; 43d Ohio; 63d Ohio; Jenks's Company, Illinois Cavalry;
   3d Battery, Michigan Light Artillery; 8th Battery, Wisconsin Light Artillery;[2]
   Battery F, 2d U.S. Artillery
Second Brigade — Col. Joseph A. Mower (w)
   26th Illinois; 47th Illinois; 5th Minnesota; 11th Missouri; 8th Wisconsin; 2d
   Battery, Iowa Light Artillery

Third Division — Brig. Gen. Charles S. Hamilton
Escort
   Company C, 5th Missouri Cavalry
First Brigade — Brig. Gen. Napoleon B. Buford
   48th Indiana; 59th Indiana; 5th Iowa; 4th Minnesota; 26th Missouri; Battery M,
   1st Missouri Light Artillery; 11th Battery, Ohio Light Artillery
Second Brigade — Brig. Gen. Jeremiah C. Sullivan (w), Col. Samuel A. Holmes
   56th Illinois; 10th Iowa; 17th Iowa; 10th Missouri; Company F, 24th Missouri;
   80th Ohio; 6th Battery, Wisconsin Light Artillery; 12th Battery, Wisconsin Light
   Artillery

Cavalry Division — Col. John K. Mizner
7th Illinois; 11th Illinois; 2d Iowa; 7th Kansas; 3d Michigan; Companies E, H, I,
and K, 5th Ohio

Unattached
64th Illinois (Yates Sharpshooters); Companies A, B, C, D, H, and I (Siege Artillery),
1st U. S. Infantry

ARMY OF WEST TENNESSEE

Second Division — Brig. Gen. Thomas A. Davies

First Brigade — Brig. Gen. Pleasant A. Hackleman (mw), Col. Thomas W. Sweeny
52d Illinois; 2d Iowa; 7th Iowa; 58th Illinois (detachment); 8th Iowa (detachment); 12th Iowa (detachment); 14th Iowa (detachment)[3]
Second Brigade — Brig. Gen. Richard J. Oglesby (w), Col. August Mersy
9th Illinois; 12th Illinois; 22d Ohio; 81st Ohio
Third Brigade — Col. Silas D. Baldwin (w), Col. John V. Du Bois
7th Illinois; 50th Illinois; 57th Illinois
Artillery — Maj. George H. Stone
Batteries D, H, I, and K, 1st Missouri Light Artillery

Unattached
14th Missouri (Western Sharpshooters)

Sixth Division — Brig. Gen. Thomas J. McKean
First Brigade — Col. Benjamin Allen, Brig. Gen. John McArthur
21st Missouri; 16th Wisconsin; 17th Wisconsin
Second Brigade — Col. John M. Oliver
Ford's Company, Illinois Cavalry; 15th Michigan; Companies A, B, C, and E, 18th Missouri; 14th Wisconsin; 18th Wisconsin
Third Brigade — Col. Marcellus M. Crocker
11th Iowa; 13th Iowa; 15th Iowa; 16th Iowa
Artillery — Capt. Andrew Hickenlooper
Battery F, 2d Illinois Light Artillery; 1st Battery, Minnesota Light Artillery; 3d, 5th, and 10th Batteries, Ohio Light Artillery

Confederate Forces

ARMY OF WEST TENNESSEE
Maj. Gen. Earl Van Dorn

Price's Corps, or Army of the West
Maj. Gen. Sterling Price

First Division — Brig. Gen. Louis Hébert, Brig. Gen. Martin E. Green
First Brigade — Col. Elijah Gates
16th Arkansas; 2d Missouri; 3d Missouri; 5th Missouri; 1st Missouri Cavalry;[4] Wade's Missouri Battery
Second Brigade — Col. W. Bruce Colbert
14th Arkansas; 17th Arkansas; 3d Louisiana; 40th Mississippi; 1st Texas Legion; 3d Texas Cavalry[5]; Clark Missouri Battery; Dawson's St. Louis Battery
Third Brigade — Brig. Gen. Martin E. Green, Col. W. H. Moore (mw)
7th Mississippi Battalion; 43d Mississippi; 4th Missouri; 6th Missouri; 3d Missouri Cavalry[6]; Guibor's Missouri Battery; Landis's Missouri Battery
Fourth Brigade — Col. John D. Martin (mw); Col. Robert McLain (w)
37th Alabama; 36th Mississippi; 37th Mississippi; 38th Mississippi; Lucas's Battery[7]

Maury's Division — Brig. Gen. Dabney H. Maury
Moore's Brigade — Brig. Gen. John C. Moore

42d Alabama; 15th Arkansas; 23d Arkansas; 35th Mississippi; 2d Texas; Bledsoe's
Missouri Battery

Cabell's Brigade — Brig. Gen. W. L. Cabell
18th Arkansas; 19th Arkansas; 20th Arkansas; 21st Arkansas; Jones's Arkansas
Battalion; Rapley's Arkansas Battalion; Appeal Battery

Phifer's Brigade — Brig. Gen. C. W. Phifer
3d Arkansas Cavalry[8]; 6th Texas Cavalry[9]; 9th Texas Cavalry[10]; Stirman's
Arkansas Sharpshooters; McNally's Battery

Cavalry[11] — Brig. Gen. F. C. Armstrong
Slemon's Regiment; Wirt Adams's (Mississippi) Regiment

Reserve Artillery
Hoxton's Tennessee Battery; Sengstak's Battery

District of the Mississippi

First Division — Maj. Gen. Mansfield Lovell
First Brigade — Brig. Gen. Albert Rust
4th Alabama Battalion; 31st Alabama; 35th Alabama; 9th Arkansas; 3d Kentucky;
7th Kentucky; Hudson's Battery

Second Brigade[12] — Brig. Gen. John Villepigue
33d Mississippi; 39th Mississippi

Third Brigade — Brig. Gen. John S. Bowen
6th Mississippi; 15th Mississippi; 22d Mississippi; Caruthers's Mississippi Battalion;
1st Missouri; Watson's Battery

Cavalry Brigade — Col. W. H. Jackson
1st Mississippi; 7th Tennessee

The Opposing Forces at the Battle of Davis Bridge

Union Forces

ARMY OF WEST TENNESSEE
Escort
Company A, 2d Illinois Cavalry

Fourth Division — Maj. Gen. Stephen A. Hurlbut
First Brigade — Brig. Gen. Jacob G. Lauman
28th Illinois; 32d Illinois; 41st Illinois; 53d Illinois; 3d Iowa; Battery C, 1st Missouri
Light Artillery; 15th Battery, Ohio Light Artillery; 1st and 2d Battalions, 5th
Ohio Cavalry

Second Brigade — Brig. Gen. James C. Veatch
14th Illinois; 15th Illinois; 46th Illinois; 25th Indiana; 53d Indiana; Battery L,
2d Illinois Light Artillery; 7th Battery, Ohio Light Artillery

Provisional Brigade — Col. Robert K. Scott
12th Michigan; 68th Ohio

Confederate Forces

Maury's Division — Brig. Gen. Dabney H. Maury

Moore's Brigade — Brig. Gen. John C. Moore
    42d Alabama; 15th Arkansas; 23d Arkansas; 35th Mississippi; 2d Texas; Bledsoe's Missouri Battery

Cabell's Brigade — Brig. Gen. W. L. Cabell
    18th Arkansas; 19th Arkansas; 20th Arkansas; 21st Arkansas; Rapley's Arkansas Battalion Sharpshooters; Jones's Arkansas Battalion; Appeal Battery

Phifer's Brigade — Col. Lawrence S. Ross
    6th Texas Cavalry[13]; 9th Texas Cavalry[14]; 3d Arkansas Cavalry[15]; Stirman's Sharpshooters; McNally's Battery

Cavalry

Wirt Adams's Mississippi Regiment

Other Artillery Engaged

Dawson's St. Louis Battery

NOTES

ADAH   Alabama Department of Archives and History, Montgomery
ArU   University of Arkansas, Fayetteville
CWMC   Civil War Miscellaneous Collection, U.S. Army Military History
         Institute, Carlisle Barracks, Pa.
CWRT   Harrisburg Civil War Round Table Collection, U.S. Army Military
         History Institute, Carlisle Barracks, Pa.
CWTI   Civil War Times Illustrated Collection, United States Army Military
         History Institute, Carlisle Barracks, Pa.
DLC   Library of Congress, Washington, D.C.
DU   Duke University, William R. Perkins Library, Durham, N.C.
HL   Huntington Library, San Marino, Calif.
HNOC   Historic New Orleans Collection, Kemp and Leila Williams
         Foundation, New Orleans, La.
IaHS   State Historical Society of Iowa, Iowa City
IaU   University of Iowa, Iowa City
ISHL   Illinois State Historical Library, Springfield
IU   Illinois Historical Survey, University of Illinois Library, Champaign-
         Urbana
KHS   Kansas State Historical Society, Topeka
MiU   University of Michigan, Michigan Historical Collections, Bentley
         Historical Library, Ann Arbor
MnHS   Minnesota Historical Society, St. Paul
MoHS   Missouri Historical Society, St. Louis
NMMA   Northeast Mississippi Museum Association, Corinth
OHS   Ohio Historical Society, Columbus
OR   *War of the Rebellion: A Compilation of the Official Records of the Union and
         Confederate Armies*, series 1. Washington, D.C., 1880–1901
SNMP   Shiloh National Military Park Library, Shiloh, Tenn.
TxU   University of Texas, Center for American History, Austin
UCLA   University of California, Los Angeles
UNC   University of North Carolina, Southern Historical Collection, Chapel
         Hill
USAMHI   United States Army Military History Institute, Carlisle Barracks, Pa.
WHMC   University of Missouri, Western Historical Manuscripts Collection,
         Columbia and St. Louis
WiHS   State Historical Society of Wisconsin, Madison
WRHS   Western Reserve Historical Society, Cleveland, Ohio

1. Castel, *Price and the War*, 84; Shalhope, *Price*, 206, 208; Connelly, *Army of the Heartland*, 142, 174–75; Duke, "Confederate Intrenchments," 1, Historical Reports, SNMP; Pope, "War Reminiscences XII"; Anderson, *Memoirs*, 191–92; Webb, *Battles and Biographies*, 292.

2. Castel, *Price and the War*, 4; Justin Smith, *War with Mexico*, 1:290, 2:266; Webb, *Battles and Biographies*, 289; Shalhope, *Price*, 75–77; Bauer, *Mexican War*, 158–59.

3. Reynolds, "Price and the Confederacy," 11, 13–14, 22, 26–29, MoHS; W. H. B., "Sterling Price," 366–67; Pope, "War Reminiscences XII"; Shalhope, *Price*, 149–50, 163–64; Malone, *Dictionary of American Biography*, 15:216; Broadhead, "Early Events," 23. For a different interpretation of Price's motives, see Castel, *Price and the War*, 12–24.

4. W. H. B., "Sterling Price," 367–69, 370; Castel, *Price and the War*, 27–29, 38–40, 44–45; *OR* 8:695–97, 702, 729; Shalhope, *Price*, 188, 192–93; Monaghan, *War on the Border*, 199–201; Reynolds, "Price and the Confederacy," 34.

5. Connelly, *Army of the Heartland*, 104; *OR* 8:730–31; Shalhope, *Price*, 194, 197; Reynolds, "Price and the Confederacy," 34.

6. Reynolds, "Price and the Confederacy," 35; Shea and Hess, *Pea Ridge*, 20; Warner, *Generals in Gray*, 133; Maury, "Recollections of Van Dorn," 191.

7. Maury, "Recollections of Van Dorn," 191, 197–98; Miller, *Soldier's Honor*, 286; "Maj. Gen. Earl Van Dorn," 385; Pope, "War Reminiscences XII"; *OR* 8:734.

8. Hartje, *Van Dorn*, 12–13, 92–93; Maury, "Recollections of Van Dorn," 197; Miller, *Soldier's Honor*, 42–53.

9. Woodworth, *Davis and His Generals*, 50–56; Archer Jones, *Confederate Strategy*, 51–52; Maury, "Recollections of Van Dorn," 192; Shea and Hess, *Pea Ridge*, 22; Castel, *Price and the War*, 68.

10. Maury, "Recollections of the Elkhorn Campaign," 181–83; Castel, *Price and the War*, 69–81; Shea and Hess, *Pea Ridge*, 237–40; W. H. B., "Sterling Price," 372; Tunnard, *Southern Record*, 188.

11. Connelly, *Army of the Heartland*, 126–44; Liddell, *Liddell's Record*, 57–58; *OR* 8:804 and 10(2):354.

12. Castel, *Price and the War*, 81–82; Shalhope, *Price*, 206–7; Gottschalk, *In Deadly Earnest*, 71–74; *OR* 8:814.

CHAPTER TWO

1. McFeely, *Grant*, 91–97; Ulysses S. Grant, *Memoirs*, 1:286–87; Cooling, *Forts Henry and Donelson*, 44–47, 224–25, 245; Connelly, *Army of the Heartland*, 137–39.

2. Stephen E. Ambrose, *Halleck*, 18–20, 33–35; Hattaway and Jones, *How the North Won*, 156.

3. McFeely, *Grant*, 104–10; Ulysses S. Grant, *Memoirs*, 1:360, 371.

4. Ulysses S. Grant, *Memoirs*, 1:371.

5. Snead, "With Price," 717–18; Ulysses S. Grant, *Memoirs*, 1:371–72, 377; Stephen E. Ambrose, *Halleck*, 41–43; Pope, "Siege of Corinth I" and "War Reminiscences XIII."

6. Wallace, *Autobiography*, 2:575–76; Chetlain, *Recollections*, 91; Ulysses S. Grant, *Memoirs*, 1:381; McFeely, *Grant*, 107.

7. Nabors, *Old Tishomingo County*, 20–21; Pope, "War Reminiscences XIII"; Archer Jones, *Confederate Strategy*, 6; Ulysses S. Grant, *Memoirs*, 1:375; Rosecrans, "Battle of Corinth," 738–39; Hattaway and Jones, *How the North Won*, 205; Lamers, *Edge of Glory*, 83.

8. Archer Jones, *Confederate Strategy*, 61.

9. Stephen E. Ambrose, *Halleck*, 5, 50–51.

10. Nabors, *Old Tishomingo County*, 9, 11–20, 36, 55; Gaither, "Back to Corinth," 13, 41; Price, "Corona College," 1, Historical Reports, SNMP; Ingersoll, *Iowa and the Rebellion*, 101; Ulysses S. Grant, *Memoirs*, 1:373; Rawlins Diary, September 12, 1862, University of North Texas.

11. Rosecrans, "Battle of Corinth," 739–40; Griffin Diary, September 16, 1862, ISHL; Ulysses S. Grant, *Memoirs*, 1:373–74; Lucius F. Hubbard, *Minnesota in the Battles of Corinth*, 6; Barron, *Lone Star Defenders*, 100–102; James F. Mohr to his brother, May 22, 1862, Mohr Papers, Filson Club; Horton, "'Sul' Ross," 6, Horton Manuscripts, NMMA; *Mankato (Minn.) Semi-Weekly Record*, September 12, 1862.

12. Gaither, "Back to Corinth," 41; Price, "Corona College," 1–2, Historical Reports, SNMP; Nabors, *Old Tishomingo County*, 20–21, 65–67; John Duckworth to William Rosser, December 25, 1862, Second Iowa Infantry File, SNMP; Alonzo L. Brown, *Fourth Minnesota*, 112–13.

13. Snead, "With Price," 719; Connelly, *Army of the Heartland*, 175–77; Keen, *Living and Fighting*, 32–33; Ulysses S. Grant, *Memoirs*, 1:374; Gottschalk, *In Deadly Earnest*, 110–11; Price, "Corona College," 3, Historical Reports, SNMP.

14. Harwell, *Journal of a Nurse*, 27–28; Duke, "Confederate Intrenchments," 1–2, Historical Reports, SNMP; Reynolds, "Price and the Confederacy," 40; Castel, *Price and the War*, 85; *OR* 17(2):628.

15. Snead, "With Price," 719–20, 723; *Dayton Weekly Journal*, June 17, 1862; *OR* 17(2):599, 604, 626, 644; Pope, "Siege of Corinth I."

16. *OR* 13:814–15, 831–32; Castel, *Price and the War*, 87; Hartje, *Van Dorn*, 207.

17. Reynolds, "Price and the Confederacy," 36, 42–44; Snead, "With Price," 723–24; Castel, *Price and the War*, 90.

18. Shalhope, *Price*, 213; Snead, "With Price," 724; *OR* 13:838; Reynolds, "Price and the Confederacy," 36–37; Castel, *Price and the War*, 89.

19. Castel, *Price and the War*, 91; Snead, "With Price," 725.

20. Lamers, *Edge of Glory*, 9–12, 17–19; Reid, *Ohio in the War*, 1:311–27.

21. Beatty, *Citizen Soldier*, 24–25, 34; Lamers, *Edge of Glory*, 20–21, 37–61, 78–86; Reid, *Ohio in the War*, 1:348; Pope, "Siege of Corinth I"; Rosecrans, "Battle of Corinth," 737.

22. McFeely, *Grant*, 106, 117; Julia Dent Grant, *Memoirs*, 99–100; Ulysses S. Grant, *Memoirs*, 1:386, 392–93; Sherman, *Memoirs*, 275–76.

CHAPTER THREE

1. Hattaway and Jones, *How the North Won*, 205–7; Stephen E. Ambrose, *Halleck*, 54; Ulysses S. Grant, *Memoirs*, 1:381; Fiske, *Mississippi Valley*, 135–37; Francis Vinson Greene, *The Mississippi*, 30–31; Woodworth, *Davis and His Generals*, 109; Kenneth P. Williams, *Lincoln Finds a General*, 4:2–3; Lamers, *Edge of Glory*, 88–89.

2. Ulysses S. Grant, *Memoirs*, 1:381.

3. Francis Vinson Greene, *The Mississippi*, 32–33; Basler, *Collected Works of Lincoln*, 5:260–61, 295.

4. Francis Vinson Greene, *The Mississippi*, 34–35; Hattaway and Jones, *How the North Won*, 207–8; Sherman, *Memoirs*, 277–78; Fiske, *Mississippi Valley*, 142–43; Stephen E. Ambrose, *Halleck*, 55–56.

5. Hattaway and Jones, *How the North Won*, 205; Throne, *Boyd Diary*, 62–63; Lamers, *Edge of Glory*, 91; Barron, *Lone Star Defenders*, 101; William Brown to his wife, September 5, 1862, Brown Letters, WiHS; Griffin Diary, September 16, 1862, ISHL; Soper, "Company D," 135; Sherman, *Memoirs*, 279; Lucius F. Hubbard, *Minnesota in the Battles of Corinth*, 6–7.

6. Comte de Paris, *Civil War*, 2:405; Ulysses S. Grant, *Memoirs*, 381–82; "Rosecrans's Campaigns," in *Report of the Joint Committee*, 3:17; Lamers, *Edge of Glory*, 92–93; Rosecrans, "Battle of Corinth," 737; Dyer, *Compendium*, 1:476.

7. Lamers, *Edge of Glory*, 93; Warner, *Generals in Blue*, 198–99.

8. Basler, *Collected Works of Lincoln*, 5:312–13; Nicolay and Hay, *Lincoln*, 5:354–56; Ulysses S. Grant, *Memoirs*, 1:392–93; McFeely, *Grant*, 119–20.

9. Ulysses S. Grant, *Memoirs*, 1:393–94; Comte de Paris, *Civil War*, 2:396.

10. McCord, "Battle of Corinth," 569–70; *OR* 17(2):142–44; Comte de Paris, *Civil War*, 2:396–97; Ulysses S. Grant, *Memoirs*, 1:394–97; Simon, *Grant Papers*, 5:255–56, 278; Shanks, *Recollections*, 122; Lamers, *Edge of Glory*, 95–96; Rosecrans, "Battle of Corinth," 740; Stanley, *Memoirs*, 107–8.

11. Rosecrans, "Battle of Corinth," 740; Comte de Paris, *Civil War*, 2:405; McCord, "Battle of Corinth," 573; Stanley, *Memoirs*, 107–8; Payne, "Sixth Missouri," 464–65; Alonzo L. Brown, *Fourth Minnesota*, 112–13; *OR* 17(1):247; Charles H. Smith, *Fuller's Ohio Brigade*, 75.

CHAPTER FOUR

1. Snead, "With Price," 725; Fauntleroy, "Elk Horn to Vicksburg," 23–24; *OR* 17(2):644–46; Castel, *Price and the War*, 92; Gottschalk, *In Deadly Earnest*, 117.

2. *OR* 17(1):382–83 and (2):683–85; Snead, "With Price," 726.

3. Anderson, *Memoirs*, 114; Bevier, *First and Second Missouri Brigades*, 124, 333; Castel, "Little Diary," 5, 41–45; Castel, *Price and the War*, 93; Maury, "Campaign against Grant," 289.

4. Tunnard, *Southern Record*, 28–29.

5. Barron, *Lone Star Defenders*, 98–99; Tunnard, *Southern Record*, 28–29; Cater, *As It Was*, 137; *Houston Tri-Weekly Telegraph*, July 16, 1862; Anderson, *Memoirs*, 212–16.

6. Connelly, *Army of the Heartland*, 187–96.

7. Snead, "With Price," 725–26; Archer Jones, *Confederate Strategy*, 63–64; William C. Davis, *Jefferson Davis*, 463; Connelly, *Army of the Heartland*, 207; Woodworth, *Davis and His Generals*, 136–37; Maury, "Campaign against Grant," 286; *OR* 17(2):626–29, 637, 654, 656.

8. Snead, "With Price," 727.

9. *OR* 17(2):663, 665.

10. Snead, "With Price," 727; *OR* 17(2):662.

11. Kenneth P. Williams, *Lincoln Finds a General*, 4:4–5, 11–13; Fiske, *Mississippi Valley*, 137–40.

12. Castel, "Little Diary," 44; *OR* 17(1):120 and (2):663–64, 667, 675–77; Snead, "With Price," 727; Hartje, *Van Dorn*, 209.

13. Maury, "Campaign against Grant," 286; *OR* 17(1):120 and (2):675–76, 681–82; Hartje, *Van Dorn*, 210–11; Snead, "With Price," 728.

14. J. P. Young, *Seventh Tennessee Cavalry*, 45; *OR* 17(1):47–48, 120, and (2):682; Comte de Paris, *Civil War*, 2:399.

15. Simon, *Grant Papers*, 5:289–90, 308, 310–11, 327–28, and 6:4; Lamers, *Edge of Glory*, 99; "Rosecrans's Campaigns," in *Report of the Joint Committee*, 3:17–18; William Starke Rosecrans to John Rawlins, August 12, 1862, Rosecrans Papers, UCLA.

16. *OR* 16(2):315 and 17(2):176–79; Simon, *Grant Papers*, 5:289–98, 302–6; Ulysses S. Grant, *Memoirs*, 1:397; Rosecrans, "Battle of Corinth," 738; Lamers, *Edge of Glory*, 98.

17. *OR* 17(1):65 and (2):189–95; Simon, *Grant Papers*, 6:5–10, 12–13; Ulysses S. Grant, *Memoirs*, 1:401.

18. Simon, *Grant Papers*, 6:10–11; J. P. Young, *Seventh Tennessee Cavalry*, 47; Throne, *Boyd Diary*, 65–66.

19. *OR* 17(1):51–52; J. P. Young, *Seventh Tennessee Cavalry*, 47.

CHAPTER FIVE

1. Hartje, *Van Dorn*, 210; *OR* (1):121 and (2):685, 687–88, 890.

2. Carlisle Diary, 2, NMMA; Castel, "Little Diary," 45; *OR* 17(1):376 and (2): 691–97; Tunnard, *Southern Record*, 181; Hartje, *Van Dorn*, 210–11.

3. Simon, *Grant Papers*; "Rosecrans's Testimony," in *Report of the Joint Committee*, 3:18; Lamers, *Edge of Glory*, 98–99; *OR* 17(2):206.

4. *OR* 17(2):694–98.

5. Mattox, "Chronicle," 208; Hubbell, "Diary," 97; Castel, "Little Diary," 45; Wright Schaumburg to his wife, October 9, 1862, Schaumburg-Wright Family Papers, HNOC.

6. Tunnard, *Southern Record*, 181; Anderson, *Memoirs*, 217; Carlisle Diary, 2, NMMA; Hubbell, "Diary," 98; Fauntleroy, "Elk Horn to Vicksburg," 29; Frederick, "Porter Diary," 302; Albert McCollom to his parents, September 24, 1862, McCollom Papers, ArU; Absalom F. Dantzler to his wife, September 18, 1862, Dantzler Letters, DU; Castel, "Little Diary," 45.

7. Simon, *Grant Papers*, 6:27–28, 30–31; "Rosecrans's Campaigns," in *Report of the Joint Committee*, 3:18; *OR* 17(2):34–35; Charles H. Smith, *Fuller's Ohio Brigade*, 76; "Correspondence of Wisconsin Volunteers," 4:115, 117, Quiner Papers, WiHS; Van Eman, "Iuka."

8. *OR* 17(2):702; Fauntleroy, "Elk Horn to Vicksburg," 29–30; Tyler Diary, September 13, 1862, WiHS; Albert McCollom to his parents, September 24, 1862, McCollom Papers, ArU; Barron, *Lone Star Defenders*, 104; Tunnard, *Southern Record*, 181–82; Castel, "Little Diary," 45–46.

9. Lamers, *Edge of Glory*, 45–46; Scott, *Four Years' Service*, 13; Barron, *Lone Star Defenders*, 104; Albert McCollom to his parents, September 24, 1862, McCollom Papers, ArU; Banks, "Letters," 143; Castel, "Little Diary," 46.

10. Anderson, *Memoirs*, 219; Robert Kepner to his sister, September 22, 1862, Mead Family Papers, IaU; "Correspondence of Wisconsin Volunteers," 4:117, Quiner Papers, WiHS.

11. Nabors, *Old Tishomingo County*, 39–40, 42; Griggs, *Opening of the Mississippi*, 105–6; Bevier, *First and Second Missouri Brigades*, 125; "Correspondence of Wisconsin Volunteers," 4:117, Quiner Papers, WiHS; map of Iuka in Hamilton Papers, IU; Ingersoll, *Iowa and the Rebellion*, 101.

12. "Correspondence of Wisconsin Volunteers," 4:115–18, Quiner Papers, WiHS; Alonzo L. Brown, *Fourth Minnesota*, 76; Kitchens, *Rosecrans Meets Price*, 41–50.

13. *OR* 17(2):702; Scott, *Four Years' Service*, 13; Solomon Scrapbook, 387, DU; Fauntleroy, "Elk Horn to Vicksburg," 30; Bevier, *First and Second Missouri Brigades*, 332; Barron, *Lone Star Defenders*, 105; Tunnard, *Southern Record*, 181.

14. William B. McGrorty to his wife, September 16, 1862, McGrorty Papers, MnHS; Anderson, *Memoirs*, 218–19; Helm, "Close Fighting," 171; Rawlins Diary, September 15, 1862, University of North Texas; Fay, *This Infernal War*, 158.

15. *Jackson Mississippian*, September 24, 1862; Simpson Diary, September 14, 1862, Mississippi Department of Archives and History; Fauntleroy, "Elk Horn to Vicksburg," 30; Tunnard, *Southern Record*, 182; Carlisle Diary, 3, NMMA; Anderson, *Memoirs*, 219; Bevier, *First and Second Missouri Brigades*, 132.

16. Castel, "Little Diary," 46; Barron, *Lone Star Defenders*, 104–5; Anderson, *Memoirs*, 219; Absalom F. Dantzler to his wife, September 18, 1862, Dantzler Letters, DU; Hubbell, "Diary," 98.

17. Kitchens, *Rosecrans Meets Price*, 60–61; Snead, "With Price," 731; Castel, *Price and the War*, 97–98; *OR* 17(2):702–3.

18. *OR* 17(2):696–98.

19. Woodworth, *Davis and His Generals*, 152–54; *OR* 17(2):698, 707; Castel, *Price and the War*, 98.

20. *OR* 17(2):214, 224; Lamers, *Edge of Glory*, 99–100.

21. *OR* 17(1):65; Edward A. Webb to his sister, September 16, 1862, Webb Papers, WRHS; Simon, *Grant Papers*, 6:38–42.

22. Simon, *Grant Papers*, 6:38, 44, 63; *OR* 17(1):65 and (2):214; William B. McGrorty to his wife, September 16, 1862, McGrorty Papers, MnHS.

23. "Rosecrans's Campaigns," in *Report of the Joint Committee*, 3:18; Throne, *Boyd Diary*, 66–68; Clark, *Downing's Diary*, 60, 68; *OR* 17(2):221.

24. Warner, *Generals in Blue*, 338; Cromwell Memoirs, 2, ISHL; John W. Fuller to David Sloane Stanley, August 23, 1887, West-Stanley-Wright Family Papers, USAMHI.

25. Warner, *Generals in Blue*, 338; John W. Fuller to David Sloane Stanley, August 23, 1887, West-Stanley-Wright Family Papers, USAMHI; Cromwell Memoirs, 2, ISHL; "Rosecrans's Campaigns," in *Report of the Joint Committee*, 3:18; Joseph A. Mower to William Starke Rosecrans, September 16, 1862, Rosecrans Papers, UCLA; Bryner, *Bugle Echoes*, 51; Tyler Diary, September 16, 1862, WiHS; Hubbell, "Diary," 98.

26. Anderson, *Memoirs*, 220; Tunnard, *Southern Record*, 182; *OR* 17(1):136.

27. *OR* 17(1):136, 223–24; Hubbell, "Diary," 98; Throne, *Boyd Diary*, 66; Chance, *Second Texas*, 58; "Rosecrans's Campaigns," in *Report of the Joint Committee*, 3:18.

28. "Rosecrans's Campaigns," in *Report of the Joint Committee*, 3:18; *OR* 17(1):117 and (2):224; Lamers, *Edge of Glory*, 101.

29. Simon, *Grant Papers*, 6:51; *OR* 17(1):65, 117–18; Comte de Paris, *Civil War*, 2:401; Ulysses S. Grant, *Memoirs*, 1:408.

30. "Rosecrans's Campaigns," in *Report of the Joint Committee*, 3:18; Lamers, *Edge of Glory*, 104; *OR* 17(1):66; Charles S. Hamilton to William Starke Rosecrans, September 18, 1862, Rosecrans Papers, UCLA; Simon, *Grant Papers*, 6:58; Ulysses S. Grant, *Memoirs*, 1:408–10; Hamilton, "Iuka," 734.

31. Lamers, *Edge of Glory*, 103; Comte de Paris, *Civil War*, 2:401; Simon, *Grant Papers*, 6:56; "Rosecrans's Campaigns," in *Report of the Joint Committee*, 3:18.

32. Throne, *Boyd Diary*, 66. Boyd incorrectly dates this entry as September 16. See also Clark, *Downing's Diary*, 69, and Tyler Diary, September 17, 1862, WiHS.

CHAPTER SIX

1. *OR* 17(1):376–77, 441, 453, and (2):701, 703–4; Hartje, *Van Dorn*, 212–13; Snead, "With Price," 732.

2. *OR* 17(2):707; Stirman, *In Fine Spirits*, 48; Hubbell, "Diary," 98; Castel, "Little Diary," 46.

3. *OR* 17(1):121 and (2):708; John Tyler to William L. Yancey, October 15, 1862, WHMC-Columbia.

4. *OR* 17(1):66; "Rosecrans's Campaigns," in *Report of the Joint Committee*, 3:18.

5. Simon, *Grant Papers*, 6:64–65; *OR* 17(1):66, 118.

6. "Rosecrans's Campaigns," in *Report of the Joint Committee*, 3:18; *OR* 17(1):66; Sanborn, *Descriptions of Battles*, 8.

7. Ulysses S. Grant, *Memoirs*, 1:411; *OR* 17(1):67; Lamers, *Edge of Glory*, 106–7; Simon, *Grant Papers*, 6:65–66.

8. Albert McCollom to his parents, September 24, 1862, McCollom Papers, ArU; Castel, "Little Diary," 46; *OR* 17(1):121–22; *Jackson Mississippian*, undated clipping in Solomon Scrapbook, 387, DU; Griscom, *Fighting with Ross' Texas Cavalry*, 43.

9. *Milwaukee Sentinel*, September 26, 1862; Charles H. Smith, *Fuller's Ohio Brigade*, 76–77; "Rosecrans's Campaigns," in *Report of the Joint Committee*, 3:18; Byers,

"Men in Battle," 11–12; *OR* 17(1):79; John R. P. Martin to his mother, September 21, 1862, Historical Files, Vicksburg National Military Park Library.

10. *Milwaukee Sentinel*, September 26, 1862; "Rosecrans's Campaigns," in *Report of the Joint Committee*, 3:18; *OR* 17(1):69, 94, 105, 116.

11. *OR* 17(1):138; Pierce, *Second Iowa Cavalry*, 30.

12. *OR* 17(1):94, 116; John Miller Diary, September 19, 1862, Miller Family Papers, Michigan State University; Sanborn, *Descriptions of Battles*, 8–9.

13. Sanborn, *Descriptions of Battles*, 9; Hamilton, "Iuka," 734; *OR* 17(1):69.

14. Alonzo L. Brown, *Fourth Minnesota*, 80; *OR* 17(1):69; "Rosecrans's Campaigns," in *Report of the Joint Committee*, 3:18–19; Theophilus S. Dickey to his wife, September 21, 1862, Wallace-Dickey Papers, ISHL.

15. Sanborn, *Descriptions of Battles*, 9; Alonzo L. Brown, *Fourth Minnesota*, 80; Hamilton, "Iuka," 734; *OR* 17(1):69, 90, 94.

16. William Starke Rosecrans to John W. Fuller, September 19, 1878, Rosecrans Papers, UCLA; "Rosecrans's Campaigns," in *Report of the Joint Committee*, 3:19.

CHAPTER SEVEN

1. Alonzo L. Brown, *Fourth Minnesota*, 80–81; *OR* 17(1):113, 116; McDonald, *Iuka's History*, 11, 15.

2. *OR* 17(1):90, 94, 116; Alonzo L. Brown, *Fourth Minnesota*, 80–81; Hamilton, "Iuka," 735; Sanborn, *Descriptions of Battles*, 9.

3. Stuart, *Iowa Colonels and Regiments*, 131; Warner, *Generals in Blue*, 315.

4. Byers, "Men in Battle," 12; Campbell Diary, September 19, 1862, WRHS; *OR* 17(1):90, 94, 99; Alonzo L. Brown, *Fourth Minnesota*, 81; Sanborn, *Descriptions of Battles*, 9; McDonald, *Iuka's History*, 15.

5. McDonald, *Iuka's History*, 15; Alonzo L. Brown, *Fourth Minnesota*, 81–82; Sanborn, *Descriptions of Battles*, 9; *Stark County Democrat*, October 8, 1862; William D. Evans to Caroline, September 26, 1862, Evans Papers, WRHS; Stirman, *In Fine Spirits*, 48; *OR* 17(1):90, 99, 103–4; Moore, *Rebellion Record*, 4:484; Dean, *Twenty-sixth Missouri*, 3; Van Eman, "Iuka."

6. Alonzo L. Brown, *Fourth Minnesota*, 82; *OR* 17(1):90; Moore, *Rebellion Record*, 5:484; Hamilton, "Iuka," 734.

7. Alonzo L. Brown, *Fourth Minnesota*, 82; Comte de Paris, *Civil War*, 2:402; J. H. Greene, *Reminiscences*, 29; *OR* 17(1):90, 97, 104; Hamilton, "Iuka," 734; Sanborn, *Descriptions of Battles*, 9; Helm, "Close Fighting," 171.

8. *OR* 17(1):122, 124; Wright Schaumburg to his wife, October 9, 1862, Schaumburg-Wright Family Papers, HNOC; Snead, "With Price," 732; McDonald, *Iuka's History*, 15.

9. *OR* 17(1):124; Carlisle Diary, 2, NMMA.

10. *Jackson Mississippian*, September 24, 1862.

11. *OR* 17(1):124, 130; Tunnard, *Southern Record*, 182; Rose, *Ross' Texas Brigade*, 70.

12. Helm, "Close Fighting," 171; Rose, *Ross' Texas Brigade*, 72, 97; Anderson, *Memoirs*, 222.

13. *OR* 17(1):130; McDonald, *Iuka's History*, 15; Bearss, *Decision in Mississippi*, 42; Anderson, *Memoirs*, 222.

14. Sanborn, *Descriptions of Battles*, 9–10; *OR* 17(1):90, 104; Dean, *Twenty-sixth Missouri*, 3; Stone, *Boomer Memoir*, 235; Hamilton, "Iuka," 735.

15. Alonzo L. Brown, *Fourth Minnesota*, 82–83; Campbell Diary, September 19, 1862, WRHS; *OR* 17(1):95; Hamilton, "Iuka," 735.

16. Sanborn, *Descriptions of Battles*, 19; Neil, *Battery at Close Quarters*, 8–9; Sears, *Eleventh Ohio Battery*, 1; Alonzo L. Brown, *Fourth Minnesota*, 83.

17. *OR* 17(1):97; Kitchens, *Rosecrans Meets Price*, 100.

18. *OR* 17(1):90, 101; Dyer, *Compendium*, 3:1, 297; McDonald, *Iuka's History*, 16; Reeves Diary, September 19, 1862, Reeves Papers, MnHS; Alonzo L. Brown, *Fourth Minnesota*, 84–86, 98–100, and "Fourth Minnesota at Iuka"; William M. Davis, "Iuka."

19. *OR* 17(1):90, 97, 102, 130; Dean, *Twenty-sixth Missouri*, 3; Moore, *Rebellion Record*, 5:484.

20. *OR* 17(1):90, 97; Campbell Diary, September 19, 1862, WRHS; Allen, "Iuka"; Hamilton, "Iuka," 735.

21. *OR* 17(1):73, 99, 105, 113; Campbell Diary, September 19, 1862, WRHS; Sanborn, *Descriptions of Battles*, 10; Bearss, *Decision in Mississippi*, 39.

22. Leonard Brown, *American Patriotism*, 178; *OR* 17(1):73, 105; Kibbe, "Iuka and Corinth"; Kenderdinn, "Fired on Friends."

23. Stuart, *Iowa Colonels and Regiments*, 215, 220; Members of the Twelfth Wisconsin Battery to Brigadier General Charles Hamilton, October 26, 1862, and M. H. Watson to Hamilton, July 18, 1862, both in Hamilton Papers, IU; Immell, "Iuka and Corinth"; *OR* 17(1):80, 107, 130; William M. Davis, "Iuka."

24. *OR* 17(1):86, 105, 111; Frost, *Tenth Missouri*, 70, 76.

25. Sanborn, *Descriptions of Battles*, 10; *OR* 17(1):95, 97; "Correspondence of Wisconsin Volunteers," 4:116, Quiner Papers, WiHS.

CHAPTER EIGHT

1. Also known as the Twenty-seventh Texas Cavalry.

2. Tunnard, *Southern Record*, 184; *OR* 17(1):125, 127, 129; Campbell Diary, September 19, 20, 1862, WRHS.

3. *OR* 17(1):122, 124, 131, and (2):708; Snead, "With Price," 732; McDonald, *Iuka's History*, 16.

4. *OR* 17(1):127, 129, 131–32, 134–35; *Jackson Mississippian*, September 24, 1862; McDonald, *Iuka's History*, 16; Tunnard, *Southern Record*, 183–84; Bearss, *Decision in Mississippi*, 42, places the entire brigade in a single line.

5. *OR* 17(1):125, 127, 133; Sanborn, *Descriptions of Battles*, 12; Smith and Smith, *Colonel Gilbert*, 172.

6. *OR* 17(1):128; Barron, *Lone Star Defenders*, 106–7; Tunnard, *Southern Record*, 184–85; Helm, "Close Fighting," 171; J. H. Greene, *Reminiscences*, 29.

7. Moore, *Rebellion Record*, 5:484; *OR* 17(1):128–29; Barron, *Lone Star Defenders*, 107; Tunnard, *Southern Record*, 183.

8. *OR* 17(1):97; Moore, *Rebellion Record*, 5:484; Alonzo L. Brown, *Fourth Minnesota*, 88–89; Barron, *Lone Star Defenders*, 107; Helm, "Close Fighting," 171; Sears, *Eleventh Ohio Battery*, 1.

9. John Risedorph to Lucy, September 26, 1862, Risedorph Papers, MnHS; *OR* 17(1):97, 128–29; Alonzo L. Brown, *Fourth Minnesota*, 89; McDonald, *Iuka's History*, 16–17; Rose, *Ross' Texas Brigade*, 95–96; Helm, "Close Fighting," 171; Sanborn, *Descriptions of Battles*, 13.

10. *OR* 17(1):100, 129; Alonzo L. Brown, *Fourth Minnesota*, 89, 91; Sanborn, *Descriptions of Battles*, 13.

11. *OR* 17(1):129; Helm, "Close Fighting," 171.

12. Stuart, *Iowa Colonels and Regiments*, 134; Campbell Diary, September 19, 1862, WRHS; Dodge, "Man of Iuka"; *OR* 17(1):99; Alonzo L. Brown, *Fourth Minnesota*, 87–88; Byers, *Iowa in War Times*, 152; Sanborn, *Descriptions of Battles*, 13; Tunnard, *Southern Record*, 185.

13. *OR* 17(1):99; Alonzo L. Brown, *Fourth Minnesota*, 88; Byers, *Iowa in War Times*, 154–55, and *With Fire and Sword*, 31–32; Sanborn, *Descriptions of Battles*, 13; Tunnard, *Southern Record*, 185; Moore, *Rebellion Record*, 5:484.

14. *OR* 17(1):102–3; Moore, *Rebellion Record*, 5:484; Alonzo L. Brown, *Fourth Minnesota*, 91; Stone, *Boomer Memoir*, 236.

15. *St. Paul Pioneer Press*, January 14, 1884; Moore, *Rebellion Record*, 5:484; *OR* 17(1):95, 103; Dean, *Twenty-sixth Missouri*, 4–6.

16. Allen, "Iuka"; Moore, *Rebellion Record*, 5:484; *OR* 17(1):103–4.

17. *OR* 17(1):95, 99, 127–28; Allen, "Hot Little Fight."

18. Beyer and Keydel, *Deeds of Valor*, 1:92–93; Neil, *Battery at Close Quarters*, 10; Sears, "11th Ohio Battery at Iuka" and *Eleventh Ohio Battery*, 3; Rose, *Ross' Texas Brigade*, 71.

19. Sears, "11th Ohio Battery at Iuka" and *Eleventh Ohio Battery*, 2–3; Neil, *Battery at Close Quarters*, 10–11; Rose, *Ross' Texas Brigade*, 71.

20. Helm, "Close Fighting," 171; Barron, *Lone Star Defenders*, 107; *OR* 17(1): 128–29.

21. Neil, *Battery at Close Quarters*, 11, 14; Sears, "11th Ohio Battery at Iuka."

CHAPTER NINE

1. *OR* 17(1):123, 127, 129; Rose, *Ross' Texas Brigade*, 71.

2. *OR* 17(1):122, 125, 136; Hamilton, "Iuka," 735; Von Phul, "Little's Burial," 213; Philip T. Tucker, *Fighting Chaplain*, 81–82; Bevier, *First and Second Missouri Brigades*, 333; Snead, "With Price," 732; Rose, *Ross' Texas Brigade*, 70; McDonald, *Iuka's History*, 16; *Jackson Mississippian*, September 24, 1862; Castel, "Little Diary," 46; Anderson, *Memoirs*, 222.

3. Castel, "Iuka," 16–17; *OR* 17(1):134–36.

4. *OR* 17(1):135–36.

5. *OR* 17(1):101; Risedorph Reminiscences, 7–8, Risedorph Papers, MnHS; Alonzo L. Brown, *Fourth Minnesota* and "Fourth Minnesota at Iuka."

6. *OR* 17(1):108–9, 134; Alonzo L. Brown, *Fourth Minnesota*, 87; Absalom F. Dantzler to his wife, September 24, 29, 1862, Dantzler Letters, DU.

7. Leonard Brown, *American Patriotism*, 178; *OR* 17(1):80, 107, 109, 134; Immell, "Iuka and Corinth"; Absalom F. Dantzler to his wife, September 29, 1862, Dantzler Letters, DU; Alonzo L. Brown, *Fourth Minnesota*, 87.

8. *OR* 17(1):106, 109–10, 112; Stuart, *Iowa Colonels and Regiments*, 313, 317–19; Kenderdinn, "Fired on Friends"; Sanders, "Iuka"; Spencer, "Baptism of Fire."

9. Stuart, *Iowa Colonels and Regiments*, 317–19; Kenderdinn, "Fired on Friends"; *OR* 17(1):110; Sanders, "Iuka."

10. *OR* 17(1):112.

11. Tunnard, *Southern Record*, 185; *OR* 17(1):106.

12. *OR* 17(1):106, 110, 129.

CHAPTER TEN

1. Alonzo L. Brown, *Fourth Minnesota*, 85–86; Pepper, *Under Three Flags*, 91; Hamilton, "Iuka," 735; Sanborn, *Descriptions of Battles*, 13–14.

2. *OR* 17(1):85, 88; Hickok, "Iuka and Corinth"; Moore, *Rebellion Record*, 5:485.

3. *OR* 17(1):81, 86, 88–89; Bryner, *Bugle Echoes*, 53; James Mellor to his wife, September 29, 1862, Croton Papers, University of Wisconsin; J. H. Greene, *Reminiscences*, 29; "Correspondence of Wisconsin Volunteers," 4:118, Quiner Papers, WiHS.

4. *OR* 17(1):86, 88, 123, 132–33; William S. Stewart to parents, September 23, 1862, Stewart Letters, WHMC-Columbia; William David Evans to Caroline, September 26, 1862, Evans Papers, and William A. Britton to Editors, September 22, 1862, Britton Letters, both in WRHS.

5. Smith and Smith, *Colonel Gilbert*, 110–11; William David Evans to Caroline, September 26, 1862, Evans Papers, WRHS; Strickling Reminiscences, 19, OHS; *OR* 17(1):84, 101; Van Eman, "Iuka"; Alonzo L. Brown, "Fourth Minnesota at Iuka"; William M. Davis, "Iuka."

6. *OR* 17(1):126; Bailey Reminiscences, 16, Bailey Papers, ArU; Fauntleroy, "Elk Horn to Vicksburg," 31; Tunnard, *Southern Record*, 183.

7. Bevier, *First and Second Missouri Brigades*, 133, 136, 333–34; Anderson, *Memoirs*, 221–22; Hubbell, "Diary," 98.

8. Smith and Smith, *Colonel Gilbert*, 111; Payne, "Sixth Missouri," 463; Tunnard, *Southern Record*, 188; *OR* 17(1):79; "Rosecrans's Campaigns," in *Report of the Joint Committee*, 3:19; Bevier, *First and Second Missouri Brigades*, 334–35; Anderson, *Memoirs*, 225; Alonzo L. Brown, *Fourth Minnesota*, 94, 96; J. H. Greene, *Reminiscences*, 30.

9. *Jackson Mississippian*, September 24, 1862; Smith and Smith, *Colonel Gilbert*, 111; Helm, "Close Fighting," 171; Anderson, *Memoirs*, 223.

10. *OR* 17(1):79; "Correspondence of Wisconsin Volunteers," 4:118, Quiner Papers, WiHS; Risedorph Reminiscences, 15–16, Risedorph Papers, MnHS.

11. Smith and Smith, *Colonel Gilbert*, 112; *Jackson Mississippian*, September 24, 1862.

CHAPTER ELEVEN

1. Von Phul, "Little's Burial," 214; Philip T. Tucker, *Fighting Chaplain*, 83–84; Snead, "With Price," 733; Maury, "Campaign against Grant," 290.

2. Maury, "Campaign against Grant," 290–91; *OR* 17(1):122; Snead, "With Price," 733; Von Phul, "Little's Burial," 214.

3. *OR* 17(1):137; Von Phul, "Little's Burial," 214–15; Snead, "With Price," 733.

4. "Rosecrans's Campaigns," in *Report of the Joint Committee*, 3:19; *OR* 17(1):67, 74; William Starke Rosecrans to Annie E. Rosecrans, September 22, Rosecrans Papers, UCLA; Hamilton, "Iuka," 735–36.

5. Van Eman, "Iuka"; Smith and Smith, *Colonel Gilbert*, 111; Strickling Reminiscences, 19, OHS.

6. Lamers, *Edge of Glory*, 113; "Rosecrans's Campaigns," in *Report of the Joint Committee*, 3:19; *OR* 17(1):70; Pierce, *Second Iowa Cavalry*, 32–33.

7. Bailey Reminiscences, 17, Bailey Papers, ArU; Tunnard, *Southern Record*, 183–84; Frederick, "Porter Diary," 303; Barron, *Lone Star Defenders*, 108.

8. Tunnard, *Southern Record*, 184, 187–88; Maury, "Campaign against Grant," 291; Bailey Reminiscences, 17, Bailey Papers, ArU; *OR* 17(1):122; Fauntleroy, "Elk Horn to Vicksburg," 32; Mattox, "Chronicle," 208.

9. "Rosecrans's Campaigns," in *Report of the Joint Committee*, 3:19; *OR* 17(1):87; Smith and Smith, *Colonel Gilbert*, 112; Strickling Reminiscences, 19, OHS; Charles H. Smith, *Fuller's Ohio Brigade*, 78; Sweet, "Civil War Experiences," 245; Alonzo L. Brown, *Fourth Minnesota*, 96–97.

10. William S. Stewart to his parents, September 23, 1862, Stewart Letters, WHMC-Columbia; Simon, *Grant Papers*, 6:72; *OR* 17(1):83; Pierce, *Second Iowa Cavalry*, 33; Smith and Smith, *Colonel Gilbert*, 112.

11. *OR* 17(1):74, 82–83; William Starke Rosecrans to John Fuller, September 19, 1878, Rosecrans Papers, UCLA; McDonald, *Iuka's History*, 19.

12. Simon, *Grant Papers*, 6:72–73; *OR* 17(1):70–71, 114, 117; Pierce, *Second Iowa Cavalry*, 33; Fox, *Seventh Kansas*, 33; William Starke Rosecrans to John W. Fuller, September 19, 1878, Rosecrans Papers, UCLA.

13. Lamers, *Edge of Glory*, 114–15; Fuller, "Our Kirby Smith," 170–71; John W. Fuller to David Sloane Stanley, August 23, 1887, West-Stanley-Wright Family Papers, USAMHI.

CHAPTER TWELVE

1. Cresap, *Appomattox Commander*, 63–64, 76–78.

2. Lamers, *Edge of Glory*, 117–18; Arthur C. Ducat to William Starke Rosecrans, April 24, 1885, Rosecrans Papers, UCLA.

3. Lamers, *Edge of Glory*, 118; *OR* 17(1):118–19.

4. Theophilus L. Dickey to his wife, September 21, 1862, Wallace-Dickey Papers, ISHL; William Starke Rosecrans to John W. Fuller, September 19, 1878, Rosecrans Papers, UCLA.

5. Lamers, *Edge of Glory*, 119; *OR* 17(1):67, 119.

6. Simon, *Grant Papers*, 6:63, 68–70; *OR* 17(1):67.

7. Simon, *Grant Papers*, 73; James C. Parrott to his wife, September 22, 1862, Parrott Papers, IaHS.

8. Lamers, *Edge of Glory*, 123.

9. Charles Brown Tompkins to his wife, September 22, 1862, Tompkins Papers, DU; James C. Parrott to his wife, September 22, 1862, Parrott Papers, IaHS; Clark, *Downing's Diary*, 69; Adams Journals, 1:28, IaU; Schilling Diary, September 19, 1862, Brown County Historical Society.

10. *OR* 17(1):79.

11. Charles Brown Tompkins to his wife, September 22, 1862, Tompkins Papers, DU; James C. Parrott to his wife, September 22, 1862, Parrott Papers, IaHS; William S. Stewart to his parents, September 23, 1862, Stewart Letters, WHMC-Columbia.

12. *OR* 17(1):64, 68, 119; "Rosecrans's Campaigns," in *Report of the Joint Committee*, 3:20.

13. Maury, "Campaign against Grant," 291; Stirman, *In Fine Spirits*, 49; *OR* 17(1): 114, 137; Pierce, *Second Iowa Cavalry*, 33; Chance, *Second Texas*, 60–61.

14. McKnight Diary, September 20, 1862, Civil War Collection, MoHS; J. H. Greene, *Reminiscences*, 30; Stanley, *Memoirs*, 107; "Rosecrans's Campaigns," in *Report of the Joint Committee*, 3:20.

15. Simon, *Grant Papers*, 6:79–80; Julia Dent Grant, *Memoirs*, 103; *OR* 17(1):71.

16. Hartje, *Van Dorn*, 212; *OR* 17(1):376; Simon, *Grant Papers*, 6:79–81.

17. *OR* 17(1):71; Simon, *Grant Papers*, 6:71; William Starke Rosecrans to Annie E. Rosecrans, September 22, 1862, Rosecrans Papers, UCLA; "Rosecrans's Campaigns," in *Report of the Joint Committee*, 3:20.

18. Hartje, *Van Dorn*, 212; *OR* 17(1):376.

19. Anderson, *Memoirs*, 226, 228; Wright Schaumburg to his wife, October 9, 1862, Schaumburg-Wright Family Papers, HNOC; *Jackson Mississippian*, September 24, 1862; Garrett, *Letters*, 65; Hubbell, "Diary," 99.

20. *OR* 17(1):72, 78–79, 126–27, 133; Kitchens, *Rosecrans Meets Price*, 187–89. From records in the National Archives, Kitchens obtained figures of 63 killed and 299 wounded for Hébert's brigade.

21. Tyler Diary, September 25–29, 1862, WiHS; Throne, *Boyd Diary*, 70–71.

CHAPTER THIRTEEN

1. John Tyler to William L. Yancey, October 15, 1862, WHMC-Columbia; Wright Schaumburg to his wife, October 9, 1862, Schaumburg-Wright Family Papers, HNOC; *OR* 17(1):377 and (2):710–12.

2. Wright Schaumburg to his wife, October 9, 1862, Schaumburg-Wright Family Papers, HNOC; Carlisle Diary, September 25, 1862, NMMA; Anderson, *Memoirs*, 229; Tunnard, *Southern Record*, 190–91; Frederick, "Porter Diary," 303; Ruyle Memoirs, 8, CWRT.

3. Castel, *Price and the War*, 104–6.

4. *OR* 17(1):378, 432, 434–35, 457–58; John Tyler to William L. Yancey, October 15, 1862, WHMC-Columbia; Payne, "Sixth Missouri," 463–64.

5. Maury, "Campaign against Grant," 293; *OR* 17(1):453.

6. Maury, "Campaign against Grant," 293; Warner, *Generals in Gray*, 194–95; Van Dorn biographical sketch, Sigel Papers, WRHS; Mansfield Lovell to his wife, October 1, 1862, Lovell Papers, HL; *OR* 17(1):440–42, 453, and (2):697, 711, 713, 716; Pope, "War Reminiscences XII"; John Tyler to William L. Yancey, October 15, 1862, WHMC-Columbia; Castel, *Price and the War*, 106–7.

7. *OR* 17(1):378, 455, and (2):704, 714–15; John Tyler to William L. Yancey, October 15, 1862, WHMC-Columbia; Hartje, *Van Dorn*, 215.

8. John Tyler to William L. Yancey, October 15, 1862, WHMC-Columbia; *Mobile Advertiser and Register*, undated clipping in Solomon Scrapbook, 396, DU; *OR* (1):385 and (2):417, 713, 718.

9. Barron, *Lone Star Defenders*, 115; Tyler Diary, September 29, 1862, WiHS; *OR* 17(1):416.

10. Anderson, *Memoirs*, 230; Barron, *Lone Star Defenders*, 115; sketch map dated March 15, 1946, Cockrell Papers, DU; J. P. Young, *Seventh Tennessee Cavalry*, 49; Ruyle Memoirs, 8, CWRT.

11. *OR* 17(1):385; Tunnard, *Southern Record*, 191; Anderson, *Memoirs*, 230; Fauntleroy, "Elk Horn to Vicksburg," 34.

12. J. P. Young, *Seventh Tennessee Cavalry*, 49; Stanley, "Corinth"; John Tyler to William L. Yancey, October 15, 1862, WHMC-Columbia; *OR* 17(1):378 and (2):717–18; Fauntleroy, "Elk Horn to Vicksburg," 34.

13. *OR* 17(1):417, 419, 421, and (2):684; Quinnie Armor to Monroe Cockrell, April 6, 1946, Cockrell Papers, DU; Henry, "Bowen," 172.

14. Hubbell, "Diary," 99; Fauntleroy, "Elk Horn to Vicksburg," 34; Keen, *Living and Fighting*, 37.

15. Kenneth P. Williams, *Lincoln Finds a General*, 4:82; Simon, *Grant Papers*, 6:83, 85–88.

16. Ulysses S. Grant, *Memoirs*, 1:414.

17. *OR* 17(1):166, 242; Simon, *Grant Papers*, 6:110; Rosecrans, "Battle of Corinth," 740–41, 743; McCord, "Battle of Corinth," 573; Stanley, "Corinth"; Warner, *Generals in Blue*, 301.

18. Rosecrans, "Battle of Corinth," 740–41, 743; "Rosecrans's Campaigns," in *Report of the Joint Committee*, 3:20; *OR* 17(1):166; Stephen E. Ambrose, *Wisconsin Boy*, 35.

19. Rosecrans, "Battle of Corinth," 743; "Rosecrans's Campaigns," in *Report of the Joint Committee*, 3:20–21; Simon, *Grant Papers*, 6:97; *OR* 17(1):166.

20. Rosecrans, "Battle of Corinth," 742; *OR* 17(1):357; Clark, *Downing's Diary*, 72; William H. Tucker, *Fourteenth Wisconsin*, 4; "Correspondence of Wisconsin Volunteers," 5:161, Quiner Papers, WiHS.

21. *OR* 17(1):178–79; Charles H. Smith, *Fuller's Ohio Brigade*, 82; Bryner, *Bugle Echoes*, 55–56; Smith and Smith, *Colonel Gilbert*, 114; Sergent Diary, October 1, 1862, NMMA. In his report, Stanley mistakenly said he had been ordered to the Tuscumbia River.

22. *OR* 17(1):352, 357; Stephen E. Ambrose, *Wisconsin Boy*, 37–38; William H. Tucker, *Fourteenth Wisconsin*, 4–5; "Correspondence of Wisconsin Volunteers," 5:161, Quiner Papers, WiHS; Dean Memoirs, 9, CWMC.

23. Simon, *Grant Papers*, 6:97–100, 110; Ulysses S. Grant, *Memoirs*, 1:415; *OR* 17(1):157.

CHAPTER FOURTEEN

1. Rosecrans, "Battle of Corinth," 743; *OR* 17(1):352, 407, 425; Stephen E. Ambrose, *Wisconsin Boy*, 38; J. P. Young, *Seventh Tennessee Cavalry*, 33–34.

2. Tunnard, *Southern Record*, 191; *OR* 17(1):407; William C. Holmes, "Corinth," 290.

3. *OR* 17(1):336, 352–53, 378, 407, 425; William H. Tucker, *Fourteenth Wisconsin*, 5–6; Stephen E. Ambrose, *Wisconsin Boy*, 38; "Correspondence of Wisconsin Volunteers," 5:162, Quiner Papers, WiHS.

4. Rosecrans, "Battle of Corinth," 743; McCord, "Battle of Corinth," 573; *OR* 17(1):282, 363–64; General Orders No. 151, dated October 25, 1862, Hamilton Papers, WiHS; Throne, *Boyd Diary*, 71; Alonzo L. Brown, *Fourth Minnesota*, 112; Sergent Diary, October 2, 1862, NMMA.

5. Rosecrans, "Battle of Corinth," 744–45; *OR* 17(1):167, 251; Stanley, "Corinth"; Jackson, *Colonel's Diary*, 68–69.

6. *OR* 17(1):205, 220, 216, 218, 226; Alonzo L. Brown, *Fourth Minnesota*, 112; Warner, *Generals in Blue*, 53–54; Simon, *Grant Papers*, 3:302; Risedorph Reminiscences, 3, Risedorph Papers, MnHS.

7. Zearing, "Letters," 162; Sergent Diary, October 3, 1862, NMMA; *OR* 17(1):251, 296, 299; Curtis Diary, 11, OHS; Daniel Leib Ambrose, *Seventh Illinois*, 91; Soper, "Company D," 134; Franklin Reed to Lemuel Reed, October 20, 1862, Reed Letters, IaHS.

8. *OR* 17(1):179, 197; Bryner, *Bugle Echoes*, 57; Smith and Smith, *Colonel Gilbert*, 114; *Stark County Democrat*, October 29, 1862; A. B. Monahan to his wife, October 12, 1862, Sixty-third Ohio Infantry File, SNMP.

9. Belknap, *Fifteenth Iowa*, 209; *OR* 17(1):336–37, 346–47, 350, 359, 362; Rosecrans, "Battle of Corinth," 745.

10. *OR* 17(1):378, 385, 404, 407, 410, and (2):717; Fauntleroy, "Elk Horn to Vicksburg," 34; John Tyler to William L. Yancey, October 15, 1862, WHMC-Columbia.

11. John Tyler to William L. Yancey, October 15, 1862, WHMC-Columbia; Anderson, *Memoirs*, 230–31; unidentified letter, October 9, 1862, McEwen Family Papers, MoHS.

12. Warner, *Generals in Blue*, 348; *OR* 17(1):354, 410; "Correspondence of Wisconsin Volunteers," 5:161–62, and Special Order of Colonel John Hancock, October

13, 1862, Quiner Papers, WiHS; Stephen E. Ambrose, *Wisconsin Boy*, 38; Christie, "Hot Times."

13. Monroe Cockrell to Edward J. East, February 25, 1946, Cockrell Papers, DU; William H. Tucker, *Fourteenth Wisconsin*, 6; *OR* 17(1):337, 354, 357; Stephen E. Ambrose, *Wisconsin Boy*, 39.

14. *OR* 17(1):354, 357; William H. Tucker, *Fourteenth Wisconsin*, 7; "Correspondence of Wisconsin Volunteers," 5:161, Quiner Papers, WiHS; John A. Duckworth to William Rosser, December 25, 1862, Second Iowa Infantry File, SNMP.

15. *OR* 17(1):344, 354, 357; Alonzo L. Brown, *Fourth Minnesota*, 117.

16. *OR* 17(1):167, 344, 354; Rosecrans, "Battle of Corinth," 746.

17. *OR* 17(1):197, 205, 251, 284, 359; Ducat, *Memoirs*, 34; McCord, "Battle of Corinth," 574–75; Curtis Diary, 11, OHS; Daniel Leib Ambrose, *Seventh Illinois*, 91–92; Chamberlin, *Eighty-first Ohio*, 25.

CHAPTER FIFTEEN

1. *OR* 17(1):378, 385–86, 389, 401; John Tyler to William L. Yancey, October 15, 1862, WHMC-Columbia; Comte de Paris, *Civil War*, 2:407.

2. Gottschalk, *In Deadly Earnest*, 138–40; *Mobile Advertiser and Register*, undated clipping in Solomon Scrapbook, 396, and William McCurdy to Mrs. A. F. Dantzler, October 11, 1862, Dantzler Letters, both at DU.

3. Alonzo L. Brown, *Fourth Minnesota*, 117; "Correspondence of Wisconsin Volunteers," 5:161, Quiner Papers, WiHS; *OR* 17(1):270, 354, 410.

4. Horton, "Battery F," 1, Horton Manuscripts, NMMA; *OR* 17(1):384, 407, 410, 412; "Correspondence of Wisconsin Volunteers," 5:161, 166, Quiner Papers, WiHS; William C. Holmes, "Corinth," 290.

5. *OR* 17(1):379; John Tyler to William L. Yancey, October 15, 1862, WHMC-Columbia.

6. *OR* 17(1):251–52, 344, 354; Anders, *Twenty-first Missouri*, 98–99; William H. Tucker, *Fourteenth Wisconsin*, 8; Chamberlin, *Eighty-first Ohio*, 26.

7. *OR* 17(1):252; Curtis Diary, 11, OHS; Chamberlin, *Eighty-first Ohio*, 26.

8. *OR* 17(1):252, 289; Daniel Leib Ambrose, *Seventh Illinois*, 92.

9. *OR* 17(1):167–68, 252.

10. Chamberlin, *Eighty-first Ohio*, 26; Curtis Diary, 11, OHS; Wright, *Corporal's Story*, 56; *OR* 17 (1):252, 282, 284, 288; Langworthy, "Corinth"; Morrison, *Ninth Illinois*, 38–39.

11. *OR* 17(1):272, 276, 280; Bell, *Tramps and Triumphs*, 12; John S. Wilcox to his wife, Wilcox Papers, ISHL; Bowen Diary, October 3, 1862, Bowen Papers, HL.

12. *OR* 17(1):205, 211, 213, 216, 226, 252, 289, 344; Cluett, *Fifty-seventh Illinois*, 39.

13. Daniel Leib Ambrose, *Seventh Illinois*, 92–93; *OR* 17 (1):252, 290; Hubert, *Fiftieth Illinois*, 131.

14. *OR* 17 (1):348–49, 407–8; "Comments about 'Lady Richardson,'" 205; Daniel Miller to his brother, October 9, 1862, Miller Papers, MoHS.

15. *OR* 17(1):296, 408–9, 418; "Comments about 'Lady Richardson,'" 205; *Corinth Weekly Herald*, September 6, 1912; Cluett, *Fifty-seventh Illinois*, 40.

16. Cluett, *Fifty-seventh Illinois*, 40; Zearing, "Letters," 162; *OR* 17(1):268, 290, 296, 408; "Comments about 'Lady Richardson,'" 205; Whitfield, "What Battery?"

17. *OR* 17(1):270; "Correspondence of Wisconsin Volunteers," 5:165, Quiner Papers, WiHS; William C. Holmes, "Corinth," 290–91.

18. Stephen E. Ambrose, *Wisconsin Boy*, 40; William C. Holmes, "Corinth," 291; H. J. Reid, "Twenty-second Mississippi," 14, Claiborne Papers, UNC.

19. *OR* 17(1):270, 357–58, 410, 412; Carter, "Lady Richardson," 306; William H. Tucker, *Fourteenth Wisconsin*, 8–9; "Correspondence of Wisconsin Volunteers," 5:161, 163–65, Quiner Papers, WiHS; William C. Holmes, "Corinth," 291; Abernathy, *Private Stockwell*, 48–49.

20. William C. Holmes, "Corinth," 291; Henry, "Bowen," 171; William H. Tucker, *Fourteenth Wisconsin*, 9; *OR* 17(1):176; Stephen E. Ambrose, *Wisconsin Boy*, 40–41.

21. William C. Holmes, "Corinth," 291; H. J. Reid, "Twenty-second Mississippi," 14, Claiborne Papers, UNC; William H. Tucker, *Fourteenth Wisconsin*, 8.

22. *OR* 17(1):293, 297; Daniel Leib Ambrose, *Seventh Illinois*, 93; Duncan, *Recollections*, 88–89; Mattox, "Chronicle," 199–200; McKinstry, "With Rogers," 221.

23. Daniel Leib Ambrose, *Seventh Illinois*, 93; *OR* 17(1):293, 344, 398, 410.

24. "Comments about 'Lady Richardson,'" 205; Henry, "Bowen," 171; *OR* 17(1): 405, 408–11, 421–22, 426; William C. Holmes, "Corinth," 292.

CHAPTER SIXTEEN

1. Wright, *Corporal's Story*, 56–57; Nelson Memoirs, 2, CWMC.

2. *OR* 17(1):268, 386; Curtis Diary, 11–12, OHS; Hubbell, "Diary," 100; John Tyler to William L. Yancey, October 15, 1862, WHMC-Columbia.

3. Warner, *Generals in Blue*, 81, 347; Trowbridge, "Saving a General," 20, 22; Chetlain, *Recollections*, 71; Cozzens, "'Poor Little Ninth,'" 34, 43–44; John Tyler to William L. Yancey, October 15, 1862, WHMC-Columbia; Chamberlin, *Eighty-first Ohio*, 27; *OR* 17(1):386.

4. Wright, *Corporal's Story*, 58; *OR* 17(1):252–53, 268, 283; Griscom, *Fighting with Ross' Texas Cavalry*, 44; Nelson Memoirs, 23, and Carlisle Journal, 81, both in CWMC; Hubbell, "Diary," 100; Anderson, *Memoirs*, 232; William McCurdy to Mrs. A. F. Dantzler, October 11, 1862, Dantzler Letters, DU.

5. *OR* 17(1):282–83; Morrison, *Ninth Illinois*, 38–39; Bevier, *First and Second Missouri Brigades*, 148–49; Fauntleroy, "Elk Horn to Vicksburg," 34; Payne, "Sixth Missouri," 464.

6. McKee Diary, October 3, 1862, CWMC; *OR* 17(1):276, 280; Bell, *Tramps and Triumphs*, 12.

7. Colwell, "Infantryman at Corinth," 10–11; *OR* 17(1):253; Chamberlin, *Eighty-first Ohio*, 27; Wright, *Corporal's Story*, 59.

8. *OR* 17(1):253–54, 386; Wright, *Corporal's Story*, 59.

9. *OR* 17(1):433–34; Fauntleroy, "Elk Horn to Vicksburg," 34; Payne, "Sixth Missouri," 464; Anderson, *Memoirs*, 232.

10. Maury, "Campaign against Grant," 295; Warner, *Generals in Gray*, 219; Ralph J.

Smith, *Reminiscences*, 10; Horton, "Battery F," 2, Horton Manuscripts, NMMA; *OR* 17(1):290, 344, 354, 357, 398.

11. *OR* 17(1):350, 354, 357, 377; Christie, "Hot Times"; "Correspondence of Wisconsin Volunteers," 5:162, Quiner Papers, WiHS.

12. *OR* 17(1):293–94, 296, 337, 344, 348–50, 355, 398; Cluett, *Fifty-seventh Illinois*, 41; Hubert, *Fiftieth Illinois*, 133; Anders, *Twenty-first Missouri*, 100; John Tyler to William L. Yancey, October 15, 1862, WHMC-Columbia; Duncan, *Recollections*, 89–90.

13. *OR* 17(1):344, 351, 398, 426; Zearing, "Letters," 162; Anders, *Twenty-first Missouri*, 101; Duncan, *Recollections*, 89–90; Hubert, *Fiftieth Illinois*, 133; Daniel Leib Ambrose, *Seventh Illinois*, 96.

14. *OR* 17(1):344, 351; Throne, *Boyd Diary*, 71; Anders, *Twenty-first Missouri*, 101.

15. Hubert, *Fiftieth Illinois*, 134; Duncan, *Recollections*, 89–90.

16. *OR* 17(1):293, 295, 297, 337–38, 344, 348; Cluett, *Fifty-seventh Illinois*, 41; Anders, *Twenty-first Missouri*, 101; Daniel Leib Ambrose, *Seventh Illinois*, 96; Adams Journals, 1:30, IaU.

17. Stuart, *Iowa Colonels and Regiments*, 255, 258, 263; Warner, *Generals in Blue*, 102.

18. Belknap, *Fifteenth Iowa*, 210; *OR* 17(1):359, 362, 364; Throne, *Boyd Diary*, 71; Adams Journals, 1:30, IaU; Parkhurst, "Attack," 173.

19. Parkhurst, "Attack," 174–75; Throne, *Boyd Diary*, 71–72; Adams Journals, 1:30, IaU; *OR* 17(1):337–38, 359.

20. *OR* 17(1):264, 394, 397; Throne, *Boyd Diary*, 72; Parkhurst, "Attack," 176–77; Duncan, *Recollections*, 91; Belknap, *Fifteenth Iowa*, 211.

21. Throne, *Boyd Diary*, 72–73; Belknap, *Fifteenth Iowa*, 211; *OR* 17(1):359, 364, 398; Horton, "Battery F," 2, Horton Manuscripts, NMMA; Adams Journals, 1:30, IaU.

22. *OR* 17(1):398, 405, 419, 426; John Tyler to William L. Yancey, October 15, 1862, WHMC-Columbia; Duncan, *Recollections*, 91; Mansfield Lovell to his wife, October 1, 1862, Lovell Papers, HL.

CHAPTER SEVENTEEN

1. William S. Rosecrans to Ulysses S. Grant, October 3, 1862, Rosecrans Papers, UCLA.

2. McKee Diary, October 3, 1862, CWMC; Bowen Diary, October 3, 1862, Bowen Papers, HL; *OR* 17(1):254–55; Curtis Diary, 12, OHS. For the sake of clarity I have referred to the two adjacent fields east of the Chewalla road as the east White House fields; that west of the road, as the west White House field.

3. *OR* 17(1):254–55, 269, 299; Bowen Diary, October 3, 1862, Bowen Papers, HL; Franklin B. Reed to Lemuel C. Reed, October 20, 1862, Reed Letters, IaHS; Wright, *Corporal's Story*, 59–60; Bell, *Tramps and Triumphs*, 12; Reed, *Twelfth Iowa*, 85; Soper, "Company D," 136–37.

4. *OR* 17(1):255, 288; Bowen Diary, October 3, 1862, Bowen Papers, HL; Nelson Memoirs, 24, CWMC; Colwell, "Infantryman at Corinth," 11.

5. Wright, *Corporal's Story*, 59–60; Bell, *Tramps and Triumphs*, 12.

6. *OR* 17(1):168, 254–57; Hamilton, "Hamilton's Division," 757.
7. Hamilton, "Hamilton's Division," 757; *OR* 17(1):205, 213, 216, 227; Alonzo L. Brown, *Fourth Minnesota*, 127.
8. Ducat, *Memoirs*, 27, 32–33; Hamilton and Ducat, *Correspondence*, 6, 10; Arthur C. Ducat to William Starke Rosecrans, December 29, 1880, Rosecrans Papers, UCLA; Alonzo L. Brown, *Fourth Minnesota*, 127.
9. "Rosecrans's Campaigns," in *Report of the Joint Committee*, 3:21; Arthur C. Ducat to William Starke Rosecrans, December 29, 1880, Rosecrans Papers, UCLA.
10. *OR* 17(1):179, 197, 183; William Starke Rosecrans to David Sloane Stanley, October 3, 1862, Rosecrans Papers, UCLA; Charles H. Smith, *Fuller's Ohio Brigade*, 82; Rosecrans, "Battle of Corinth," 746; Bryner, *Bugle Echoes*, 58.
11. Arthur C. Ducat to William Starke Rosecrans, December 29, 1880, Rosecrans Papers, UCLA; Ducat, *Memoirs*, 32–33; *OR* 17(1):168; Hamilton and Ducat, *Correspondence*, 11; "Rosecrans's Campaigns," in *Report of the Joint Committee*, 3:21.
12. Cozzens, *No Better Place to Die*, 16–17; *OR* 17(1):213.
13. Hamilton, "Hamilton's Division," 757; Ducat, *Memoirs*, 33; Arthur C. Ducat to William Starke Rosecrans, December 29, 1880, Rosecrans Papers, UCLA; Rosecrans, "Battle of Corinth," 747; *Chicago Tribune*, October 25, 1882.

CHAPTER EIGHTEEN

1. Maury, "Campaign against Grant," 295; *OR* 17(1):379, 394, 401, 433; John Tyler to William C. Yancey, October 15, 1862, WHMC-Columbia.
2. Bevier, *First and Second Missouri Brigades*, 149; Hubbell, "Diary," 100; *OR* 17(1):389–90.
3. *OR* 17(1):255, 269, 390; Payne, "Sixth Missouri," 464; Chamberlin, *Eighty-first Ohio*, 27; Bowen Diary, October 3, 1862, Bowen Papers, HL; *Missouri Republican*, June 19, 1886.
4. *Missouri Republican*, June 19, 1886; Brown Recollections, 103–4, OHS; Bevier, *First and Second Missouri Brigades*, 149–50; John Tyler to William L. Yancey, October 15, 1862, WHMC-Columbia; Bowen Diary, October 3, 1862, Bowen Papers, HL; Payne, "Sixth Missouri," 464; *OR* 17(1):255, 269, 272, 289, 390. Estimates of the length of the cannonade varied from twenty to forty-five minutes.
5. Thomas W. Sweeny to W. B. Bodye, October 20, 1862, Sweeny Papers, HL; *OR* 17(1):256, 272–73; John S. Wilcox to his wife, October 9, 1862, Wilcox Papers, ISHL; *Missouri Republican*, June 19, 1886; Bell, *Tramps and Triumphs*, 12; Payne, "Sixth Missouri," 464; John Tyler to William L. Yancey, October 15, 1862, WHMC-Columbia.
6. Stuart, *Iowa Colonels and Regiments*, 61; *OR* 17(1):256, 273; Payne, "Sixth Missouri," 464; Wright, *Corporal's Story*, 60.
7. Thomas W. Sweeny to W. B. Bodye, October 20, 1862, Sweeny Papers, HL; *OR* 17(1):256, 273, 276.
8. Stuart, *Iowa Colonels and Regiments*, 59, 61–64; *OR* 17(1):256, 273; Payne, "Sixth Missouri," 464; John Tyler to William L. Yancey, October 15, 1862, WHMC-Columbia.

9. Chetlain, *Recollections*, 95; Brown Recollections, 92, OHS; Trowbridge, "Saving a General," 22; *OR* 17(1):256, 273.

10. Franklin B. Reed to Lemuel C. Reed, October 20, 1862, Reed Letters, IaHS; *OR* 17(1):299; Benner, *Sul Ross*, 84; Griscom Diary, October 3, 1862, TxU; Soper, "Company D," 137; Myron Underwood to his wife, October 19, 1862, Underwood Papers, IaU.

11. *OR* 17(1):198–99, 256.

12. *OR* 17(1):256, 273; Wright, *Corporal's Story*, 60; Chamberlin, *Eighty-first Ohio*, 27–28; Nelson Memoirs, 25, and Carlisle Journal, 82, both in CWMC; Chetlain, *Recollections*, 95.

13. *OR* 17(1):197, 201; McLain, "Old Abe," 407; Fowler, *Old Abe*, 10–12.

14. *OR* 17(1):201, 207, 256–57, 273, 299; Bell, *Tramps and Triumphs*, 12–13; Phillips Memoirs, 22–23, CWTI; Bryner, *Bugle Echoes*, 58.

15. Gilbert Certificate, CWMC; *OR* 17(1):197, 201.

16. Hubbell, "Diary," 100; Kavanaugh Memoirs, 44, WHMC-Columbia; Anderson, *Memoirs*, 233; Ruyle Memoirs, 9, CWRT; Fauntleroy, "Elk Horn to Vicksburg," 34; *OR* 17(1):389.

17. *OR* 17(1):197, 199, 201, 203–4, 394; Maury, "Campaign against Grant," 295; Payne, "Sixth Missouri," 464; Griscom, *Fighting with Ross' Texas Cavalry*, 45; Bryner, *Bugle Echoes*, 58.

18. Bryner, *Bugle Echoes*, 59; *OR* 17(1):197–99, 201, 204; McKinney Diary, October 3, 1862, CWTI.

19. "Correspondence of Wisconsin Volunteers," 4:123, 127, Quiner Papers, WiHS; *OR* 17(1):203; J. H. Greene, *Reminiscences*, 31.

20. Bradley, *Confederate Mail Carrier*, 72.

21. Van Eman, "Bloody Robinett"; Charles H. Smith, *Fuller's Ohio Brigade*, 82; *OR* 17(1):247.

22. *OR* 17(1):179–83, 188, 197, 204, 401; Benner, *Sul Ross*, 85; William C. Holmes, "Corinth," 292; Anderson, *Memoirs*, 234; A. B. Monahan to his wife, October 12, 1862, Sixty-third Ohio Infantry File, SNMP; Fuller, "Our Kirby Smith," 82; Smith and Smith, *Colonel Gilbert*, 114.

23. John Tyler to William L. Yancey, October 15, 1862, and Kavanaugh Memoirs, 45, both in WHMC-Columbia.

24. John Tyler to William L. Yancey, October 15, 1862, WHMC-Columbia; *OR* 17(1):433–35; Maury, "Campaign against Grant," 295; Bevier, *First and Second Missouri Brigades*, 337–38.

25. Anderson, *Memoirs*, 233.

26. *OR* 17(1):399, 404, 419, 427; William C. Holmes, "Corinth," 292.

27. *OR* 17(1):405; Judge Morse to Mansfield Lovell, October 27, 1862, Lovell Papers, HL; Maury, "Campaign against Grant," 297–98.

28. Ducat, *Memoirs*, 33; Rosecrans, "Battle of Corinth," 747; William Starke Rosecrans to Charles S. Hamilton, October 3, 1862, Rosecrans Papers, UCLA.

29. Hamilton, "Hamilton's Division," 758; Hamilton and Ducat, *Correspondence*, 11; Ducat, *Memoirs*, 34.

30. Ducat, *Memoirs*, 33–34; Hamilton, "Hamilton's Division," 758; *OR* 17(1):205, 207, 227, 230; Alonzo L. Brown, *Fourth Minnesota*, 119.

31. Hamilton, "Hamilton's Division," 758; *OR* 17(1):205, 216; *Missouri Republican*, June 19, 1886; Risedorph Reminiscences, 3–4, Risedorph Papers, MnHS; Hamilton and Ducat, *Correspondence*, 11.

32. Alonzo L. Brown, *Fourth Minnesota*, 127.

33. *OR* 17(1):218–19, 222; Alonzo L. Brown, *Fourth Minnesota*, 127; Hamilton, "Hamilton's Division," 758.

34. Leonard Brown, *American Patriotism*, 179; *OR* 17(1):230, 234, 237.

35. Bevier, *First and Second Missouri Brigades*, 150; *Missouri Republican*, June 19, 1886; *OR* 17(1):227, 235, 394, 401.

36. Rosecrans, "Battle of Corinth," 747.

CHAPTER NINETEEN

1. Thomas W. Sweeny to his daughter, October 13, 1862, Sweeny Papers, HL; *OR* 17(1):257, 419; John S. Wilcox to his wife, October 9, 1862, Wilcox Papers, ISHL; Van Eman, "Bloody Robinett"; Hovey Cadman to his wife, October 13, 1862, Cadman Papers, UNC; *Daily Toledo Blade*, October 14, 1862; Bell, *Tramps and Triumphs*, 13; Colwell, "Infantryman at Corinth," 11–12; Kavanaugh Memoirs, 46, WHMC-Columbia; Hittle, "Division of Davies"; Byers, *With Fire and Sword*, 33–34.

2. *OR* 17(1):177, 257; Trowbridge, "Saving a General," 22; Chamberlin, *Eighty-first Ohio*, 30.

3. "Correspondence of Wisconsin Volunteers," 5:163, Quiner Papers, WiHS; Colwell, "Infantryman at Corinth," 12; Byers, *With Fire and Sword*, 34.

4. *OR* 17(1):457.

5. Hamilton and Ducat, *Correspondence*, 4–5; Hamilton, "Hamilton's Division," 758; Alonzo L. Brown, *Fourth Minnesota*, 128.

6. "Rosecrans's Campaigns," in *Report of the Joint Committee*, 3:21; Rosecrans, "Battle of Corinth," 748; *OR* 17(1):170; General Order 151, dated October 25, 1862, Hamilton Papers, IU.

7. *OR* 17(1):257, 273, 297, 355; Bell, *Tramps and Triumphs*, 13; Hubert, *Fiftieth Illinois*, 137; William H. Tucker, *Fourteenth Wisconsin*, 10.

8. *OR* 17(1):179, 184, 189; Van Eman, "Bloody Robinett"; Charles H. Smith, *Fuller's Ohio Brigade*, 82–83; Fuller, "Our Kirby Smith," 172; Smith and Smith, *Colonel Gilbert*, 114.

9. *OR* 17(1):198–200; Gilmore, "63d Ohio"; Lucius F. Hubbard, *Minnesota in the Battles of Corinth*, 15.

10. Fuller, "Our Kirby Smith," 162–68, 172; Stanley, *Memoirs*, 112–13; Charles H. Smith, *Fuller's Ohio Brigade*, 83.

11. Alonzo L. Brown, *Fourth Minnesota*, 120; *OR* 17(1):169, 206, 215, 219, 220, 225, 230; Bowen Diary, October 4, 1862, Bowen Papers, HL; McCord, "Battle of Corinth"; Milburn, "Reminiscences."

12. Rosecrans, "Battle of Corinth," 748; J. B. Rogers, *War Pictures*, 195; *OR* 17(1):169, 249, 258–59, 266–67, 273, 291, 297; John A. Duckworth to William Rosser, December 25, 1862, Second Iowa Infantry File, SNMP; Thomas W.

Sweeny to W. B. Bodye, October 8, 1862, Sweeny Papers, HL; Hittle, "Division of Davies"; Du Bois, "Journal and Letters," 50; McCord, "Battle of Corinth," 577.

13. William H. Tucker, *Fourteenth Wisconsin*, 10–11; *OR* 17(1):355; Burge Memoirs, 3, CWMC.

14. *OR* 17(1):422, 431, 433, and (2):719; Gottschalk, *In Deadly Earnest*, 147; Anderson, *Memoirs*, 235; Griscom Diary, October 4, 1862, TxU; W. T. Young to Thomas M. Owen, February 12, 1909, Forty-second Alabama Infantry Papers, ADAH.

15. *OR* 17(1):436, 444, 457.

16. *OR* 17(1):379 and (2):719–20; Maury, "Campaign against Grant," 295; Bevier, *First and Second Missouri Brigades*, 150; Kavanaugh Memoirs, 46, WHMC-Columbia.

17. Maury, "Campaign against Grant," 295–96; *OR* 17(1):379, 405, 422.

18. Comte de Paris, *Civil War*, 2:410; *OR* 17(1):394, 405, 422, and (2):720; Jackson, *Colonel's Diary*, 85.

19. *OR* 17(1):158; Simon, *Grant Papers*, 6:103–4; Lamers, *Edge of Glory* 144–45; Ulysses S. Grant, *Memoirs*, 1:416.

20. Simon, *Grant Papers*, 6:104–6, 111–13; *OR* 17(1):154, 267, 305, 370; Ulysses S. Grant, *Memoirs*, 1:415–16; Morris, *Thirty-first Regiment Volunteers*, 49.

CHAPTER TWENTY

1. Anonymous soldier's letter, October 13, 1862, Cadman Papers, UNC; Minturn, "Corinth"; *OR* 17(1):184, 192.

2. *OR* 17(1):184, 192, 394, 398; McKinstry, "With Rogers"; McCord, "Battle of Corinth."

3. *Iowa State Register*, October 29, 1862; Rosecrans, "Battle of Corinth," 748–49; Colwell, "Infantryman at Corinth," 12; Adams, "Battery Robinett"; Strickling Reminiscences, 21, and Kelly Diary, 12, both in OHS; Lucius F. Hubbard, *Minnesota in the Battles of Corinth*, 20; Kate McConnell to Monroe Sword, October 21, 1862, Sword Papers, DU; *OR* 17(1):275.

4. Bryner, *Bugle Echoes*, 63; Smith and Smith, *Colonel Gilbert*, 115; Stanley, *Memoirs*, 109–10; *OR* 17(1):184, 192, 194, 247, 394; Maury, "Campaign against Grant," 295; McCord, "Battle of Corinth," 578; Arnold, "Corinth," 199; *Daily Toledo Blade*, October 14, 1862.

5. "Correspondence of Wisconsin Volunteers," 4:122, Quiner Papers, WiHS; Ray Reminiscences, 25, Ray Papers, ArU; *OR* 17(1):110, 169; Wright, *Corporal's Story*, 63; Soper, "Company D," 138; Stanley, *Memoirs*, 110.

6. *OR* 17(1):390; Columbus Sykes to his wife, August 31, 1862, Sykes Letters, CWMC; Bevier, *First and Second Missouri Brigades*, 235–36.

7. *OR* 17(1):379, 387; Hubbell, "Diary," 100; Fauntleroy, "Elk Horn to Vicksburg," 35; Anderson, *Memoirs*, 235–36.

8. *OR* 17(1):180, 184–85, 198; Jackson, *Colonel's Diary*, 70; Arnold, "Corinth," 199; Van Eman, "Bloody Robinett"; John W. Fuller to David Sloane Stanley, August

23, 1887, West-Stanley-Wright Family Papers, USAMHI; Maury, "Campaign against Grant," 296.

9. *OR* 17(1):291, 390–91, 401; Kavanaugh Memoirs, 46, WHMC-Columbia.

10. Bevier, *First and Second Missouri Brigades*, 341; Hubbell, "Diary," 101; *OR* 17(1):401.

11. *OR* 17(1):281; F. W. Dunn to his father, October 14, 1862, Dunn Letters, MiU; Childress Diary, 31, ISHL; *OR* 17(1):249–50; Bargus Diary, October 4, 1862, CWTI.

12. Payne, "Sixth Missouri," 465.

13. Kavanaugh Memoirs, 46–48, 53, WHMC-Columbia; McCord, "Battle of Corinth," 580; Bailey Reminiscences, 17, Bailey Papers, ArU; *OR* 17(1):239–41, 259.

14. *OR* 17(1):267, 291; Daniel Leib Ambrose, *Seventh Illinois*, 99; Reed, *Twelfth Iowa*, 86; Bradley, *Confederate Mail Carrier*, 73; Nelson Memoirs, 27, CWMC.

15. Thomas W. Sweeny to W. B. Bodye, October 8, 1862, Sweeny Papers, HL; John S. Wilcox to his wife, October 9, 1862, Wilcox Papers, ISHL; Bowen Diary, October 4, 1862, Bowen Papers, HL; *OR* 17(1):259, 274, 277, 285, 289; Langworthy, "Corinth"; Colwell, "Infantryman at Corinth," 13.

16. *OR* 17(1):259, 269–71; McCord, "Battle of Corinth," 580; notes on Federal artillery dispositions, Cockrell Papers, DU.

17. *OR* 17(1):391; Bevier, *First and Second Missouri Brigades*, 156, 341; John S. Wilcox to his wife, October 9, 1862, Wilcox Papers, ISHL; Ruyle Memoirs, 9, CWRT; Hubbell, "Diary," 101–2; Anderson, *Memoirs*, 236–37.

18. *OR* 17(1):231, 235, 239–40; Hood, "Sixth Wisconsin Battery"; Bartholomew, "Another Defense"; Immell, "Twelfth Wisconsin Battery at Corinth."

19. *OR* 17(1):259–60; Horton, "Batteries Powell and Williams," 1–2, Horton Manuscripts, NMMA; Griggs, *Opening of the Mississippi*, 113; "Correspondence of Wisconsin Volunteers," 8:36, Quiner Papers, WiHS.

20. Daniel Leib Ambrose, *Seventh Illinois*, 99–100; *OR* 17(1):291, 293; Bell, *Tramps and Triumphs*, 13.

21. Stanley, *Memoirs*, 110; Charles H. Smith, *Fuller's Ohio Brigade*, 85–86; Bevier, *First and Second Missouri Brigades*, 341–42; Absalom Dyson to his wife, October 20, 1862, Dyson–Bell–Sans Souci Papers, WHMC–St. Louis.

22. Will Snyder to his parents, November 25, 1862, NMMA; Fauntleroy, "Elk Horn to Vicksburg," 35.

23. *OR* 17(1):233, 238; Raum, "Corinth"; Frost, *Tenth Missouri*, 81; Morgan, "Once Was Enough"; Immell, "Twelfth Wisconsin Battery Takes Advantage"; "Correspondence of Wisconsin Volunteers," 8:336, Quiner Papers, WiHS.

24. Immell, "Twelfth Wisconsin Battery Takes Advantage."

25. "Correspondence of Wisconsin Volunteers," 8:36, Quiner Papers, WiHS; Frost, *Tenth Missouri*, 81–82; Raum, "Corinth."

26. Bevier, *First and Second Missouri Brigades*, 154, 342; Hubbell, "Diary," 101; Anderson, *Memoirs*, 237; *OR* 17(1):231, 260, 270, 274, 277; Corsair, "Corinth"; John S. Wilcox to his wife, October 9, 1862, Wilcox Papers, ISHL; Thomas W. Sweeny to W. B. Bodye, October 8, 1862, Sweeny Papers, HL; Raum, "Corinth."

27. *OR* 17(1):215, 219, 220, 222, 225, 391; Tunnard, *Southern Record*, 192; Neil, *Battery at Close Quarters*, 17, 19; Byers, "Men in Battle," 12–13.

28. Neil, *Battery at Close Quarters*, 19–20; Alonzo L. Brown, *Fourth Minnesota*, 124; *St. Paul Pioneer Press*, January 14, 1884; *OR* 17(1):226; Reeves Diary, October 4, 1862, Reeves Papers, MnHS.

29. William McCurdy to Mrs. A. F. Dantzler, October 11, 1862, Dantzler Letters, DU; Andrew J. Patrick to his father, October 15, 1862, Patrick Letters, University of Southern Mississippi; Tunnard, *Southern Record*, 192; *OR* 17(1):223, 226.

30. *OR* 17(1):215–20, 224.

31. Franklin B. Reed to Lemuel C. Reed, October 20, 1862, Reed Letters, IaHS; Hubert, *Fiftieth Illinois*, 145–47; Rosecrans, "Battle of Corinth," 753.

CHAPTER TWENTY-ONE

1. Maury, "Campaign against Grant," 298; *OR* 17(1):398; Duncan, *Recollections*, 94; McKinstry, "With Rogers," 221.

2. Jackson, *Colonel's Diary*, 85.

3. William P. Rogers, "Diary and Letters," 260–61, 267, 276–77, 289, 295; Lawrence S. Ross to his wife, August 25, 1862, Ross Family Papers, Baylor University; Cockrell, *Lost Account*, 51–52; *Houston Tri-Weekly Telegraph*, July 16, 1862; William P. Rogers to his wife, August 21, 1862, Rogers Papers, TxU; McGinnis, "Roger's Death"; McKinstry, "With Rogers," 221.

4. *OR* 17(1):379, 339; Maury, "Campaign against Grant," 298.

5. Jackson, *Colonel's Diary*, 85; Duncan, *Recollections*, 94; Hawke, *Corinth*, 45; *Houston Tri-Weekly Telegraph*, November 5, 1862; *OR* 17(1):185, 398; Griscom, *Fighting with Ross' Texas Cavalry*, 46.

6. Jackson, *Colonel's Diary*, 71; Minturn, "Corinth"; *OR* 17(1):185; Gilmore, "63d Ohio"; Fuller, "Our Kirby Smith," 174; Hatheway, "Fierce Fighting"; Lybarger, "Corinth."

7. *OR* 17(1):394, 402; Bevier, *First and Second Missouri Brigades*, 154; Gottschalk, *In Deadly Earnest*, 150.

8. *OR* 17(1):238, 402; Raum, "Corinth."

9. *OR* 17(1):238, 260–61, 391, 402; Raum, "Corinth"; Martin, "Reminiscences," 69.

10. Jackson, *Colonel's Diary*, 71, 81.

11. Ibid., 72; *Harper's Weekly*, November 1, 1862; Minturn, "Corinth"; George H. Cadman to his wife, October 13, 1862, Cadman Papers, UNC; *OR* 17(1):196; Duncan, *Recollections*, 96; *Corinth Herald*, February 20, 1902.

12. Jackson, *Colonel's Diary*, 72, 85–86; Duncan, *Recollections*, 97; McKinstry, "With Rogers," 221; *OR* 17(1):192, 398–99; Strickling Reminiscences, 21, OHS; Charles H. Smith, *Fuller's Ohio Brigade*, 86; *Battle of Corinth Souvenir Program*, 11.

13. Cockrell, *Lost Account*, 30–31; *OR* 17(1):185, 398; "Col. William P. Rogers," 57; Stirman, *In Fine Spirits*, 52.

14. Charles H. Smith, *Fuller's Ohio Brigade*, 86; Strickling Reminiscences, 21, OHS; *OR* 17(1):192; *Battle of Corinth Souvenir Program*, 11; Arnold, "Corinth," 199.

15. *OR* 17(1):181, 185–92; Fuller, "Our Kirby Smith," 174–76; Jackson, *Colonel's Diary*, 72; Bradney, "Saw the 63d."

16. *OR* 17(1):185, 192; *Battle of Corinth Souvenir Program*, 11; Jackson, *Colonel's Diary*, 72; *Harper's Weekly*, November 1, 1862; A. B. Monahan to his wife, October 12, 1862, Sixty-third Ohio Infantry File, SNMP.

17. Duncan, *Recollections*, 97; Arnold, "Corinth," 199; *Mobile Advertiser and Register*, undated clipping in Solomon Scrapbook, DU; Mayo, "Colonel Rogers," 57–58; *Galveston Daily News*, March 8, 1896; Jackson, *Colonel's Diary*, 78.

18. *OR* 17(1):185; Bradney, "Saw the 63d"; Fuller, "Our Kirby Smith," 175.

19. Jackson, *Colonel's Diary*, 86; McKinstry, "With Rogers," 221; *Corinth Herald*, February 20, 1902.

20. Arnold, "Corinth," 199; *Battle of Corinth Souvenir Program*, 11; Budd, "Colonel Rogers."

21. Minturn, "Corinth"; Hatheway, "Fierce Fighting"; Lybarger, "Corinth."

22. *OR* 17(1):192; *Battle of Corinth Souvenir Program*, 11; Jackson, *Colonel's Diary*, 74–76.

23. *OR* 17(1):192, 202; *Battle of Corinth Souvenir Program*, 11; McNeal, "Charge of the Texans"; Henney, "11th Mo. at Corinth"; Frank T. Gilmore to David Sloane Stanley, January 13, 1897, and John W. Fuller to David Sloane Stanley, November 1, 1887, both in West-Stanley-Wright Family Papers, USAMHI.

24. Fuller, "Our Kirby Smith," 175; "What One Comrade Knows"; Stanley, *Memoirs*, 113–14; *OR* 17(1):181; *Dayton Weekly Journal*, November 4, 1862; Frank T. Gilmore to David Sloane Stanley, January 13, 1897, West-Stanley-Wright Family Papers, USAMHI.

25. William S. Stewart to his mother, October 12, 1862, Stewart Letters, WHMC-Columbia; Henney, "The Eleventh Missouri"; *OR* 17(1):202.

26. Arnold, "Corinth," 199; Jackson, *Colonel's Diary*, 86; McKinstry, "With Rogers," 222; Ray Reminiscences, 26, Ray Papers, ArU; Wade Diary, October 4, 1862, CWMC.

27. *Battle of Corinth Souvenir Program*, 11; Arnold, "Corinth," 199; "Death of Colonel Rogers"; "Death of Colonel Rogers at Corinth"; Budd, "Colonel Rogers"; Adams, "Battery Robinett"; Duncan, *Recollections*, 99–100; Ray Reminiscences, 26, Ray Papers, ArU; Fuller, "Our Kirby Smith," 174; Daniel H. J. Cobb to his sister, November 1, 1862, Cobb Letters, MiU; John A. Duckworth to William Rosser, December 25, 1862, Second Iowa Infantry File, SNMP; McGinnis, "Corinth to Vicksburg"; Lindsey, "At Corinth"; Pomeroy Diary, October 4, 1862, KHS; *Houston Tri-Weekly Telegraph*, November 5, 1862. There are nearly as many versions of Colonel Rogers's death as there were witnesses to it. Some say he died on horseback carrying the colors; others say he was atop the parapet of Battery Robinett on foot, waving either his handkerchief, his sword, his pistol, or a flag; still others claim he was dismounted in the ditch. I have tried to reconstruct the moment from the most plausible accounts.

28. "Col. William P. Rogers," 58; Jackson, *Colonel's Diary*, 87–88; J. B. Rogers, *War Pictures*, 202.

29. Bryner, *Bugle Echoes*, 63; *OR* 17(1):185, 191; *Daily Toledo Blade*, October 14, 1862; *Utica Evening Telegraph*, October 22, 1862; *Stark County Democrat*, October 29, 1862; Frank T. Gilmore to David Sloane Stanley, January 13, 1897, West-Stanley-Wright Family Papers, USAMHI; Minturn, "Corinth"; Stanley, *Memoirs*, 113.

30. Strickling Reminiscences, 22, OHS; Cockrell, *Lost Account*, 31; William D. Evans to Caroline, October 8, 1862, Evans Papers, WRHS; Benner, *Sul Ross*, 86; Scott, *Four Years' Service*, 15.

31. Graves Diary, October 4, 1862, TxU; H. S. Halbert to Green W. Kerr, June 20, 1881, NMMA.

32. Charles H. Smith, *Fuller's Ohio Brigade*, 88–89, 101; *OR* 17(1):186, 188; Gilmore, "63d Ohio"; *Daily Toledo Blade*, October 20, 1862; Moore, *Civil War in Song and Story*, 463; Charles W. Smith, "Address," 2, OHS; Frank T. Gilmore to David Sloane Stanley, January 13, 1897, West-Stanley-Wright Family Papers, US-AMHI.

33. Hawke, *Corinth*, 47; Cockrell, *Lost Account*, 33; Stirman, *In Fine Spirits*, 52; *OR* 17(1):180; Daniel Leib Ambrose, *Seventh Illinois*, 101; Du Bois, "Journal and Letters," 48.

34. Stanley, "Corinth"; *OR* 17(1):180, 443; Pomeroy Diary, October 4, 1862, KHS; Hubert, *Fiftieth Illinois*, 143; Du Bois, "Journal and Letters," 48.

35. *Corinth Herald*, January 3, 1895; *OR* 17(1):209, 236–37; Ingersoll, *Iowa and the Rebellion*, 294–95; Reeves Diary, October 4, 1862, Reeves Papers, MnHS; Raum, "Corinth"; Stirman, *In Fine Spirits*, 52; "Correspondence of Jeremiah C. Sullivan," 140.

36. Lucius F. Hubbard to his aunt, October 13, 1862, Hubbard Papers, MnHS; William B. McGrorty to his wife, October 7, 1862, McGrorty Papers, MnHS; *Minnesota in the Wars*, 263–64; Jack Arkins to his brother, October 25, 1862, Arkins Papers, MnHS; Lucius F. Hubbard, *Minnesota in the Battles of Corinth*, 15–16, 21–22.

37. *OR* 17(1):180; Gottschalk, *In Deadly Earnest*, 157; Bevier, *First and Second Missouri Brigades*, 155.

CHAPTER TWENTY-TWO

1. *OR* 17(1):408; William C. Holmes, "Corinth," 292; G. A. Foote to his father, October 26, 1862, Foote Family Papers, Filson Club; John Tyler to William L. Yancey, October 15, 1862, WHMC-Columbia.

2. *OR* 17(1):408, 412–13; Henry, "Bowen," 171; Hirsh, "Shot Through," 505–6; H. J. Reid, "Twenty-second Mississippi," 15, Claiborne Papers, UNC; William C. Holmes, "Corinth," 292.

3. Throne, *Boyd Diary*, 75.

4. *OR* 17(1):380.

5. Maury, "Campaign against Grant," 297–99; J. H. Greene, *Reminiscences*, 32.

6. "Correspondence of Wisconsin Volunteers," 8:36–37, Quiner Papers, WiHS; *Alexandria Democrat*, October 22, 1862; Maury, "Campaign against Grant," 297; Rose, *Ross' Texas Brigade*, 73; *OR* 17(1):380, 411; *Grenada Appeal*, October 14, 1862.

7. *OR* 17(1):181, 345; Hubert, *Fiftieth Illinois*, 159; Rosecrans, "Battle of Corinth," 752; Montgomery, *Reminiscences*, 92–93; Stanley, *Memoirs*, 114; Adams, "Battery Robinett"; McCord, "Battle of Corinth," 581–82.

8. Rosecrans, "Battle of Corinth," 752–53; *OR* 17(1):170; "Rosecrans's Campaigns," in *Report of the Joint Committee*, 3:21; Fox, "Seventh Kansas Cavalry," 35; "Correspondence of Wisconsin Volunteers," 4:128, Quiner Papers, WiHS; *OR* 17(1):170; Dean Memoirs, 10, CWMC.

9. McCord, "Battle of Corinth," 581–82; Chetlain, "Corinth," 379–80; Hubert, *Fiftieth Illinois*, 159.

10. Bryner, *Bugle Echoes*, 64–65; Adams, "Battery Robinett"; "Correspondence of Wisconsin Volunteers," 4:124, Quiner Papers, WiHS; J. B. Rogers, *War Pictures*, 203; Chetlain, "Corinth," 380.

11. Fuller, "Our Kirby Smith," 176–77.

12. "Rosecrans's Campaigns," in *Report of the Joint Committee*, 3:22; Morris, *Thirty-first Regiment Volunteers*, 49; Bedford Memoirs, 25, Bedford Papers, DLC; Rosecrans, "Battle of Corinth," 753; *OR* 17(1):154–55, 305, 367; Simon, *Grant Papers*, 6:111–15.

13. *OR* 17(1):170, 246, 370; "Rosecrans's Campaigns," in *Report of the Joint Committee*, 3:22; Simon, *Grant Papers*, 6:115; Rosecrans, "Battle of Corinth," 753.

14. Castel, "Corinth," 21; Maury, "Recollections of Van Dorn," 195.

15. Fay, *This Infernal War*, 165; Hubbell, "Diary," 102; Anderson, *Memoirs*, 239.

16. Hartje, *Van Dorn*, 234–35; John Tyler to William L. Yancey, October 15, 1862, WHMC-Columbia; Maury, "Recollections of Van Dorn," 195.

17. *OR* 17(1):380 and (2):720; John Tyler to William L. Yancey, October 15, 1862, WHMC-Columbia; Hartje, *Van Dorn*, 234–35; Maury, "Recollections of Van Dorn," 196.

18. Maury, "Recollections of Van Dorn," 196; Castel, *Price and the War*, 120; Hartje, *Van Dorn*, 235; Cockrell, *Lost Account*, 72.

19. *OR* 17(1):380; Cockrell, *Lost Account*, 73; Maury, "Campaign against Grant," 302; Hartje, *Van Dorn*, 235.

20. Maury, "Campaign against Grant," 302–3; *OR* 17(1):380; Hartje, *Van Dorn*, 235.

21. Maury, "Campaign against Grant," 303.

CHAPTER TWENTY-THREE

1. Warner, *Generals in Blue*, 244–45; Lash, "Hurlbut," v–vii, 135–41, 146–47; 150–52.

2. Ainsworth Diary, October 4, 1862, Ainsworth Papers, University of Southern Mississippi; *OR* 17(1):155, 302, 305, 320, 321, 333; Comte de Paris, *Civil War*, 2:415–16; Quinnie Armour to Monroe Cockrell, April 6, 1946, and Fred Smith to Cockrell, February 18, 1946, both in Cockrell Papers, DU.

3. *OR* 17(1):302, 305, 315, 320–21, 325, 328–29; Robert L. Smith, "Hurlbut's Men"; Dugan, *Hurlbut's Fighting Fourth*, 185.

4. *OR* 17(1):322, 325, 328–29, 331–32; Dugan, *Hurlbut's Fighting Fourth*, 185; McDaniel, *Davis Bridge*, 5.

5. Maury, "Campaign against Grant," 303; McDaniel, *Davis Bridge*, 5.

6. Stuart, *Iowa Colonels and Regiments*, 166–67; *OR* 17(1):317, 392, 399; Wilder Reminiscences, 30, DLC; Maury, "Campaign against Grant," 303.

7. McDaniel, *Davis Bridge*, 27; Dugan, *Hurlbut's Fighting Fourth*, 174.

8. *OR* 17(1): 309, 310, 321–22, 399; Jacob Brunner to his wife, October 13, 1862, Brunner Letters, OHS; Wilder Reminiscences, 30, DLC; Robert L. Smith, "Hurlbut's Men"; Quinnie Armour to Monroe Cockrell, April 6, 1946, Cockrell Papers, DU; Maury, "Campaign against Grant," 303.

9. *OR* 17(1): 310, 325, 327; Dugan, *Hurlbut's Fighting Fourth*, 175.

10. Dugan, *Hurlbut's Fighting Fourth*, 175; *OR* 17(1): 322, 327, 329, 333, 393; Albert McCollom to his brother, December 5, 1862, McCollom Papers, ArU; *Daily Toledo Blade*, October 17, 1862; Wilder Reminiscences, 30–31, DLC; Thomas B. Jones, *Forty-sixth Illinois*, 223–24; Camm, "Diary," 911; Jacob Brunner to his wife, October 13, 1862, Brunner Letters, OHS.

11. Grebe Memoirs, 14, Grebe Collection, DLC; Albert McCollom to his wife, October 13, 1862, McCollom Papers, ArU; Keen, *Living and Fighting*, 39; *OR* 17(1): 323, 325, 400; Camm, "Diary," 911; Barber, *Army Memoirs*, 82.

12. Dugan, *Hurlbut's Fighting Fourth*, 179–80.

13. Maury, "Campaign against Grant," 303; *OR* 17(1): 132, 383, 400; *Mobile Advertiser and Register*, undated clipping in Solomon Scrapbook, 400, DU.

14. Benner, *Sul Ross*, 87; Sid S. Johnson, *Texans Who Wore the Gray*, 94–95; Scott, *Four Years' Service*, 17; "General and Governor Ross," 169; Rose, *Ross' Texas Brigade*, 165–66; Griscom, *Fighting with Ross' Texas Cavalry*, 47; Sparks, *War between the States*, 194; *OR* 17(1): 388, 393, 395, 400; Maury, "Campaign against Grant," 303–4.

15. *OR* 17(1): 302, 306, 313, 323; Cockrell, *Lost Account*, 73; Stuart, *Iowa Colonels and Regiments*, 167.

16. Camm, "Diary," 911; *OR* 17(1): 323–31; McDaniel, *Davis Bridge*, 27; Dugan, *Hurlbut's Fighting Fourth*, 176; Barber, *Army Memoirs*, 82–83.

17. Stuart, *Iowa Colonels and Regiments*, 163; Willit S. Haynes to his sisters, October 8, 1862, Haynes Letters, Bradley University; *OR* 17(1): 312–13, 316; Mark Bassett to Edward Lecour, May 27, 1896, Fifty-third Illinois Infantry File, SNMP.

18. Stirman, *In Fine Spirits*, 63; *OR* 17(1): 403.

19. Camm, "Diary," 911; Stuart, *Iowa Colonels and Regiments*, 167; *OR* 17(1): 323.

20. *OR* 17(1): 306, 308; Stuart, *Iowa Colonels and Regiments*, 167.

21. Stuart, *Iowa Colonels and Regiments*, 167; *OR* 17(1): 306, 308, 312, 317, 323.

CHAPTER TWENTY-FOUR

1. Maury, "Campaign against Grant," 304; Barron, *Lone Star Defenders*, 121–22; Fred B. Smith to Monroe Cockrell, February 18, 1946, Cockrell Papers, DU; *OR* 17(1): 380, 396, 399, 403; Fauntleroy, "Elk Horn to Vicksburg," 35.

2. *OR* 17(1): 306–8, 383, 388, 391–92, 394–95, 403; Maury, "Campaign against Grant," 304; Bevier, *First and Second Missouri Brigades*, 344.

3. Simon, *Grant Papers*, 6: 116–18; *OR* 17(1): 158; Stanley, *Memoirs*, 114–15.

4. Daniel Leib Ambrose, *Seventh Illinois*, 109; Chamberlin, *Eighty-first Ohio*, 34; Stanley, *Memoirs*, 115.

5. Lamers, *Edge of Glory*, 160; *OR* 17(1): 182, 246, 266.

6. *OR* 17(1):210; "Correspondence of Wisconsin Volunteers," 8:36, Quiner Papers, WiHS.

7. Lamers, *Edge of Glory*, 160–61; Stanley, *Memoirs*, 115; Hubert, *Fiftieth Illinois*, 162; *OR* 17(1):162, 210; "Correspondence of Wisconsin Volunteers," 8:36, Quiner Papers, WiHS.

8. *OR* 17(1):182, 263; Lamers, *Edge of Glory*, 161; Rosecrans, "Battle of Corinth," 754.

9. *OR* 17(1):161, 246, 345–46, 367–73, 413, 423; "Rosecrans's Campaigns," in *Report of the Joint Committee*, 3:22; Dean Memoirs, 10–11, CWMC.

10. *OR* 17(1):368, 413, 423; Morris, *Thirty-first Regiment Volunteers*, 49; Henry, "Bowen," 171.

11. *OR* 17(1):161, 345–46; "Rosecrans's Campaigns," in *Report of the Joint Committee*, 3:22; Cockrell, *Lost Account*, 71.

12. Rosecrans, "Battle of Corinth," 754.

13. *OR* 17(1):161–62 and (2):267; Ulysses S. Grant, *Memoirs*, 1:416, 418.

14. Barron, *Lone Star Defenders*, 122; *Corinth Herald*, November 17, 1904.

15. Carlisle Journal, 8, CWMC.

16. Anderson, *Memoirs*, 243–44; Fauntleroy, "Elk Horn to Vicksburg," 35.

17. Carlisle Journal, 8, CWMC; *OR* 17(1):721; Scott, *Four Years' Service*, 20; Anderson, *Memoirs*, 241.

18. Anderson, *Memoirs*, 241; Barron, *Lone Star Defenders*, 123.

19. Barron, *Lone Star Defenders*, 123–24; Anderson, *Memoirs*, 244; John M. Hubbard, *Notes of a Private*, 36; Tunnard, *Southern Record*, 193.

20. Fay, *This Infernal War*, 167; Anderson, *Memoirs*, 244; *OR* 17(1):417–18, 431, 436, 439; Scott, *Four Years' Service*, 18; Montgomery, *Reminiscences*, 94.

21. *OR* 17(1):423; Fauntleroy, "Elk Horn to Vicksburg," 35; Tunnard, *Southern Record*, 193.

22. "Rosecrans's Campaigns," in *Report of the Joint Committee*, 3:22; Lamers, *Edge of Glory*, 164; *OR* 17(1):338.

23. Clark, *Downing's Diary*, 74; Charles Cody to his family, October 16, 1862, Cody Papers, IaU; William B. Britton to "Messrs. Editors," October 9, 1862, Britton Letters, WRHS; Bedford Memoirs, 27, Bedford Papers, DLC; Throne, *Boyd Diary*, 76–77.

24. "Rosecrans's Campaigns," in *Report of the Joint Committee*, 3:22; Simon, *Grant Papers*, 6:126–27; *OR* 17(1):162–63, 368, 371, 373.

25. Simon, *Grant Papers*, 6:129; *OR* 17(1):162; "Rosecrans's Campaigns," in *Report of the Joint Committee*, 3:22.

26. *OR* 17(1):306; Simon, *Grant Papers*, 6:127; Lash, "Hurlbut," 139–40.

27. *OR* 17(1):381, 417–18, 423–25, 429; Tunnard, *Southern Record*, 193.

28. Cole, "War Experiences," 262; Tunnard, *Southern Record*, 193; *OR* 17(1):721.

29. Hubert, *Fiftieth Illinois*, 163; *OR* 17(1):183, 210; Daniel Leib Ambrose, *Seventh Illinois*, 111; Alonzo L. Brown, *Fourth Minnesota*, 134.

30. Lamers, *Edge of Glory*, 166–67; "Rosecrans's Campaigns," in *Report of the Joint Committee*, 3:23; Simon, *Grant Papers*, 6:131–32.

31. Lamers, *Edge of Glory*, 167; *OR* 17(1):133, 158–59; Ulysses S. Grant, *Memoirs*, 1:419; Simon, *Grant Papers*, 6:138.

32. *OR* 17(2):269; Simon, *Grant Papers*, 6:130, 133–34, 142; "Rosecrans's Campaigns," in *Report of the Joint Committee*, 3:23; Rosecrans, "Battle of Corinth," 755.

33. "Rosecrans's Campaigns," in *Report of the Joint Committee*, 3:23; John J. Bennett to his brother, October 7, 1862, Bennett Family Papers, Louisiana State University; Hubert, *Fiftieth Illinois*, 164; Rosecrans, "Battle of Corinth," 755–56.

CHAPTER TWENTY-FIVE

1. Rosecrans, "Battle of Corinth," 751; Bowen Diary, October 4, 1862, Bowen Papers, HL; Ferguson, "Annals of the War."

2. *OR* 17(1):382–85, 395–97; Livermore, *Numbers and Losses*, 94.

3. Wright Schaumburg to his wife, October 9, 1862, Schaumburg-Wright Family Papers, HNOC; *OR* 17(1):395–96.

4. Bevier, *First and Second Missouri Brigades*, 154; Stirman, *In Fine Spirits*, 52; Maury, "Campaign against Grant," 298.

5. Mansfield Lovell to his daughter, October 22, 1862, CWTI, and to his wife, October 13, 1862, Lovell Papers, HL.

6. Octavia Sulivane to her sister, February 5, 1863, Sulivane Letters, Smith Collection, USAMHI; Pollard, *Lost Cause*, 337; *Mobile Advertiser and Register* and *Atlanta Southern Confederacy*, undated clippings in Solomon Scrapbook, DU; *Daily Toledo Blade*, October 22, 1862.

7. *OR* 17(1):788–89.

8. Earl Van Dorn to Emily Van Dorn, October 14, 1862, Van Dorn Papers, ADAH.

9. Hartje, *Van Dorn*, 247–48; *OR* 17(1):381–82 and (2):726–28; Octavia Sulivane to her sister, February 5, 1863, Sulivane Letters, Smith Collection, USAMHI.

10. Earl Van Dorn to T. N. Waul, November 29, 1862, and to Emily Van Dorn, October 30, 1862, both in Van Dorn Papers, ADAH; Octavia Sulivane to her sister, February 5, 1863, Sulivane Letters, Smith Collection, USAMHI.

11. Hartje, *Van Dorn*, 248–49; *OR* 17(1):415–46.

12. *OR* 17(1):415–16; Octavia Sulivane to her sister, February 5, 1863, Sulivane Letters, Smith Collection, USAMHI.

13. *OR* 17(1):418–20.

14. *OR* 17(1):444–45, 458–59; Octavia Sulivane to her sister, February 5, 1863, Sulivane Letters, Smith Collection, USAMHI.

15. Octavia Sulivane to her sister, February 5, 1863, Sulivane Letters, Smith Collection, USAMHI; Hartje, *Van Dorn*, 250; Francis Vinson Greene, *The Mississippi*, 56–58.

16. Francis Vinson Greene, *The Mississippi*, 62; Maury, "Campaign against Grant," 304–5; Hartje, *Van Dorn*, 250–51; Tunnard, *Southern Record*, 212; *OR* 17(2):901.

17. *OR* 17(2):902–3; Tunnard, *Southern Record*, 216; Maury, "Campaign against Grant," 304–5; Francis Vinson Greene, *The Mississippi*, 62–64; Octavia Sulivane to her sister, February 5, 1863, Sulivane Letters, Smith Collection, USAMHI; Hartje, *Van Dorn*, 253.

18. William C. Davis, *Jefferson Davis*, 475–78; Hartje, *Van Dorn*, 254–55; Rose, *Ross' Texas Brigade*, 131; Octavia Sulivane to her sister, February 5, 1863, Sulivane Letters, Smith Collection, USAMHI; Maury, "Recollections of Van Dorn," 197.

19. Lamers, *Edge of Glory*, 168–69; Ducat, *Memoirs*, 28.

20. Lamers, *Edge of Glory*, 172; *OR* 17(1):176.

21. *OR* 17(1):172–73.

22. Lamers, *Edge of Glory*, 172–73.

23. Simon, *Grant Papers*, 6:146–47, 166–67; Lamers, *Edge of Glory*, 172–73.

24. Lamers, *Edge of Glory*, 172–74; John W. Fuller to David Sloane Stanley, August 23, 1887, West-Stanley-Wright Family Papers, USAMHI; Ulysses S. Grant, *Memoirs*, 1:417–18; William Starke Rosecrans to his wife, September 20, 1874, Rosecrans Papers, UCLA; Sherman, *Memoirs*, 284.

25. Simon, *Grant Papers*, 6:163–66.

26. *OR* 17(2):286–87.

27. Julia Dent Grant, *Memoirs*, 104.

28. Simon, *Grant Papers*, 6:182, 185.

29. Ibid., 6:185; Julia Dent Grant, *Memoirs*, 104–5; Ulysses S. Grant, *Memoirs*, 1:420.

CHAPTER TWENTY-SIX

1. Francis Vinson Greene, *The Mississippi*, 54; Sherman, *Memoirs*, 284; Stanley, *Memoirs*, 115; Chetlain, "Corinth," 381–82.

2. Nelson Memoirs, 29, CWMC; Lamers, *Edge of Glory*, 178–79; Arthur C. Ducat to William Starke Rosecrans, April 24, 1885, Rosecrans Papers, UCLA.

3. "Correspondence of Wisconsin Volunteers," 4:126, 128, Quiner Papers, WiHS; A. B. Monahan to his wife, October 12, 1862, Sixty-third Ohio Infantry File, SNMP; Clark, *Downing's Diary*, 75–76; Hubert, *Fiftieth Illinois*, 164; Daniel Leib Ambrose, *Seventh Illinois*, 112–13; John S. Wilcox to his wife, October 16, 1862, Wilcox Papers, ISHL.

4. Lamers, *Edge of Glory*, 168; Maury, "Campaign against Grant," 304–5.

5. Connelly, *Army of the Heartland*, 238, 270.

6. Von Clausewitz, *On War*, 535.

7. Throne, "Iowa Doctor," 99; *OR* 17(2):270; Roberts, "Caring for the Wounded," 328; Simon, *Grant Papers*, 6:145.

8. Trowbridge, "Saving a General," 22; Simon, *Grant Papers*, 6:145.

9. Simon, *Grant Papers*, 6:143.

10. Trowbridge, "Saving a General," 22–25.

11. Jackson, *Colonel's Diary*, 77–81.

12. Castel, *Price and the War*, 133–34, 139; Shalhope, *Price*, 225–32; Gottschalk, *In Deadly Earnest*, 176–77.

13. Hartje, *Van Dorn*, 262–65.

14. Ibid., 290, 296.

15. Pope, "War Reminiscences XII"; Hartje, *Van Dorn*, 308–9.

16. Ibid., 310–13; Van Dorn biographical sketch, Sigel Papers, WRHS; Sparks, *War between the States*, 226–27.

17. Hartje, *Van Dorn*, 318–19.

APPENDIX

1. Not shown in table of organization found in *OR* 17(1):374–75; however, Armstrong mentions this regiment as belonging to his brigade. See *OR* 17(2):675.

2. One section only.

3. Detachments of the 58th Illinois, the 8th Iowa, the 12th Iowa, and the 14th Iowa constituted the organization known as the Union Brigade.

4. Serving as infantry.

5. Serving as infantry.

6. Serving as infantry.

7. Not identified as such, but probable.

8. Serving as infantry.

9. Serving as infantry.

10. Serving as infantry.

11. Probably incomplete.

12. Probably incomplete.

13. Serving as infantry.

14. Serving as infantry.

15. Serving as infantry.

BIBLIOGRAPHY

MANUSCRIPTS

Alabama Department of Archives and History, Montgomery
  Forty-second Alabama Infantry Papers
  Earl Van Dorn Papers
Arkansas History Commission, Little Rock
  W. L. Skaggs Collection
Baylor University, Texas Collection, Waco, Tex.
  Ross Family Papers
Bradley University, Peoria Historical Society Collection, Peoria, Ill.
  Willit Samuel Haynes Letters
Brown County Historical Society, New Ulm, Minn.
  August Schilling Diary
Chicago Historical Society, Chicago, Ill.
  Joseph H. Barrett Collection
  John A. Rawlins Papers
Duke University, William R. Perkins Library, Durham, N.C.
  Monroe F. Cockrell Papers
  Absalom F. Dantzler Letters
  M. J. Solomon Scrapbook
  Mary E. Sword Papers
  Ella Gertrude (Clanton) Thomas Journal
  Charles Brown Tompkins Papers
Emory University, Robert W. Woodruff Library, Atlanta, Ga.
  Honnell Family Papers
Filson Club, Louisville, Ky.
  Ben Albert Miscellaneous Papers
  Foote Family Papers
  James F. Mohr Papers
Historic New Orleans Collection, Kemp and Leila Williams Foundation,
    New Orleans, La.
  Schaumburg-Wright Family Papers
Huntington Library, San Marino, Calif.
  Edwin A. Bowen Papers
  Mansfield Lovell Papers
  Thomas W. Sweeny Papers
Illinois Historical Survey, University of Illinois Library, Champaign-Urbana
  Charles S. Hamilton Papers
Illinois State Historical Library, Springfield
  George Lemon Childress Diary
  John N. Cromwell Memoirs
  John A. Griffin Diary

Charles S. Hamilton Papers
Robert Ingersoll Papers
Lease Family Papers
Wallace-Dickey Papers
John S. Wilcox Papers
Indiana Historical Society, Indianapolis
　Civil War File of James C. Veatch
Kansas State Historical Society, Topeka
　Fletcher Pomeroy Diary
Library of Congress, Washington, D.C.
　Wilmer Bedford Papers
　Alpheus S. Bloomfield Papers
　John N. Ferguson Diary
　Balzar Grebe Collection
　Hiram B. Howe Papers
　Joseph Lester Collection
　Bela T. St. John Papers
　Oliphant M. Todd Diary
　John Whitten Diary
　William F. Wilder Reminiscences
Louisiana State University, Hill Memorial Library, Baton Rouge
　Bennett Family Papers
Michigan State University, East Lansing
　Chamberlain Family Papers
　Crosby Family Papers
　Miller Family Papers
Minnesota Historical Society, St. Paul
　William Arkins and Family Papers
　Lucius F. Hubbard and Family Papers
　William Bernard McGrorty and Family Papers
　Richard S. Reeves Papers
　John Risedorph Papers
Mississippi Department of Archives and History, Jackson
　Hunter, Joseph J. "A Sketch of the History of the Noxubee Troopers,
　1st Mississippi Cavalry"
　A. W. Simpson Diary
Missouri Historical Society, St. Louis
　Civil War Collection
　　Thomas Hogan Letter
　　David McKnight Diary
　McEwen Family Papers
　Herman B. Miller Papers
　Reynolds, T. C. "General Sterling Price and the Confederacy"
Northeast Mississippi Museum Association, Corinth
　James Newton Carlisle Diary
　H. S. Halbert Letter

Hugh Horton Manuscripts
C. J. Sergent Diary
Will Snyder Letter
Ohio Historical Society, Columbus
Edwin Brown Recollections
Jacob Brunner Letters
Jacob Crouse Letters
Edwin Curtis Diary
John Kelly Diary
Smith, Charles W. "Address to Comrades of Fuller's Ohio Brigade, September 6, 1888"
Joseph Strickling Reminiscences
Oregon Historical Society, Portland
Samuel Arnold Randle Collection
Shiloh National Military Park Library, Shiloh, Tenn.
Historical Reports
Duke, Cecil A. "Confederate Intrenchments"
Price, May. "Story of Corona College"
Fifty-third Illinois Infantry File
Second Iowa Infantry File
Sixty-third Ohio Infantry File
Sioux City Public Museum, Sioux City
Henry Clay McNeil Papers
State Historical Society of Iowa, Iowa City
James C. Parrott Papers
Franklin B. Reed Letters
State Historical Society of Wisconsin, Madison
William Brown Letters
Charles S. Hamilton Papers
E. B. Quiner Papers
James Monroe Tyler Diary
Tennessee State Library and Archives, Nashville
George Elliott Journal
Joseph R. Mothershead Journal
United States Army Military History Institute, Carlisle Barracks, Pa.
Leslie Anders Collection
Nehemiah Davis Starr Letters
Civil War Miscellaneous Collection
William Burge Memoirs
Hugh T. Carlisle Journal
Edward Dean Memoirs
James M. Gilbert Certificate
John J. McKee Diary
Joseph K. Nelson Memoirs
Columbus Sykes Letters
William Wade Diary

Civil War Times Illustrated Collection
   George Bargus Diary
   Henry Lewis Little Diary
   Mansfield Lovell Letter
   David McKinney Diary
   Lewis F. Phillips Memoirs
Harrisburg Civil War Round Table Collection
   William A. Ruyle Memoirs
Murray J. Smith Collection
   Octavia Sulivane Letters
West-Stanley-Wright Family Papers
University of Arkansas, Fayetteville
   Joseph M. Bailey Papers
   James Henderson Berry Papers
   Robert and Sephronia Clark McCollom Papers
   William Stephen Ray Papers
University of California, Los Angeles
   William Starke Rosecrans Papers
University of Iowa, Iowa City
   Henry Clay Adams Civil War Journals
   Charles Cody Papers
   George Hall Family Papers
   Mead Family Papers
   Myron Underwood Papers
University of Michigan, Michigan Historical Collections, Bentley Historical Library,
   Ann Arbor
   Donald H. J. Cobb Letters
   Newell Ranson Dunn Letters
University of Missouri, Western Historical Manuscripts Collection, Ellis Library,
   Columbia
   Francis Marion Cockrell Scrapbooks
   J. B. Courts Letter
   William H. Kavanaugh Memoirs
   Thomas Pollock Letters
   Avington Wayne Simpson Diary
   I. V. Smith Memoirs
   William S. Stewart Letters
   Benjamin F. Sweet Memoirs
   John Tyler Letter
University of Missouri, Western Historical Manuscripts Collection, St. Louis
   Dyson–Bell–Sans Souci Papers
University of North Carolina, Southern Historical Collection, Chapel Hill
   George Hovey Cadman Papers
   J. F. H. Claiborne Papers
   Louis Hébert Autobiography
University of North Texas Archives, Denton

F. A. Rawlins Diary
University of Southern Mississippi, McLain Library, Hattiesburg
   Garrett Smith Ainsworth Papers
   Andrew J. Patrick Letters
University of Texas, Center for American History, Austin
   L. H. Graves Diary
   George L. Griscom Diary
   William P. Rogers Papers
University of Wisconsin, La Crosse
   Edward Croton Papers
Vicksburg National Military Park Library, Historical Files, Vicksburg, Miss.
   John R. P. Martin Letter
Western Reserve Historical Society, Cleveland, Ohio
   William A. Britton Letters
   John Q. A. Campbell Diary
   William David Evans Papers
   Franz Sigel Papers
   Edward A. Webb Papers
Wright State University Archives, Dayton, Ohio
   James F. Overholser Memoirs

NEWSPAPERS

Alexandria (La.) Democrat
Atlanta Southern Confederacy
Chicago Tribune
Cincinnati Daily Commercial
Corinth Herald
Corinth Weekly Herald
Daily Toledo Blade
Dayton Weekly Journal
Detroit Advertiser and Tribune
Galveston Daily News
Grenada Appeal
Harper's Weekly
Houston Tri-Weekly Telegraph
Iowa State Register (Des Moines)

Jackson Mississippian
Kankakee Daily Journal
Mankato (Minn.) Semi-Weekly Record
Mankato Semi-Weekly Union
Memphis Appeal
Milwaukee Sentinel
Missouri Republican (St. Louis)
Mobile Advertiser and Register
New Orleans Picayune
St. Paul (Minn.) Pioneer Press
Stark County (Ohio) Democrat
Utica (N.Y.) Evening Telegraph
Weekly Enterprise (Enterprise, Ala.)

NATIONAL TRIBUNE

Adams, W. W. "Battery Robinett." May 5, 1898.
Allen, John H. "Battle of Iuka." April 27, 1911.
———. "A Hot Little Fight." March 17, 1927.
"Attack on Corinth." January 3, 1884.

Bartholomew, W. A. "Another Defense of the Sixth Wisconsin Battery." September 20, 1883.

"Battle of Corinth From the Diary of a Confederate Soldier." February 22, 1883.

Bifram, Samuel. "At the Battle of Corinth." June 30, 1910.

Bledsoe, B. H. "With the 11th Iowa." March 17, 1898.

"Bloody Repulse at Corinth." April 8, 1882.

"The Bloody Repulse at Corinth." April 29, 1882.

Boynton, S. B. "The Boy Who Saved Corinth." August 15, 1901.

Bradney, Thomas J. "Saw the 63d Fight." February 23, 1899.

Brown, Alonzo L. "The Fourth Minnesota at Iuka." December 11, 1884.

———. "Iuka. The 48th Indiana's Conduct and the 11th Ohio Battery's Losses." October 2, 1884.

Budd, Charles H. "Colonel Rogers. His Death at Corinth — Map of the Fatal Spot." November 12, 1885.

Campbell, William. "Iuka. Other Testimony." October 2, 1884.

Christie, Thomas. "Hot Times on the Chewalah [sic] Road." November 26, 1914.

———. "Old Abe at Corinth." November 19, 1914.

Conn, Charles M. "On the Hatchie. The 12th Mich. the First to Cross the Bridge." April 22, 1886.

Corsair, David. "Fighting at Corinth." July 21, 1898.

Davis, William M. "Iuka." January 8, 1885.

"Death of Colonel Rogers." April 22, 1886.

"Death of Colonel Rogers at Corinth." October 8, 1885.

De Daines, James. "Desperate Fighting There." August 18, 1898.

Dodge, Alf. "The Man of Iuka." April 4, 1904.

Gilmore, F. T. "Defending the 63d Ohio." February 9, 1899.

Hatheway, James. "Fierce Fighting at Robinett." August 31, 1899.

Henney, George W. "The 11th Mo. at Corinth." June 28, 1894.

Hickok, A. D. "Iuka and Corinth: The Eagle Regiment's Share in the Great Victories." January 7, 1909.

Hittle, W. H. "The Division of Davies." March 21, 1901.

Holmes, George W. "At Battery Robinett." May 26, 1898.

Hood, N. R. "The Sixth Wisconsin Battery at Corinth." August 30, 1883.

Immell, L. D. "Iuka and Corinth." August 6, 1903.

———. "The Twelfth Wisconsin Battery at Corinth." February 16, 1888.

———. "The Twelfth Wisconsin Battery Takes Advantage of a Crisis." January 10, 1889.

Kenderdinn, H. M. "Fired on Friends. Tight Place in which the 17th Iowa Found Itself at Iuka." January 9, 1896.

Kibbe, A. R. "Iuka and Corinth. Commander of the 12th Wisconsin Battery." September 24, 1903.

Langworthy, L. L. "The Battle of Corinth. A Claim That the 12th Ill. Had a Strong Hand in Saving the Day." January 24, 1901.

Lindsey, W. J. "At Corinth." March 1, 1906.

Lybarger, E. L. "The Battle of Corinth." November 26, 1908.

McGinnis, J. I. "Corinth to Vicksburg." June 28, 1894.

———. "Roger's Death." November 1, 1894.

McNeal, James. "Charge of the Texans." January 5, 1899.

Manley, A. J. "The 14th Wisconsin at Corinth." January 15, 1885.

Mayo, J. L. "Colonel Rogers at Corinth." October 15, 1885.

Milburn, Henry. "A Round of Reminiscences. Boy Lieutenant Writes of Corinth."
March 7, 1901.

Minturn, W. H. H. "Battle of Corinth. Fuller's Ohio Brigade and the Part It Took in
This Engagement." May 23, 1901.

Morgan, Uriah. "Once Was Enough." March 7, 1912.

Pope, John. "Siege of Corinth . . . Part I." May 17, 1888.

———. "War Reminiscences . . . Part XII." February 26, 1891.

———. "War Reminiscences . . . Part XIII." March 5, 1891.

Raum, Green B. "The Battle of Corinth." December 20, 1900.

Russell, Q. O. "Iuka." October 2, 1884.

Sanders, J. L. "Battle of Iuka." August 29, 1918.

Sears, Cyrus. "The 11th Ohio Battery at Iuka." November 6, 1884.

Shaw, R. K. "The 63d Ohio at Corinth." October 29, 1885.

Shuman, J. T. "The 63d Ohio at Robinett." March 2, 1899.

Smith, Robert L. "Hurlbut's Men." July 18, 1896.

Spencer, Dan. "The Battle of Iuka. An Iowa Boy's Baptism of Fire." June 27, 1918.

Stanley, David S. "Corinth." February 8, 1894.

Van Eman, J. H. "Battle of Iuka. Experiences of a Fuller's Brigade Veteran." May 29,
1902.

———. "Bloody Robinett." May 24, 1894.

"What One Comrade Knows about Battery Robinett." March 15, 1894.

Whitefield, W. G. "What Battery Was This?" July 11, 1895.

OFFICIAL DOCUMENTS

Report of the Joint Committee on the Conduct of the War. 10 vols. Washington, D.C.,
1863–66.

*The War of the Rebellion: A Compilation of the Official Records of the Union and
Confederate Armies.* Washington, D.C., 1880–1901.

ADDRESSES, ARTICLES, AND ESSAYS

Arnold, T. B. "The Battles around Corinth, Miss." *Confederate Veteran* 5, no. 5 (May
1897): 199–200.

Banks, Robert W. "The Civil War Letters of Robert W. Banks." *Journal of Mississippi
History* 5 (1943): 141–54.

Booton, W. W. "A Federal Boy Soldier at Corinth." *Confederate Veteran* 5, no. 1
(January 1897): 6.

Broadhead, James O. "Early Events of the War in Missouri." In *War Papers and
Personal Reminiscences, 1861–1865. Read before the Commandery of the State of*

Missouri, Military Order of the Loyal Legion of the United States, 1:1–28. St. Louis, 1892.

Byers, Samuel H. M. "How Men Feel in Battle." *American History Illustrated*, April 1987, 11–17.

Camm, William. "Diary of Colonel William Camm, 1861 to 1865." *Journal of the Illinois State Historical Society* 28, no. 4 (January 1926): 793–969.

Carter, Frank. "Lady Richardson." *Confederate Veteran* 3, no. 10 (October 1895): 306.

Castel, Albert. "Battle without a Victor . . . Iuka." *Civil War Times Illustrated* 11, no. 6 (October 1972): 12–18.

———. "Victory at Corinth." *Civil War Times Illustrated* 17, no. 6 (October 1978): 12, 16–22.

———, ed. "The Diary of Gen. Henry Little." *Civil War Times Illustrated* 11, no. 6 (October 1972): 4–11, 41–47.

Chetlain, Augustus. "The Battle of Corinth." In *Military Essays and Recollections. Papers Read before the Commandery of the State of Illinois, Military Order of the Loyal Legion of the United States*, 2: 373–82. Chicago, 1894.

Cole, C. M. "Vivid War Experiences at Ripley, Miss." *Confederate Veteran* 8, no. 6 (June 1905): 262–64.

"Col. H. J. Reid." *Confederate Veteran* 14, no. 3 (March 1906): 132.

Colwell, Charles. "An Infantryman at Corinth." *Civil War Times Illustrated* 13, no. 7 (November 1974): 11–14.

"Col. William P. Rogers." *Confederate Veteran* 4, no. 2 (February 1896): 57–58.

"Comments about 'Lady Richardson.'" *Confederate Veteran* 3, no. 7 (July 1895): 205–6.

"Correspondence of Jeremiah C. Sullivan." *Annals of the State Historical Society of Iowa*, July 1863, 140.

Cozzens, Peter. "'My Poor Little Ninth': The Ninth Illinois at Shiloh." *Illinois Historical Journal* 83 (Spring 1990): 31–44.

Crane, John. "Letter to Mrs. Bownson, Chairman of the William P. Rogers Monument Committee." *Confederate Veteran* 15, no. 6 (June 1907): 245–46.

"Dabney Herndon Maury." *Southern Historical Society Papers* 17 (1899–1900): 335–45.

Du Bois, John. "Journal and Letters of Col. John Du Bois." *Missouri Historical Review* 61, no. 1 (October 1966): 22–50.

Fauntleroy, James H. "Elk Horn to Vicksburg: James H. Fauntleroy's Diary for the Year 1862." *Civil War History* 2, no. 1 (March 1956): 7–44.

Ferguson, P. C. "Annals of the War. Iuka and Corinth." *Philadelphia Weekly Times*, April 1, 1882.

Fox, Simeon M. "The Early History of the Seventh Kansas Cavalry." *Collections of the Kansas State Historical Society, 1909–1910* 11 (1910): 238–53.

Frederick, J. V., ed. "An Illinois Soldier in North Mississippi." *Journal of Mississippi History* 1 (1939): 182–94.

———. "War Diary of William C. Porter." *Arkansas Historical Quarterly* 11 (1952): 286–314.

Fuller, John W. "Our Kirby Smith." In *Sketches of War History, 1861–1865. Papers Read before the Ohio Commandery of the Military Order of the Loyal Legion of the United States, 1886–1888*, 2:161–79. Cincinnati, 1888.

Gaither, Rice. "Back to Corinth." *New York Times Magazine*, June 22, 1947, 13, 41–44.

"General and Governor Ross, of Texas." *Confederate Veteran* 2, no. 5 (May 1894): 169.

"Gen. W. L. Cabell." *Confederate Veteran* 2, no. 2 (February 1894): 67–68.

Gunn, Jack W. "The Battle of Iuka." *Journal of Mississippi History* 24 (1962): 142–57.

Hamilton, Charles S. "The Battle of Iuka." In *Battles and Leaders of the Civil War*, edited by Robert Underwood Johnson and Clarence Clough Buel, 2:734–35. New York, 1887.

———. "Hamilton's Division at Corinth." In *Battles and Leaders of the Civil War*, edited by Robert Underwood Johnson and Clarence Clough Buel, 2:757–58. New York, 1887.

Hamilton, Charles S., and Arthur C. Ducat. *Correspondence in Regard to the Battle of Corinth, Miss., October 3d and 4th, 1862.* Chicago, 1882.

Helm, W. P. "Close Fighting at Iuka, Miss." *Confederate Veteran* 19, no. 11 (April 1911): 171.

Henry, Pat. "Maj. Gen. John S. Bowen, C.S.A." *Confederate Veteran* 22, no. 4 (April 1914): 171–72.

Hirsh, I. E. "Shot Through by a Cannon Ball." *Confederate Veteran* 11, no. 11 (November 1903): 505.

Holmes, William C. "The Battle of Corinth." *Confederate Veteran* 28, no. 8 (August 1919): 290–92.

Hubbard, Lucius F. *Minnesota in the Battles of Corinth, May–October 1862.* St. Paul, 1907.

Hubbell, Finley L. "Diary of Lieut. Col. Hubbell." *Land We Love* 6, no. 2 (December 1868): 97–105.

Inge, F. A. "Corinth, Miss., in War Times." *Confederate Veteran* 23, no. 9 (September 1915): 412–13.

Johnson, Frederick Marion. "Diary of a Soldier in Grant's Rear Guard." *Journal of Mississippi History* 45 (1983): 194–204.

Jordan, Weymouth T., ed. "Matthew Andrew Dunn Letters." *Journal of Mississippi History* 1 (1939): 110–27.

"Letter of Lieut. Col. Wm. E. Small." *Annals of the State Historical Society of Iowa*, April 1863, 87.

Lybarger, E. L. *A Paper read at a Reunion of Fuller's Brigade, Held at Marietta, Ohio, September 10, 1885.* N.p., n.d.

McCord, W. B. "Battle of Corinth." In *Glimpses of the Nation's Struggle. Fourth Series. Papers Read before the Minnesota Commandery of the Military Order of the Loyal Legion of the United States, 1892–1897*, 567–84. St. Paul, 1898.

McKinstry, J. A. "With Col. Rogers When He Fell." *Confederate Veteran* 4, no. 7 (July 1896): 220–22.

McLain, David. "The Story of Old Abe." *Wisconsin Magazine of History* 8, no. 4 (June 1925): 407–14.

"Maj. Gen. Earl Van Dorn." *Confederate Veteran* 19, no. 8 (April 1911): 385–87.

Martin, R. T. "Reminiscences of an Arkansan." *Confederate Veteran* 17, no. 2 (February 1908): 69–70.

Mattox, Henry E., ed. "Chronicle of a Mississippi Soldier." *Journal of Mississippi History* 52 (1990): 199–211.

Maury, Dabney. "Grant's Campaign in North Mississippi." *Southern Magazine* 13 (July–December 1873): 410–17.

———. "Recollections of Campaign against Grant in North Mississippi, 1862–1863." *Southern Historical Society Papers* 13 (January–December 1885): 285–310.

———. "Recollections of Earl Van Dorn." *Southern Historical Society Papers* 19 (January–December 1891): 191–201.

———. "Recollections of the Elkhorn Campaign." *Southern Historical Society Papers* 2 (January–December 1876): 180–92.

Neil, Henry M. *A Battery at Close Quarters. A Paper Read before the Ohio Commandery of the Loyal Legion, October 6, 1909.* Columbus, 1909.

Parkhurst, Clint. "The Attack on Corinth." *Palimpsest* 3, no. 6 (June 1922): 169–91.

Payne, James E. "The Sixth Missouri at Corinth." *Confederate Veteran* 36, no. 12 (December 1928): 462–65.

"Reminiscent Paragraphs." *Confederate Veteran* 1, no. 9 (September 1893): 267.

Roberts, J. C. "Caring for the Wounded at Iuka." *Confederate Veteran* 2, no. 10 (October 1894): 328.

Rogers, William P. "The Diary and Letters of William P. Rogers." *Southwestern Historical Quarterly* 32, no. 4 (April 1929): 259–99.

Rosecrans, William S. "The Battle of Corinth." In *Battles and Leaders of the Civil War*, edited by Robert Underwood Johnson and Clarence Clough Buel, 2:737–56. New York, 1887.

Sanborn, John B. *Some Descriptions of the Battles in which the Commands of Gen. John B. Sanborn, of St. Paul, Minnesota, Participated.* St. Paul, 1900.

Sears, Cyrus. *Eleventh Ohio Battery at Iuka.* Cincinnati, 1898.

Snead, Thomas L. "With Price East of the Mississippi." In *Battles and Leaders of the Civil War*, edited by Robert Underwood Johnson and Clarence Clough Buel, 2:717–34. New York, 1887.

Soper, E. B. "A Chapter from the History of Company D, Twelfth Iowa Infantry Volunteers." In *War Sketches and Incidents as Related by the Companions of the Iowa Commandery Military Order of the Loyal Legion of the United States*, 2:129–42. Des Moines, 1898.

Stanley, David S. "The Battle of Corinth." In *Personal Recollections of the War of the Rebellion. Addresses Delivered before the Commandery of the State of New York, Military Order of the Loyal Legion of the United States*, 2d series, 267–79. New York, 1897.

Stewart, William S. "William S. Stewart Letters, January 13, 1861, to December 4, 1862," pt. 3. *Missouri Historical Review* 61 (July 1967): 463–88.

Sweet, Benjamin F. "Civil War Experiences." *Missouri Historical Review* 43 (October 1948–July 1949): 237–47.

Throne, Mildred, ed. "An Iowa Doctor in Blue: The Letters of Seneca B. Thrall, 1862–1864." *Iowa Journal of History* 58, no. 2 (April 1960): 97–188.

Trowbridge, Silas T. "Saving a General." *Civil War Times Illustrated* 11, no. 6 (July 1972): 20–25.

Tucker, William H. *The Fourteenth Wisconsin Vet. Vol. Infantry at the Battle of Corinth.* Indianapolis, 1893.

Von Phul, Frank. "General Little's Burial." *Southern Historical Society Papers* 29 (January–December 1901): 213–15.

Welch, William W. "Annals of the War. Battle of Hatchie River." *Philadelphia Weekly Times*, March 15, 1884.

W. H. B. "Maj. Gen. Sterling Price." *Land We Love* 1 (September 1866): 364–74.

Wood, William. "General Sterling Price: The New Mexico Insurrection, 1846–47." *Magazine of American History* 18 (December 1887): 333–35.

Zearing, James R. "Civil War Letters of Major James Roberts Zearing, M.D., 1861–1865." *Transactions of the Illinois State Historical Society* 28 (1922): 150–203.

AUTOBIOGRAPHIES, COLLECTED WORKS, DIARIES, LETTERS,
MEMOIRS, AND PERSONAL NARRATIVES

Abernathy, Byron R., ed. *Private Elisha Stockwell, Jr., Sees the Civil War*. Norman, Okla., 1956.

Ambrose, Stephen E., ed. *A Wisconsin Boy in Dixie: The Selected Letters of James K. Newton*. Madison, Wis., 1961.

Army Life and Stray Shots. Memphis, Tenn., 1863.

Barber, Lucius W. *Army Memoirs of Lucius W. Barber, Company "D," Fifteenth Illinois Volunteer Infantry*. Chicago, 1894.

Barrett, J. O. *The Soldier Bird. "Old Abe": The Live War-Eagle of Wisconsin*. Madison, Wis., 1876.

Basler, Roy P., ed. *The Collected Works of Abraham Lincoln*. 10 vols. New Brunswick, N.J., 1953.

Beatty, John. *The Citizen Soldier; or, Memoirs of a Volunteer*. Cincinnati, 1879.

Bradley, James. *The Confederate Mail Carrier*. Mexico, Mo., 1894.

Byers, Samuel H. M. *With Fire and Sword*. New York, 1911.

Cater, Douglas John. *As It Was: Reminiscences of a Soldier of the Third Texas Cavalry and the Nineteenth Louisiana Infantry*. Austin, 1911.

Chetlain, Augustus. *Recollections of Seventy Years*. Galena, Ill., 1899.

Clark, Olynthus B., ed. *Downing's Civil War Diary by Sergeant Alexander G. Downing*. Des Moines, 1916.

Ducat, Arthur C. *Memoirs of Gen. A. C. Ducat*. Chicago, 1897.

Duncan, Thomas D. *Recollections of Thomas D. Duncan*. N.p., n.d.

Fay, Edwin. *This Infernal War: The Confederate Letters of Sgt. Edwin H. Fay*. Edited by Bell I. Wiley. Austin, 1958.

Garrett, David R. *The Civil War Letters of David R. Garrett*. Edited by Max Lale. Marshall, Tex., 1964.

Goodloe, Albert T. *Confederate Echoes*. Nashville, 1907.

Grant, Julia Dent. *The Personal Memoirs of Julia Dent Grant*. Edited by John Y. Simon. Carbondale, Ill., 1975.

Grant, Ulysses S. *Personal Memoirs of U. S. Grant*. 2 vols. New York, 1885–86.

Greene, J. H. *Reminiscences of the War*. Medina, Ohio, 1886.

Griggs, George. *Opening of the Mississippi*. Madison, Wis., 1864.

Harwell, Richard, ed. *Kate: The Journal of a Confederate Nurse*. Baton Rouge, 1959.

Hubbard, John M. *Notes of a Private*. Memphis, Tenn., 1909.

Jackson, Oscar L. *The Colonel's Diary*. Sharon, Pa., 1922.

Jones, Jenkin L. *An Artilleryman's Diary*. Madison, Wis., 1914.

Keen, Newton A. *Living and Fighting with the Texas Sixth Cavalry*. Gaithersburg, Md., 1986.

Liddell, St. John Richardson. *Liddell's Record*. Dayton, Ohio, 1985.

Maury, Dabney H. *Recollections of a Virginian*. New York, 1894.

Montgomery, Frank A. *Reminiscences of a Mississippian in Peace and War*. Cincinnati, 1901.

Moore, Frank, ed. *The Rebellion Record*. 11 vols. New York, 1861–68.

Pepper, George W. *Under Three Flags; or, The Story of My Life*. Cincinnati, 1899.

Rogers, J. B. *War Pictures. Experiences and Observations of a Chaplain*. Chicago, 1863.

Rowland, Dunbar, ed. *Jefferson Davis, Constitutionalist: His Letters, Papers, and Speeches*. 10 vols. Jackson, Miss., 1923.

Scott, Joseph M. *Four Years' Service in the Confederate Army*. Mulberry, Ark., 1897.

Shanks, William F. G. *Personal Recollections of Distinguished Generals*. New York, 1866.

Sherman, William T. *Memoirs of General W. T. Sherman*. New York, 1990.

Simon, John Y., ed. *The Papers of Ulysses S. Grant*. 17 vols. Carbondale, Ill., 1967– .

Smith, Ralph J. *Reminiscences of the Civil War and Other Sketches*. Waco, Tex., 1962.

Sparks, A. W. *The War between the States as I Saw It*. Longview, Tex., 1987.

Stanley, David S. *Personal Memoirs of Major-General D. S. Stanley, U.S.A.* Cambridge, Mass., 1917.

Stirman, Ras. *In Fine Spirits: The Civil War Letters of Ras Stirman*. Fayetteville, Ark., 1986.

Stone, Mary Amelia, ed. *Memoir of George Boardman Boomer*. Boston, 1864.

Throne, Mildred, ed. *The Civil War Diary of C. F. Boyd*. Iowa City, 1953.

Wallace, Lewis. *Lew Wallace, an Autobiography*. 2 vols. New York, 1906.

Wright, Charles. *A Corporal's Story. Experiences in the Ranks of Company C, Eighty-first Ohio Vol. Infantry*. Philadelphia, 1887.

UNIT HISTORIES

Ambrose, Daniel Leib. *History of the Seventh Regiment Illinois Volunteer Infantry*. Springfield, Ill., 1868.

Anders, Leslie. *The Eighteenth Missouri*. Indianapolis, 1968.

———. *The Twenty-first Missouri: From Home Guard to Union Regiment*. Westport, Conn., 1975.

Anderson, Ephraim McD. *Memoirs: Historical and Personal; Including the Campaigns of the First Missouri Confederate Brigade*. St. Louis, 1868.

Barron S. B. *The Lone Star Defenders: A Chronicle of the Third Texas Cavalry, Ross' Brigade*. New York, 1908.

Belknap, William. *History of the Fifteenth Regiment, Iowa Veteran Volunteer Infantry*. Keokuk, Iowa, 1887.

Bell, John T. *Tramps and Triumphs of the Second Iowa Infantry, Briefly Sketched*. Omaha, Nebr., 1886.

Bevier, R. S. *History of the First and Second Missouri Confederate Brigades, 1861–1865.* St. Louis, 1879.

Brown, Alonzo L. *History of the Fourth Regiment of Minnesota Infantry Volunteers.* St. Paul, 1892.

Bryner, Cloyd. *Bugle Echoes: The Story of Illinois Fortyseventh.* Springfield, Ill., 1905.

Chamberlin, W. H. *History of the Eighty-first Regiment Ohio Infantry Volunteers.* Cincinnati, 1865.

Chance, Joseph E. *The Second Texas Infantry: From Shiloh to Vicksburg.* Austin, 1984.

Cluett, William W. *History of the Fifty-seventh Regiment Illinois Volunteer Infantry.* Princeton, Ill., 1886.

Dean, Benjamin D. *Recollections of the Twenty-sixth Missouri Infantry.* Lamar, Mo., 1892.

Dugan, James. *History of Hurlbut's Fighting Fourth Division.* Cincinnati, 1863.

Fox, Simeon M. *The Seventh Kansas Cavalry.* Topeka, Kans., 1908.

Frost, M. O. *Regimental History of the Tenth Missouri Volunteer Infantry.* Topeka, Kans., 1892.

Gottschalk, Philip. *In Deadly Earnest: The History of the First Missouri Brigade, C.S.A.* Columbia, Mo., 1991.

Griscom, George L. *Fighting with Ross' Texas Cavalry Brigade, C.S.A.* Hillsboro, Tex., 1976.

Hubert, Charles F. *History of the Fiftieth Regiment Illinois Volunteer Infantry.* Kansas City, Mo., 1894.

Jones, Thomas B. *Complete History of the Forty-sixth Regiment Illinois Volunteer Infantry.* Freeport, Ill., 1907.

Morris, William S. *History of the Thirty-first Regiment Volunteers.* Evansville, Ind., 1902.

Morrison, Marion. *A History of the Ninth Regiment Illinois Volunteer Infantry.* Monmouth, Ill., 1864.

Pierce, Lyman B. *History of the Second Iowa Cavalry.* Burlington, Iowa, 1865.

Reed, David. *Campaigns and Battles of the Twelfth Regiment Iowa Veteran Volunteer Infantry.* Evanston, Ill., 1903.

Rood, H. H. *History of Company A, Thirteenth Iowa Veteran Infantry.* Cedar Rapids, Iowa, 1889.

Rose, Victor M. *Ross' Texas Brigade.* Louisville, 1881.

Smith, Charles H. *The History of Fuller's Ohio Brigade, 1861–1865.* Cleveland, 1909.

Tunnard, William H. *A Southern Record, the History of the Third Regiment Louisiana Infantry.* Baton Rouge, 1866.

Weiss, Enoch. *The Forty-eighth Regiment Indiana Veteran Volunteer Infantry in the Civil War.* N.p., n.d.

Williams, John M. *"The Eagle Regiment," Eighth Wis. Inf'ty Vols.* Belleville, Wis., 1890.

Young, J. P. *The Seventh Tennessee Cavalry. (Confederate). A History.* Nashville, 1890.

OTHER SECONDARY SOURCES

Ambrose, Stephen E. *Halleck: Lincoln's Chief of Staff.* Baton Rouge, 1962.

The Battle of Corinth. 125th Anniversary Official Souvenir Program. Corinth, Miss., 1987.

Bauer, K. Jack. *The Mexican War*. New York, 1974.

Bearss, Edwin C. *Decision in Mississippi*. Jackson, Miss., 1962.

Benner, Judith. *Sul Ross: Soldier, Statesman, Educator*. College Station, Tex., 1983.

Bettersworth, John K., ed. *Mississippi in the Confederacy as They Saw It*. 2 vols. Jackson, Miss., 1976.

Beyer, W. F., and O. F. Keydel, eds. *Deeds of Valor*. 2 vols. Detroit, 1905.

Brown, Leonard. *American Patriotism*. Des Moines, 1869.

Byers, Samuel H. M. *Iowa in War Times*. Des Moines, 1888.

Castel, Albert. *General Sterling Price and the Civil War in the West*. Baton Rouge, 1968.

Cockrell, Monroe F. *The Lost Account of the Battle of Corinth*. Jackson, Tenn., 1955.

Comte de Paris, Phillipe Albert d'Orleans. *History of the Civil War in America*. 4 vols. Philadelphia, 1885–86.

Connelly, Thomas. *Army of the Heartland*. Baton Rouge, 1967.

Cooling, Benjamin Franklin. *Forts Henry and Donelson: The Key to the Confederate Heartland*. Knoxville, Tenn., 1987.

Cozzens, Peter. *No Better Place to Die: The Battle of Stones River*. Urbana, Ill., 1990.

Cresap, Bernar. *Appomattox Commander: The Story of General E. O. C. Ord*. San Diego, 1981.

Davis, William C. *Jefferson Davis: The Man and His Hour*. New York, 1991.

Dyer, Frederick H. *A Compendium of the War of the Rebellion*. Des Moines, 1909. Reprint. 3 vols. New York, 1959.

Fiske, John. *The Mississippi Valley in the Civil War*. Boston, 1902.

Fowler, Frank A. *Old Abe, the Eighth Wisconsin War Eagle*. Madison, Wis., 1885.

Gentry, Claude. *The Battle of Corinth*. Baldwyn, Miss., 1976.

Greene, Francis Vinson. *The Mississippi*. New York, 1882.

Hartje, Robert. *Van Dorn, Confederate General*. Nashville, 1967.

Hattaway, Herman, and Archer Jones. *How the North Won: A Military History of the Civil War*. Urbana, Ill., 1983.

Hawke, Paul. *The Siege and Battle of Corinth: A Strategy for Preservation, Protection, and Interpretation Prepared for the Citizens of Corinth*. Atlanta, 1991.

Ingersoll, Lurton. *Iowa and the Rebellion*. Philadelphia, 1866.

Johnson, Sid S. *Texans Who Wore the Gray*. Tyler, Tex., 1907.

Jones, Archer. *Confederate Strategy from Shiloh to Vicksburg*. Baton Rouge, 1991.

Kitchens, Ben Earl. *Rosecrans Meets Price: The Battle of Iuka, Mississippi*. Florence, Ala., 1987.

Lamers, William M. *The Edge of Glory: A Biography of General William S. Rosecrans*. New York, 1961.

Lash, Jeffry N. "Stephen Augustus Hurlbut: A Military and Diplomatic Politician." Ph.D. diss., Kent State University, 1980.

Livermore, Thomas. *Numbers and Losses in the Civil War in America*. Boston, 1901.

McDaniel, Robert W. *Battle of Davis Bridge*. Bolivar, Tenn., n.d.

McDonald, Lyla Merrill. *Iuka's History, Embodying Dudley's Battle of Iuka*. Iuka, Miss., n.d.

McFeely, William S. *Grant: A Biography*. New York, 1981.

Malone, Dumas, ed. *Dictionary of American Biography*. 20 vols. New York, 1928–36.

Miller, Emily Van Dorn. *A Soldier's Honor, with Reminiscences of Major-General Earl Van Dorn*. New York, 1902.

Minnesota in the Civil and Indian Wars. St. Paul, 1891.

Monaghan, Jay. *Civil War on the Western Border*. Lincoln, 1955.

Moore, Frank. *The Civil War in Song and Story*. New York, 1889.

Nabors, S. M. *History of Old Tishomingo County*. Corinth, Miss., 1980.

Nicolay, John, and John Hay. *Abraham Lincoln: A History*. 10 vols. New York, 1890.

Pollard, Edward A. *The Lost Cause*. New York, 1866.

Reid, Whitelaw. *Ohio in the War*. 2 vols. Cincinnati, 1868.

Rogers, Margaret G. *Civil War Corinth*. Corinth, Miss., 1989.

Shalhope, Robert E. *Sterling Price, Portrait of a Southerner*. Columbia, Mo., 1971.

Shanks, William F. G. *Personal Recollections of Distinguished Generals*. New York, 1866.

Shea, William L., and Earl J. Hess. *Pea Ridge: Civil War Campaign in the West*. Chapel Hill, N.C., 1992.

Smith, Justin. *The War with Mexico*. 2 vols. New York, 1919.

Smith, Ophia D., and William E. Smith. *Colonel A. W. Gilbert, Citizen Soldier of Cincinnati*. Cincinnati, 1934.

Stuart, A. A. *Iowa Colonels and Regiments*. Des Moines, 1865.

Tucker, Philip T. *The Confederacy's Fighting Chaplain: Father John B. Bannon*. Tuscaloosa, Ala., 1992.

Von Clausewitz, Carl. *On War*. New York, 1970.

Warner, Ezra. *Generals in Blue: Lives of the Union Commanders*. Baton Rouge, 1964.

———. *Generals in Gray: Lives of the Confederate Commanders*. Baton Rouge, 1959.

Webb, W. L. *Battles and Biographies of Missourians*. Kansas City, Mo., 1900.

Williams, Kenneth P. *Lincoln Finds a General*. 6 vols. New York, 1949–58.

Woodworth, Steven E. *Jefferson Davis and His Generals*. Lawrence, Kans., 1990.

Young, Kevin. *To the Tyrants Never Yield*. Corpus Christi, Tex., 1992.

district commander, 42, 45; command relations of, with Bragg, 43; command relations of, with Price, 43–46, 51, 318; actions of, during Iuka campaign, 59–60, 66, 132; plans attack on Corinth, 136–37, 140–41; actions of, during march on Corinth, 150, 155; battle actions of, at Corinth (4 October), 159, 161, 193, 203, 216, 217; believes Federals reinforced, 228–29; plans of, for 4 October, 229–30, 233, 251; battle actions of, at Corinth (4 October), 235, 252, 272–73; quixotic behavior of, during retreat from Corinth, 277–79, 282, 291, 294, 298, 300, 306; excoriated for defeat, 306–8; depressed state of mind of, 307–8; court of inquiry against, 308–9; final days of, 310, 322–24

Van Dorn, Emily, 307–8, 323

Van Eman, Pvt. J. H., 113

Vaughan, Capt. Levi, 160, 170, 181

Veatch, Brig. Gen. James, 281–83, 287–88, 290, 328

Vicksburg, Miss., 44, 300, 303, 307, 309, 316, 322

Villepigue, Brig. Gen. John, 142, 155, 174, 193, 272–73, 328

Wade's Missouri Battery (C.S.), 176, 205, 220, 325, 327

Washburne, Elihu, 16

Watson (Louisiana) Artillery, 272, 328

Wayne County, Ohio, 264

Weber, Maj. Andrew, 110, 112

Welch, Thad, 297

Western Sharpshooters, 327

White, Maj. Robert, 266

White House, 195, 198–99, 210, 212

White House fields, 195–96, 201, 204–5, 210–11, 213–14, 348 (n.2)

Whitfield, Col. John, 86, 88, 91, 99

Whitfield, Pvt. W. G., 167–68

Wilcox, Lt. Col. John, 205, 243

Willcox, Capt. Lyman, 71, 73–74

Williams, Brig. Gen. Alpheus, 226

Williams, Capt. George A., 37, 234

Wilson, Hunt, 219–20

Wilson, Sgt. T. P., 82

Wilson house, 248

Wilson's Creek, Battle of, 4

Wounded: suffering and treatment of, 116–17, 124, 171–72, 214, 221–22, 225, 234, 276, 308–9, 319–21

Wright, Cpl. Charles, 164, 175, 178, 180, 197, 206, 208

Yallabusha River, 310

Yates Sharpshooters. *See* Sixty-fourth Illinois Infantry

Young, Capt. John, 104–5, 108

Young, Pvt. John, 149

Young's Bridge, 291, 295, 299

Yow house, 102–3

Zearing, Capt. James, 168